The Making of Europe's Critical Infrastruc

The Making of Europe's Critical Infrastructure

Common Connections and Shared Vulnerabilities

Edited by

Per Högselius
KTH Royal Institute of Technology, Stockholm, Sweden

Anique Hommels
Maastricht University, The Netherlands

Arne Kaijser
KTH Royal Institute of Technology, Stockholm, Sweden

and

Erik van der Vleuten
Eindhoven University of Technology, The Netherlands

Selection, introduction and editorial matter © Per Högselius,
Anique Hommels, Arne Kaijser and Erik van der Vleuten 2013, 2016
Individual chapters © Respective authors 2013, 2016

All rights reserved. No reproduction, copy or transmission of this publication may be made without written permission.

No portion of this publication may be reproduced, copied or transmitted save with written permission or in accordance with the provisions of the Copyright, Designs and Patents Act 1988, or under the terms of any licence permitting limited copying issued by the Copyright Licensing Agency, Saffron House, 6–10 Kirby Street, London EC1N 8TS.

Any person who does any unauthorized act in relation to this publication may be liable to criminal prosecution and civil claims for damages.

The authors have asserted their rights to be identified as the authors of this work in accordance with the Copyright, Designs and Patents Act 1988.

First published 2013
Published in paperback 2016 by
PALGRAVE MACMILLAN

Palgrave Macmillan in the UK is an imprint of Macmillan Publishers Limited, registered in England, company number 785998, of Houndmills, Basingstoke, Hampshire RG21 6XS.

Palgrave Macmillan in the US is a division of St Martin's Press LLC,
175 Fifth Avenue, New York, NY 10010.

Palgrave Macmillan is the global academic imprint of the above companies and has companies and representatives throughout the world.

Palgrave® and Macmillan® are registered trademarks in the United States, the United Kingdom, Europe and other countries.

ISBN 978–1–137–35872–1 hardback
ISBN 978–1–137–58098–6 paperback

A catalogue record for this book is available from the British Library.

A catalog record for this book is available from the Library of Congress.

Contents

List of Figures and Tables   vii

General Introduction

1. Europe's Critical Infrastructure and Its Vulnerabilities – Promises, Problems, Paradoxes   3
   Erik van der Vleuten, Per Högselius, Anique Hommels, and Arne Kaijser

Part I   Connecting a Continent

Introduction   23

2. Natural Gas in Cold War Europe: The Making of a Critical Infrastructure   27
   Per Högselius, Anna Åberg, and Arne Kaijser

3. Inventing Electrical Europe: Interdependencies, Borders, Vulnerabilities   62
   Vincent Lagendijk and Erik van der Vleuten

Part II   Negotiating Neighbors

Introduction   105

4. An Uneasy Alliance: Negotiating Infrastructures at the Finnish-Soviet Border   108
   Karl-Erik Michelsen

5. Bulgarian Power Relations: The Making of a Balkan Power Hub   131
   Ivan Tchalakov, Tihomir Mitev, and Ivaylo Hristov

6. Border-Crossing Electrons: Critical Energy Flows to and from Greece   157
   Aristotle Tympas, Stathis Arapostathis, Katerina Vlantoni, and Yiannis Garyfallos

Part III   Coping with Complexity

Introduction   187

7. Eurocontrol: Negotiating Transnational Air Transportation in Europe   191
   Lars Heide

8 Connections, Criticality, and Complexity: Norwegian Electricity in
   Its European Context   213
   Lars Thue

9 In Case of Breakdown: Dreams and Dilemmas of a Common
   European Standard for Emergency Communication   239
   Anique Hommels and Eefje Cleophas

Conclusion

10 Europe's Infrastructure Vulnerabilities: Comparisons and
   Connections   263
   Anique Hommels, Per Högselius, Arne Kaijser, and Erik van der Vleuten

Acknowledgments   278

Notes on Contributors   280

Bibliography   283

Index   304

Figures and Tables

Tables

2.1	Western European gas imports in 1982, by exporting country (bcm)	40
2.2	Western European gas imports in 2005, by exporting country (bcm)	41
5.1	Installed capacity in the Bulgarian electricity system, 1955–65 (MW)	137
5.2	Specific consumption of electric power in Bulgaria, Greece, and Yugoslavia, 1939 and 1970 (kWh/person)	144
5.3	Installed capacity in the Bulgarian electricity system, 1970–5 (MW)	144
5.4	Installed electric power and electricity consumption in Bulgaria and other countries, 1989	145
5.5	Installed capacity in the Bulgarian electricity system, 1985–95 (MW)	151
5.6	Electricity consumption in Bulgaria, 1960–94, by sector (percentage of total)	152
6.1	Sources of electricity in Greece, 1961–2005 (percentages)	159
6.2	Share of oil and coal imported to Greece from the Eastern bloc, 1953–66 (percentages)	169
6.3	Electricity production in Greece and electricity exchanges between Greece and other countries, 1953–2009	170
6.4	Annual electricity exchanges between Greece and its neighbors, by country (GWh), 1961–2009	172
8.1	Periods of technological and institutional development	217

Figures

2.1	The natural gas export infrastructure with transnational pipelines from the Netherlands to West Germany, Belgium, and France, as of 1968, with plans for 1969	31
2.2	Vision of Soviet gas exports to Western Europe as of 1967	35
2.3	Possible natural gas flows through the Nordic countries, as envisaged by the Swedish Natural Gas Committee in 1974	38
2.4	A Soviet stamp from 1983, issued to commemorate the completion of the "Transcontinental Export Pipeline" from Siberia to Western Europe	40
2.5	European mesoregions for natural gas, by dominant source of supply	56
3.1	Early cross-border microregional and national electricity systems	69

3.2	Early proposals for a European "supergrid"	76
3.3	The United Power Systems by the early 1980s	81
3.4	Electrical Europe by 1976 and in the early 1990s	83
5.1	United electricity systems of the European COMECON countries	149
5.2	Bulgarian imports and exports of electricity, 1975–2005 (percentage of total production)	152
6.1	Projected electricity network of Greece, 1943/44	161
6.2	Diagram showing how a 1996 import of electricity from the interconnected Balkan system saved the Bulgarian network during anomalies at Kozloduy that resulted in a drop in nuclear power generation	178
6.3	Diagram showing how a 1997 drop in frequency at the Greek power network due to anomalies at the Cerna Voda nuclear plant of Romania was offset by positive changes in the exchange of electricity between Greece, Albania, Bulgaria, and Yugoslavia	179
8.1	The NorNed link between Kvinesdal, Norway, and Eemshaven, the Netherlands, being loaded onto the cable-laying vessel	214
8.2	A model of influences in transnational connections	217
8.3	The dispatchers at the control center of the Norwegian Power Pool in the 1930s	223
8.4	Statnett's national control center coordinates the operations of all players involved in the Norwegian main grid with its international connections	230
9.1	European public safety networks	251

General Introduction

1
Europe's Critical Infrastructure and Its Vulnerabilities – Promises, Problems, Paradoxes

Erik van der Vleuten, Per Högselius, Anique Hommels, and Arne Kaijser

Prologue

> Critical infrastructure (CI) can be damaged, destroyed, or disrupted by deliberate acts of terrorism, natural disasters, negligence, accidents, computer hacking, criminal activity, and malicious behaviour. To save the lives and property of people at risk in the EU [European Union]...any disruptions or manipulations of CI should, to the extent possible, be brief, infrequent, manageable, geographically isolated...The recent terrorist attacks in Madrid and London have highlighted the risk of terrorist attacks against European infrastructure. The EU's response must be swift, coordinated, and efficient.[1]

With these opening words, the European Commission, the executive body of the EU, urged an EU-wide program for the protection of critical infrastructure in 2005. Several types of events – in this book we shall speak of "critical events" – triggered this new sense of infrastructure vulnerability and risk. The ball started rolling in the United States in the mid-1990s. Hackers had just used the Internet to rob Citibank of $10 million, email-bombed the Internet service provider America Online, and broken into computer systems at the Department of Justice, the CIA, and the US Air Force. During 1995 no fewer than 250,000 attempts to hack Department of Defense computer files were registered, most of which were successful. President Bill Clinton then set up a commission on critical infrastructure – that is, infrastructure critical to the economy, society, and administration. Such infrastructure urgently needed protection, for "a satchel of dynamite and a truckload of fertilizer and diesel fuel are known terrorist tools. Today, the right command sent over a network to a power generating station's control computer could be just as devastating." Worse, "we found all our infrastructures increasingly dependent on information and communications systems that criss-cross the nation and span the globe. That dependence is the source of rising vulnerabilities...The capability to do harm...is real; it is growing at an alarming rate; and we have little defense against it."[2]

Several high-profile terrorist attacks involving public infrastructure further boosted the debate. In the United States the attacks of September 11, 2001 showed that infrastructure was vulnerable not only to cyber threats; four commercial airliners were hijacked and used as weapons against the Pentagon and New York City's World Trade Center, killing about 3000 people. In the aftermath, local power, communication, and transport systems broke down as well and severely hampered emergency response efforts. The event triggered a range of security policies in the United States, Europe, and elsewhere.[3] Subsequent attacks on ground transportation on European soil showed still other possibilities to harm societies through their infrastructure. In the morning of March 11, 2004, terrorist bombs hit four commuter trains in Madrid, killing almost 200 and wounding another 1800. In the rush hour of July 7 and 8, 2005, four suicide bombers attacked the London subway system and a double-decker bus, killing over 50 and injuring around 700. EU policy-makers needed to respond. They, too, established a critical infrastructure protection program.

By then the ground for such a program in Europe had already been prepared by a third type of event – a series of internal infrastructure disruptions. EU policy-makers were particularly shocked by the so-called Italian blackout of September 28, 2003, in which the failure of a Swiss-Italian power line during an Alpine storm eventually plunged the entire Italian peninsula and some 56 million people into darkness. The blackout trapped some 30,000 people in railway and metro cars, and hundreds in elevators. It disrupted road traffic due to traffic-light failure, interrupted the water supply, spoiled refrigerated foodstuffs, and halted industry's continuous production lines. That same year, other blackouts struck southern Sweden, eastern Denmark, London, Helsinki, and Athens. Europe's infrastructure vulnerability was perhaps most persistently demonstrated by the 2006 European blackout, in which a power failure in northern Germany instantly turned off lights in countries as far away as Croatia and Portugal. This rolling blackout even cascaded into northern Africa through the Spain–Morocco submarine cable. Repeated Russia-Ukraine gas crises, too, caused energy-supply problems throughout much of Europe; even Italian and French consumers were shown to depend significantly on Russian gas exports via Ukraine.[4]

To EU ministers and commissioners the lesson was clear: Europe was at risk and an "all-hazards approach" was needed, addressing terrorist attacks, natural disasters, and technical malfunctions. Moreover, in Europe the problem had a particularly transnational character since the "damage or loss of a piece of infrastructure in one M[ember] S[tate] may have negative effects on several others and on the European economy as a whole...This means that a common level of protection may be necessary."[5] By 2006, some common legislation was in place, though negotiations delayed the most visible piece of EU legislation – the critical infrastructure directive – until 2008.

Introduction

The ongoing policy debate about European critical infrastructure introduces the topic of this book in several ways. Clearly, infrastructure vulnerability and its governance are pressing issues today in Europe, the United States, and many other places in the world. They concern politicians of many flavors, a host of sector and civil society organizations, citizens, and scholars: critical infrastructure features prominently among the vulnerabilities of modern technological culture that are in urgent need of a better understanding.[6] Moreover, leading critical infrastructure researchers argue that infrastructure vulnerabilities reside particularly in their transborder character. Yet the vast majority of existing critical infrastructure studies take a (sub)national perspective and leave transborder dynamics poorly understood.[7] Those few studies that do address cross-border vulnerabilities tend to follow the EU policy trajectory, but rarely ask how and why infrastructure connects some regions, countries, and peoples more than others, or how such asymmetries affect the geography of Europe's infrastructure vulnerability.[8] To do so is a major purpose of this book.

In order to open up this topic, we delve below the surface of our introductory story and place the emergence and governance of critical infrastructure in a historical perspective. Europe's infrastructure vulnerabilities and coping strategies did not fall from the sky, nor did they emerge mysteriously from a vague and abstract process called "globalization." Instead our present infrastructure has a long, very concrete, and traceable history, and so do the vulnerabilities and governance responses that they evoked.[9] Take, for example, the case of Russian natural gas exports. In the later decades of the Cold War, Russian gas proved a welcome and reliable relief for strained energy supply systems in much of Central Europe, and pipelines came to stretch from Siberian gas fields to German, French, and Italian consumers. Decades later, in a very different political context, the EU regards the very same supply lines to be one of Europe's major energy vulnerabilities. Or consider fundamental computer protocols, such as the two-digit date representation of years. This was designed when the Internet was a mere dream and the new millennium still in the distant future. Yet in the late 1990s it took a panic and great effort to mitigate the cascading effects of the so-called Millennium Bug that threatened individual computers as well as the financial, military, and other sectors relying on networked information and communication technology (ICT) services.[10] Infrastructure, it has been said, develops in historical time, which transcends individual, political, and media time.[11] Accordingly, to understand its dynamics, vulnerabilities, and governance, we need to revisit the concerns, priorities, choices, and conflicts of its makers. We need to engage with history.

Such a historical perspective does more than track down the emergence and governance of Europe's transnational infrastructure vulnerabilities. It also brings into view remarkable long-term ironies that deserve serious consideration and reflection. Consider the very notion of critical infrastructure. We know from previous research that over the last two centuries many individuals, groups, and international organizations have eagerly stimulated transnational infrastructure development. Protagonists in the League of Nations, the International Labour Organization, the United Nations (UN), the Organisation for Economic Co-operation and Development (OECD), and a host of dedicated transport, communication, and energy organizations argued that infrastructure integration would boost Europe's prosperity by creating larger markets. It would also invoke peace on a war-prone continent by fostering economic interdependencies and mutual understanding. Also, the EU's founding document, the 1992 Maastricht Treaty, obliges the EU to stimulate Trans-European Networks as levers for economic and social cohesion.[12] The term "critical infrastructure" that is currently in vogue confirms the success of this collective effort. It underscores the fact that infrastructure has indeed become omnipresent and critical to the functioning of modern economies, societies, and administrations. Simultaneously, the term signifies a major downside of Europe's infrastructure transition: precisely because modern societies have become (inter)dependent on cheap and steady infrastructure services, they are also vulnerable to infrastructure abuse and disruption. When infrastructure became omnipresent, Europe faced new risks. As we shall see repeatedly in this book, this profound historical irony shows up in many forms. Even the very security arrangements of the past might produce new vulnerabilities later on. To mention just one example here, about a century ago, power companies started to connect electricity grids across borders to increase mutual system stability and enable mutual support in case of breakdown, in short, to reduce the risk of blackouts. Simultaneously, such connections introduced the historically novel vulnerability of failures cascading across borders, as demonstrated in recent transnational blackouts. Yesterday's solutions can cause today's problems. Such ironies underline the paramount importance of a careful and reflective historical examination of critical infrastructure vulnerabilities.

Still we need to delve deeper. So far we have talked about "Europe" as a fixed container taken for granted for the purpose of our historical inquiry into transnational infrastructure vulnerabilities. Worse, our opening example tacitly equates "Europe" with "European Union." Further historical scrutiny of our introductory story suggests that this will not do. Consider that for half a century the EU and its forerunners advocated a particular version of "Europe" in terms of geography, governance model, and values. Yet political and popular support failed this project repeatedly. In the last decade, EU analysts observe, the European Commission has discovered the common enemies of transnational health, environment, and security threats as a promising way to bypass such stasis and resistance: the EU project took a qualitative and quantitative leap when the EU developed a new "security identity" and successfully claimed the governance of transboundary threats from food safety, avian influenza, and natural disasters to emergency response, terrorism, and critical infrastructure.[13] When French and Dutch voters turned down the European constitution in 2006, the EU responded with a new charm initiative that placed center stage the fight against climate change – the ultimate "common threat" demanding a "common approach."[14]

What, then, is EU-critical infrastructure protection about? Is it about protecting infrastructure, about protecting the EU version of European integration, or both? This question becomes even more compelling when we take into account resistance to the program. The electric power sector and the financial sector, for instance, found their infrastructure sufficiently protected, not at risk (of technical or terrorist breakdown at least!), well on track in terms of European cooperation outside EU institutions and transcending EU territory, and certainly not in need of the EU substituting these sector's own transnational governance arrangements. Indeed, most sectors managed to steer clear of the EU-critical infrastructure directive, which, after protracted negotiations, came to apply only to energy and transport – out of 11 sectors originally proposed.[15] Sector negotiators resisted making their own sector collaborations subordinate to the particular form of Europeanization that the EU-critical infrastructure program represented. The lesson for historians is that they should not treat the EU as a natural, self-evident container for the history of Europe's critical infrastructure vulnerability and governance.[16] Rather, stakeholders in the critical infrastructure playing field were (re)negotiating the very meaning of European integration – its territoriality, core values, governance modes – for their particular sectors.

This leads us to the broader historical question of what kind of "Europe" was built in the sphere of critical infrastructure vulnerabilities, most of which emerged before the EU could exert any significant influence on infrastructure matters.[17] Did the geography, governance forms, and priorities of this "critical infrastructure Europe" differ from, contribute to, or simply reflect "political Europe" so familiar from history class, with its quarrels between nation-states, its prominent Cold War division, and most recently the increasing momentum of the EU project? What is needed is a transnational history that actively inquires about the emergence and governance of infrastructure vulnerability entwined with the territorial and political shaping of modern Europe. Such is the thrust of this book. In the following sections we shall sharpen the conceptual tools that inform our study. Next we will briefly sketch the logic and structure of the book. The individual chapters are not introduced in this General Introduction but in the introductions to the three main parts. The Conclusion (Chapter 10) picks up the threads laid out in both the general and the part introductions and it reflects on our findings.

Infrastructure's paradox

How can we unravel the historical coevolution of critical infrastructure, its vulnerabilities and governance responses, and contemporary Europe? An important premise of this book is that notions of infrastructure, vulnerability, and Europe and the relations between them cannot be understood straightforwardly in terms of univocal correlations between well-defined variables. Rather, these concepts themselves had ambivalent meanings for different stakeholders, and became entangled in different constellations in a variety of historical processes. This is why stakeholders might persistently disagree about the appropriate interpretations of, and governance responses to, infrastructure vulnerabilities.[18] To appreciate the historical and social latitude of our key concepts, we prefer to think of them not as sharply defined variables but as paradoxes that carry potentially conflicting meanings.[19] These paradoxes, in turn, set the stage for our investigation of which and whose meanings of infrastructure, vulnerability, and "Europe" were prioritized in historical processes.

Consider, to start with, the concept of infrastructure. Late nineteenth-century railway builders invented the term to denote the structure supporting the rails (embankments, cuttings, bridges and so on), but NATO and others redefined it in the 1940s as the structures underlying modern organizations or societies.[20] Since then the term has been used in broader and narrower meanings. Some people associated infrastructure with what we today call the network industries, providing transport, energy, communication, and water services. Others expanded the term to cover all sorts of basic facilities, including education, financial services, and healthcare. At times, infrastructure carried the additional connotation of natural monopoly or public good, which made the concept highly contested when the boundaries between public and private were redrawn in the neoliberal era. All of these meanings, however, shared a common view of infrastructure as technologies of connection that played a constitutive and integrative role in economies and societies. The more recent concept of "critical infrastructure" underscores the "underlying structure that keeps society together" aspect.[21] Related terms such as "networks" or "large technical systems" likewise share connotations of connectivity and an integrative foundation for modern societies.[22]

It is important to realize, however, that this connective meaning of infrastructure echoes the rhetoric of its historical protagonists. The concept is laden with stakeholder ideology, which preceded the term infrastructure itself: already in 1852, Michel Chevalier claimed that "Railways have more in common with the religious spirit than we think; never has there existed an instrument of such power to link together scattered peoples."[23] In capacities that varied from imprisoned social reform thinker to French government advisor and senator, Chevalier articulated, developed, and promoted his vision that modern means of communication, such as railways, telegraphy, and steam navigation, would overcome barriers of nature and space, improve the human condition, and propel economic prosperity and equality across national and class borders. The religion metaphor may have disappeared since, but each later generation of infrastructure proponents seems to have revived Chevalier's old promise of connectivity and socioeconomic leverage. Nineteenth-century railway and telegraphy pamphlets, interwar proposals for European electricity, telephone, aviation and motorway infrastructure, postwar visions of TV broadcasting and the "global village," 1990s celebrations of the Internet, and present-day social media marketeers share and highlight this one assumption: infrastructure connects.

We do not dispute infrastructure's connective qualities. Instead we wish to point out that by foregrounding connections, other infrastructure features fade from view. If critical scholars call connectivity claims "the myth of the network," it is because these very claims obscure how infrastructure connections for some peoples or territories often imply the non-connection or even disconnection of others.[24] Examples abound of new canals, railways, and motorways that literally cut local communities in two. Even high-profile international infrastructure connection projects might involve disconnection: the famous Gotthardt railway line and tunnel, opening in 1882, was a symbol of the penetration of the Alps to integrate Northern and Southern Europe. At the same time it bypassed the Gotthard pass above the tunnel, disconnecting from transnational trade flows one of its most prominent hubs for millennia, condemning its expensive road infrastructure to disuse and its mail-coach system to nostalgic memory.[25] In other cases, infrastructure was designed to bypass. The business model of the Great Nordic Telegraph Company's huge telegraph network, stretching from London via the Copenhagen hub across Russia to China and Japan by the 1870s, was to connect East and West while bypassing imperial Germany. Britain's submarine telegraph network that spanned the globe by 1900, too, served to bypass land-based telegraph systems on territories outside British control.[26] The incorporation of the Baltic states into the Soviet Union after the Second World War involved cutting telephone connections to the West and rerouting all telephone traffic through Moscow. We could go on and on.[27] In terms of access, too, some social groups were "more equal than others": since 1992 the Channel Tunnel has served the free flow of businessmen, tourists, and freight, while at the same time much effort was dedicated to prevent those flows of illegal immigrants and asylum-seekers that the tunnel also attracted.[28] This is our first paradox: the very same infrastructure can at once connect and fragment.[29]

A paradox, of course, is not a contradiction. It is an apparent contradiction that serves stylistic or, in our case, analytical purposes. Our infrastructure paradox of connection and rupture reminds us not to take at face value the myth of ever-increasing connectivity, and instead appreciate the latitude of historical agents (as well as present-day ones) to employ infrastructure as tools that not only connect but also create difference. Infrastructure's paradox thus translates into the historical research question regarding when, by whom and for what reasons infrastructure was made to connect or splinter. We ask why historical agents chose some infrastructure-development trajectories and rejected alternatives, who they sought to connect and who they bypassed, and how they dealt with borders in the age of connectivity.[30] Such historical choices, we suspect, were important constituents of Europe's infrastructure vulnerabilities and the governance responses that these evoked.

Ambivalent vulnerabilities

What, then, is infrastructure vulnerability about? Notions of infrastructure vulnerability and criticality are relatively recent additions to the thinking about risks of technological systems. Charles Perrow's famous work on living with high-risk technologies from the 1980s illustrates the tone of this debate well. For him, modern technology had become so complex, with many causal feedback loops and possibilities for cascading failure, that it had become susceptible to breakdowns that cannot be predicted, anticipated, or managed: small disturbances may cause unexpected chains of events that lead to bigger failures, especially in tightly coupled systems where such processes happen very quickly and cannot be halted. Prominent examples include failures in nuclear power plants, chemical plants, air-traffic control, and electric power systems. In his Normal Accident Theory, Perrow saw breakdown and accident as an inherent, "normal condition" of such systems.[31]

Like infrastructure, the terms "risk," "vulnerability," and "criticality" have been subject to considerable interpretive flexibility. Many quantitative risk studies define risk narrowly as the probability of an unwanted event multiplied by its impact. In social theory, by contrast, the terms risk and risk society have become encompassing. Ulrich Beck argues that modern societies are increasingly organized in response to human-made technological risks that spur doubts about the present course of modernization and affect social structure.[32] Vulnerability, too, is used in narrow and broad meanings. It may refer to specific people, organizations, or places but also to technological systems and even technological cultures that are susceptible to harm, and their ability to anticipate, resist, cope with, and recover from events that could impede their functioning.[33] The notion of vulnerability in critical infrastructure discourses sometimes refers to disturbance or breakdown of the infrastructure system itself, and at other times to the consequences for households, industries, and other users of infrastructure services. In this book we will speak of "system vulnerabilities" and "user vulnerabilities" to distinguish between these two kinds of vulnerability. For all of these nuances and overlaps, however, these related concepts share one dominant message foreshadowed by Perrow: harm is coming our way.

As in the case of infrastructure connectivity, we should not take this key message of harmful infrastructure vulnerability at face value. For starters we should recognize our own bias when discussing risk and vulnerability. The psychology of fear tells us that the human mind is wired to foreground threat at the expense of opportunity, and routinely defies rational choice and behavior. Scores of experiments confirm that humans are notoriously poor at estimating risk. A morbid yet telling example is the estimated 1500 additional road accident deaths in the year following 9/11, when Americans massively substituted plane travel with interstate highway travel.[34] Also, human-made technological risks outside our direct personal control (nuclear accidents, terrorist attacks on infrastructure, gas import disruption) tend to trigger our sense of vulnerability more than natural risk (including earthquakes but also major health killers, such as diabetes, cardiovascular disease, or asthma) or technology-related risks that we claim to control (think road accidents). This mental vulnerability bias is further amplified in our contemporary culture of fear, which, according to sociologists, has emerged since the 1970s and skyrocketed after 9/11.[35] The point here is not, of course, that risk and vulnerability do not exist. Rather, the associated fear is simultaneously real and hyped for psychological, commercial, political, and media reasons. A recent OECD study observes that the improvement of cybersecurity is important; yet at the same time public cybersecurity debates suffer from "exaggerated language," "sensationalism," and "grossly misleading conclusions." Cyber espionage, password phishing, or hacktivist attacks are regularly interpreted as threats to (inter)national security or even signs of an emerging cyber war, rather than innovative forms of old social practices, such as spying, theft, and public protest.[36] Another commentator in the policy debate on European critical infrastructure protection – a security sector entrepreneur – observed that fears of an "electronic 9/11" are deliberately overstated: "Nobody is getting blown to bits. It's not real terrorism. But if you add 'terrorism' to things you get more budget."[37]

As in the case of infrastructure connectivity and rupture, the sheer observation that our sense of harmful infrastructure vulnerability is psychologically, historically, and socially situated leads us to more fundamental paradoxes. Consider, for instance, that similar critical events have historically provoked opposite reactions. In New York the Great Northeast Blackout of 1965 was celebrated for stimulating local neighborhood solidarity with candlelight dinners and street dances. Yet in a context of social unrest, the New York blackout of 1977 sparked looting, arson, and violence.[38] The same type of infrastructure breakdown, it seems, may turn out to be both hopeful and harmful for users. Harmful user vulnerability should not be assumed by definition; instead we should investigate the historical processes that produced these different outcomes. A similar argument applies to system vulnerability. Infrastructure vulnerability and breakdown should not be treated as inherently malicious; they may also create hope and opportunity for much-needed innovation and change – for instance, in the direction of increased sustainability or democratic control of modern technological systems.[39] Indeed, hopes for a sustainability transition in energy or mobility depict the crisis of present-day fossil-fuel-based infrastructure as well as the ongoing financial crisis as windows of opportunity for radical change.[40]

Next to this paradox of harm and hope, stakeholders may disagree completely about whether or not an infrastructure is vulnerable to begin with. To EU policy-makers the transnational blackouts of 2003 and 2006 suggested instant vulnerability and alarm. Yet, as we shall see in this book, the electric power sector saw the very same events as confirmation that Europe's electricity supply was reliable, secure, and well organized. Paradox again. We cannot reduce this paradox of simultaneous vulnerability and reliability to the issue of "who is right," for both parties had good arguments. EU policy-makers saw how local incidents could instantly ignite economic and social disruption thousands of kilometers away, and made it their job to address such long-distance vulnerabilities. The electric power sector, by contrast, had identified the possibility of rolling blackouts long ago, defined its task in terms of anticipating and mitigating such cascading vulnerabilities, and now found its security arrangements tested and working well: the lights stayed on for the great majority of the population and the entire system was repaired quickly, mostly within half an hour. For power sector spokespersons, the daily reliability gains of cross-border grid connection greatly outweighed incidental and rapidly contained cross-border failures.[41] This paradox of simultaneous vulnerability and reliability resonates in academic vulnerability scholarship itself: Perrow's Normal Accident Theory, which we mentioned above, triggered the emergence of High Reliability Theory, to study why supposedly inherently vulnerable technological systems in fact rarely break down. Normal Accident Theory uses electric power systems and air-traffic control as examples of inherently vulnerable technologies. High Reliability Theory highlights the same infrastructure as illustrations of high reliability.[42] Present-day critical infrastructure studies still echo and reproduce this ambivalence.[43]

As in the case of infrastructure connectivity and rupture, the paradoxes of simultaneous harm and hope and of simultaneous vulnerability and reliability force us to acknowledge the latitude of stakeholders to interpret and anticipate infrastructure vulnerability and its implications in radically different ways. Again, these paradoxes translate into historical research questions. How did stakeholders come to assess, prioritize, and anticipate vulnerabilities and their implications in concrete historical processes of transnational infrastructure development?[44] As we shall see in this book, historical agents time and again were confronted with different infrastructure design options that might have conflicting implications in terms of opportunity and harm, and of reliability and vulnerability. Hence they needed to weigh and trade different vulnerabilities against each other. Gas imports from Russia could solve Bavaria's threatening energy shortages but would create new import dependencies; Bulgarian nuclear power projects would make its electricity production less dependent on Russian coal but introduce the new risk of nuclear accident; Greek power authorities feared the risks of nuclear power and accepted dependence on power imports and polluting domestic lignite power stations. ICT would improve the precision of air-traffic control but make air travel vulnerable to ICT failures; standards for emergency communication should condemn endemic miscommunication during disasters to oblivion but might also be a source of new communication problems themselves. These were all complex issues. As the parties involved disagreed, negotiated, and struck compromises, they inscribed hope as well as harm, increased reliability, and potential new vulnerabilities in the design of transnational infrastructure. This book studies which and whose vulnerabilities were prioritized in such historical processes, and how these came to make up Europe's nascent infrastructure vulnerability geography.

Finding Europe

What does it mean, finally, to study Europe's infrastructure vulnerabilities? Given our discussion so far, the reader will not be surprised to find that we reject an ahistorical up-front definition of "Europe" as a stable container for our inquiry. Instead we set out to inquire how contemporary Europe itself was shaped in the processes of emerging and governing infrastructure vulnerabilities.

Of course, the historical and social variability of the term "Europe" is much better known than the ambivalences of infrastructure and vulnerability that we discussed above. Europe has always been a highly contested political project. Already the boom in political projects for a united Europe in the 1920s and 1930s made clear that historical agents did not see Europe as an invariable entity "out there" but as something to create, build, and work hard for. These projects also revealed substantial disagreement about how this Europe should look in terms of external reach (Should Britain, Russia, Turkey, and the colonies be included?), internal structure (Should regions, countries, or new empires be the main building blocks?), and governance (Should national autonomy be mitigated by intergovernmental, (con)federal, supranational, or non-governmental decision- and rule-making?).[45] Postwar projects for European integration show similar discrepancies. The UN worked hard for all-European economic integration from Ireland to Soviet Russia. It even breached its core principle of universal organization and set up its first regional commission, the UN Economic Commission for Europe (UNECE), to help forge an all-European economy. But the new organization competed with other Europe-builders that worked on a much smaller territorial scale. The European Communities, forerunner to the EU, involved only six states in the 1950s and 1960s, and welcomed only six more in the 1970s and 1980s. Gunnar Myrdal, the first UNECE secretary general, loathed such claims to Europe for the happy few: "I always reacted...to the increasingly common application of the term 'Europe' to that narrow strip of our Continent and the term 'European' to its subregional organizations. This type of propagandistic terminology...indicates a deeper inclination which is intensely inimical to the work governments are trying to do in this [United Nations] Commission".[46] After 1989, controversy about the meaning of Europe remained. EU membership quickly increased and the term "Europe" was increasingly associated with EU territory, polity, and values – partly following the deliberate EU cultural policy to forge a common European identity.[47] Yet when the European Commission heralded the Channel Tunnel between France and England as a sign of successful EU-led European integration, Eurosceptics revolted: "If one were to judge by the Commission's report...cross-border transport and free movement of goods in Europe could not exist without the E.U. Needless to say, governments are capable of freely cooperating...without needing to surrender their powers to an unelected, supranational authority."[48]

Such persistent and highly politicized disagreement about the meaning of Europe in terms of territory, governance, and values once more forces us to treat Europe not as a fixed concept but instead to inquire how Europe historically took shape in relation to our topic – the emergence and governance of transnational infrastructure vulnerabilities. Again, teasing out a few paradoxes helps us on our way. First, consider the following paradox of European integration. The Turkish government applied for full political membership of the European Communities in 1987, but negotiations regarding EU membership remain troublesome today. The last decade even witnessed an increase in popular and political resistance to Turkish inclusion in the EU. By contrast, in that very same decade, Turkey was fully integrated into Europe's largest electric power collaboration: in 2000, Turkish electric power authorities applied for integration into the Trans-European Synchronously Interconnected System (currently called the Continental European Synchronized Area), and since the summer of 2010, Turkish electrical machines, motors, and consumer appliances move at exactly the same frequency, in tune, and in immediate interdependency (and joint vulnerability), with their German, French, Dutch or Portuguese counterparts. This feat is even more remarkable when we consider that power authorities and companies in Britain and most of Scandinavia had chosen not to join "Europe's electrical heartbeat" in the 1960s, and still today cooperate with continental European partners in an asynchronous, and thus less immediate, mode than Turkey does.[49]

This apparent contradiction of electrical integration and political non-integration begs the broader historical question of what kind of Europe was built in the realm of infrastructure and its vulnerabilities, as compared with the dynamics of political Europe. In contrast with the formal political integration process, infrastructure counts as a major arena for Europe's so-called informal or hidden integration.[50] Indeed, such organizations as the European Broadcasting Union (1950), the European Conference for Ministers of Transport (1954), the European Conference for Post and Telecommunications (1957), and EUROCONTROL (1963) all explicitly claimed to build infrastructure for "Europe." Yet all worked outside the formal EU framework and built "Europes" that differed vastly in geographical coverage and governance modes. For instance, most Europeans do not question the inclusion of Israel in the European Broadcasting Union's Eurovision network and its annual song contests since the early 1970s.[51] This book similarly queries Europe's hidden integration and fragmentation in terms of the historical emergence and governance of infrastructure vulnerabilities. Who was connected in common vulnerability to whom?

Two further paradoxes provide important clues about where to look for such a hidden Europe of infrastructure vulnerabilities. First, we are used to thinking of European integration in terms of delegating tasks to a higher authority and the associated weakening of the nation-state. Yet the postwar era of European integration also witnessed the rise of the nation-state to unprecedented power, budgets, staffing, popular identification, and sociocultural integration. In recent decades even the (micro)region and the city have experienced a revival in terms of citizen identities, and social and economic activity.[52] The age of European integration, paradoxically, is also the era of the nation-state, the region, and the city. In the infrastructure realm, too, we see simultaneous construction and governance activity on these different scales.[53] We suspect that the same may apply to infrastructure vulnerabilities. Hence we examine whether, in the age of international gas crises and blackouts, the national and local remain important units of experiencing, infuencing, and governing vulnerability, and how vulnerabilities at very different scales of experience and power coevolved or competed.[54] Unlike much international history, then, our transnational history of Europe's infrastructure vulnerability needs to embrace, not neglect, these important national and local engines of change. For this purpose we juxtapose studies of transnational infrastructure vulnerability from pan-European, national, and microregional/local perspectives, and examine how these scales historically have coevolved.

Finally, we take a special interest in the dynamics of borders. Here is the paradox: Europe is characterized as often by its many borders on a relatively small territory as by its transactions and collaborations that transcend borders – represented most prominently today by the EU. Translated to infrastructure: on the one hand, infrastructure often serves to transcend Europe's natural and political, internal, and external borders. As argued above, railroads, pipelines, transmission lines, and telecommunication cables cross national and EU borders as well as the Alps, the Urals, the Black Sea, the Bosporus, and the Mediterranean, challenging geographical distinctions between Europe, Asia, and Africa. On the other hand, just as often, infrastructure followed and reinforced such borders. Parallel electric power lines run on each side of the French-German border, and even in 2011 there is still only a single bridge across the 470 km stretch of Danube River border between Bulgaria and Romania (while there are nine bridges across the Danube in the

Hungarian capital of Budapest alone). The Iron Curtain also became an "Electric Curtain," separating Cold War collaborations in electricity and telecommunications on either side. These are just a few instances that illustrate how political borders and priorities were inscribed into Europe's infrastructure geography. This book investigates the contradictory role of borders in the historical shaping of the geography of Europe's infrastructure vulnerability, taking the most prominent border in modern European history – the Iron Curtain – as its main case.

Structure of the book

We can now lay out our project. Above we translated the important present-day policy issue of Europe's infrastructure vulnerability and its governance into a set of historical questions. The book studies how and why historical agents interpreted, negotiated, built, and governed infrastructure connections and ruptures; how they anticipated and prioritized vulnerabilities, opportunities, and reliabilities; and, while doing so, how they produced a hidden geography of European vulnerability that both overlapped with and deviated from Europe's political geography, paying particular attention to the mutual shaping of pan-European, regional, national, and local scales of vulnerability and organization, and the role of political borders.

Given the vast nature of this subject matter, we need to narrow down our inquiry. First we decided to focus on the most critical of all critical infrastructure. As noted above, EU policy documents on critical infrastructure protection identified about 11 sectors that qualified as critical infrastructure, including food, banking, health, water supply, and space infrastructure. Other agencies might work with different lists. There is a remarkable consensus, however, that energy and ICT infrastructure count as the most critical of all. They top the list in EU policy documents. Attempts at the quantitative determination of society's most critical infrastructure, using theoretical models or real-life data, arrive at similar conclusions. In a study for the UK Cabinet Office, consultancy firm Ernst & Young deconstructed 11 key sectors underpinning the modern economy into their elements and assessed their mutual dependencies, determining telecommunications and electricity supply to be the infrastructure most frequently entangled with basic operations in the economy. A study based on a database of 2517 serious critical infrastructure incidents worldwide, as reported by news media, found an overwhelming role for energy and telecommunications in failures that cascaded across infrastructure boundaries.[55] This book, accordingly, focuses on energy and ICT infrastructure.

Narrowing down our inquiry still a bit further, we take these two critical infrastructure sectors to illustrate two different sorts of infrastructure vulnerability in contemporary Europe. The European Commission observed how "Europe's critical infrastructures are highly connected and highly interdependent," which made them "vulnerable to disruption."[56] "Connectedness" here refers to cross-border connections across national borders, which make failures difficult to contain geographically, such as energy-related crises (e.g. international gas crises and rolling blackouts). "Interdependency" refers to the situation where a given type of infrastructure not only crosses political or geographical borders but also intertwines with other infrastructure. In the nineteenth century, railroads were dependent on telegraphic communication; today, all of society's infrastructure has become entwined with information and communication systems and is thus vulnerable to ICT failure.[57]

Energy infrastructure (particularly the supply of gas, electricity, and nuclear power) and its cross-border interdependencies and vulnerabilities occupy center stage in parts I and II of the book. In Part I, "Connecting a Continent," we study the emergence and governance of transnational infrastructure vulnerability from a transcontinental perspective. This part of the book discusses how Europe's vast infrastructure traverses and transcends the Continent and has produced asymmetrical long-distance vulnerabilities. This perspective also forces us to pay ample attention to Central and Eastern Europe and to avoid implicitly equating "Europe" with Western Europe, as much European historiography once did.[58] We focus on electricity and natural gas because these systems are so closely intertwined with economic and social activities of all kinds. Today the everyday life of almost all European households is dependent on the uninterrupted flows of these commodities across borders, which are taken for granted until a blackout or "gas crisis" forces them out of complacency.

In Part II, "Negotiating Neighbors," we continue to investigate the connectedness of Europe's critical energy infrastructure. However, here we zoom in on the role of nation-states in shaping and governing both domestic and cross-border infrastructure vulnerabilities. As neighboring countries negotiated their infrastructure connections and tried to anticipate the implications for vulnerability, they built the very bricks that came to make up Europe's wider geography of infrastructure vulnerability. We selected case studies from countries situated at Europe's most prominent internal political border in contemporary history, the Iron Curtain. Studying Finnish, Bulgarian, and Greek infrastructure priorities, concerns, internal struggles, and negotiations with their neighbors brings into view the complex historical choices and processes that produced Europe's vulnerability asymmetries and the ambivalent role of the Iron Curtain as a major European border.

In Part III, "Coping with Complexity," we focus on ICT and its interdependencies with other selected infrastructure, such as air-traffic control, electricity supply, and emergency services. Again, we investigate the vulnerability and governance implications of these processes from different perspectives, including the perspectives of international organization, national concerns, and bilateral negotiations, and cooperation on the microregional scale in cross-border (micro)regions that served as primary organizational units for emergency response.

We conclude this general introduction with one final paradox. The shorthand name for the research program behind this book was "Europe Goes Critical." We were well aware that most readers would interpret this title, at first, to mean European susceptibility to harm following the proliferation of transnational infrastructure. There is, however, a second meaning. In nuclear power engineering a reactor goes critical when it becomes operational (when each nuclear reaction produces sufficient neutrons to trigger a next reaction and keep the nuclear fission process going). Contemporary Europe too, we argue, became possible and operational in the wake of transnational infrastructure, the enormous flow of goods, ideas, energy, information, value, and people that it carries, and the vulnerabilities that it implies for better or for worse. A study of Europe's infrastructure vulnerabilities should indeed embrace both messages in order to capture the width and depth of this remarkable and important historical phenomenon.

Notes

1. Commission of the European Communities 2005, p. 2.
2. Presidential Commission on Critical Infrastructure Protection 1997, pp. x and i. See United States General Accounting Office 1996, p. 2.
3. For the United States, see Perrow 2007.
4. On these blackouts, see Lagendijk and Van der Vleuten, this volume. On gas crises, see Högselius et al. this volume. For evidence that the 2006 events were drawn into ongoing critical infrastructure policy-making, see Directorate General for Energy and Transport 2007, and Hämmerli and Renda 2010.
5. Commission of the European Communities 2006, p. 3. The proposal includes a list of already existing legislation.
6. Bijker 2006 and 2009; Kaijser 2011.
7. Transnational vulnerabilities are highlighted in Gheorghe and Vamanu 2005, p. 218, and Gheorghe et al. 2007, p. 6. Compare the practice of CI studies in, for example, Goossens 2004; Gheorghe et al. 2006; Murray and Grubesic 2007, and the *International Journal of Critical Infrastructures* 2004–2010.
8. Gheorghe et al. 2006 and 2007; Burgess 2007; Fritzon et al. 2007.
9. For this argument, see Van der Vleuten and Lagendijk 2010a and 2010b.
10. Edwards 1998.
11. Edwards 2003.
12. "Treaty on European Union," published in *Official Journal of the European Union* C 191 (July 29, 1992). Compare Van der Vleuten et al. 2007 and Schot 2010.
13. Boin, Ekengren, and Rhinard 2006.
14. Associated Press, "Low-Carbon Economy Proposed for Europe. Eyeing Warming and Volatility, EU Leaders Expected to Approve it in March," January 10, 2007.
15. The proposal included energy, the nuclear industry, ICT, water, food, health, financial services, transport, chemical industry, space, and research facilities. Commission of the European Communities 2006, Annex 1. On the position of the financial sector, see Financial Services Authority, "Protection of Critical Infrastructure Dossier," available at http://www.fsa.gov.uk/pages/About/What/International/pdf/DOECIP.pdf (accessed on February 22, 2011). On the electric power sector, see Van der Vleuten and Lagendijk 2010b.
16. Compare Cole and Ther 2010.
17. This happened from the 1980s. For an overview, see Schipper and Van der Vleuten 2008.
18. This applies to technological risk in general. See Jasanoff 2002.
19. Inspired by Scranton 2011. On paradox as an analytic strategy in organization science, see Scott Poole and Van de Ven 1989.
20. Van Laak 1999.
21. Boin and McConnel 2007, p. 1; Egan 2007, pp. 4–5.
22. For a discussion, see Van der Vleuten and Kaijser 2006, p. 6.

23. Chevalier cited in Mattelart 1996, p. 103. For a vivid argument, compare Chevalier 1832, pp. 35–39 and 41.
24. Badenoch 2010, especially pp. 52–57.
25. Schueler 2006 and 2008.
26. Headrick 1991.
27. Högselius 2005.
28. Van der Vleuten and Kaijser 2006, pp. 2–3.
29. Graham and Marvin 2001; Coutard 2005.
30. Here we pick up on the Large Technical Systems research tradition that sprung from Hughes 1983. For a review, see Van der Vleuten 2006; for its transnational turn, see Van der Vleuten and Kaijser 2005. For other infrastructure paradoxes, compare Edwards 2003.
31. Perrow 1984 and 2007.
32. Beck 1992, p. 21.
33. Cutter 1993; Sarewitz et al. 2003; Leach 2008; Bijker 2006.
34. For the psychology of fear we rely on Gardner 2009. On post-9/11 road accidents, see Gigerenzer 2006.
35. For the culture of fear debates, see, for example, Furedi 2006 and 2007, and Pain 2009. For fear in connection with critical infrastructure, see, for example, Williams 2002 and Burgess 2007. Observe that psychological fear research and its experiments also stem largely from this period, and may itself be an expression of this cultural phenomenon.
36. Sommer and Brown 2011, p. 80.
37. Bruce Schneier of the UK security firm Counterpane cited in "Critical Infrastructure", *Euractive.com*, January 26, 2009.
38. Nye 2010.
39. Bijker 2006.
40. This insight informs ongoing research into sustainability transitions. For example, Grin, Rotmans and Schot 2010; Verbong and Loorbach 2012.
41. Van der Vleuten and Lagendijk 2010b; Lagendijk and Van der Vleuten, this volume.
42. Roberts 1990; La Porte and Consolini 1991; La Porte 1996; Rochlin 1996.
43. Compare the vulnerability talk of Gheorghe et al. 2006 and 2007 with the reliability studies of Schulman, Roe and Van Eeten 2004; De Bruijne 2006, and De Bruijne and Van Eeten 2007. On an even more generic level, several authors highlighted the paradoxical coexistence of our present-day preoccupation with risk and vulnerability in such different realms as terrorism, ecology, infrastructure, and parenting on one hand, and historically unprecedented levels of personal, economic and social security on the other – at least in the developed world. Furedi 2007, p. ix. Gardner 2009, p. 10 speaks of the "greatest paradox of our time".
44. Here we turn to the constructivist tradition in risk studies represented by, for example, Jasanoff 1998, and Summerton and Berner 2003. For an extension to critical infrastructure studies, see Van der Vleuten and Lagendijk 2010b.
45. For a general overview, see Wilson and Van der Dussen 1995, pp. 83ff. Schot and Lagendijk 2008; Kaiser and Schot in press.
46. Myrdal 1968, p. 626. Compare Berthelot 2004.
47. For EU cultural policy, see Shore 2000.
48. Jeffrey Titford (EDD), Debates of the European Parliament, May 30, 2002. Compare Van der Vleuten and Kaijser 2006.
49. Lagendijk and Van der Vleuten, this volume; Högselius, Kaijser and Van der Vleuten in press.
50. For notions of informal and hidden integration, see Wallace 1990, pp. 8ff., and Misa and Schot 2005. For a research agenda on Europe's hidden infrastructure integration, see Van der Vleuten and Kaijser 2005.
51. For this line of research, see the collected essays in Van der Vleuten and Kaijser 2006; Schot 2007; Badenoch and Fickers 2010, and monographs such as Lagendijk 2008; Schipper 2008; Anastasiadou 2011; Lommers 2012; and Janáč 2012.

52. For example, Milward 2000; Applegate 1999; Storm 2003; and Saunier 2002.
53. On the role of state governments in international infrastructure associations, see, for example, Henrich-Franke 2007 and 2008. For national infrastructure building and regulation, see, for example, Millward 2005; for cities as a unit of infrastructure activity, see Tarr and Dupuy 1988; Graham and Marvin 2001; and Hård and Misa 2008.
54. This is the new transnational history proposed by Tyrell 1991 and 2009. For different forms of transnational history, see Van der Vleuten 2008 and Saunier 2009. On this point we deviate from the transnational histories inspired by the transnational turn in political science in the 1970s, which brought non-state actors into international relations and thereby a priori focuses on the international. This work includes studies of transnational networks by infrastructure experts; see Schot and Lagendijk 2008 and Schot and Schipper 2011.
55. Lukasik 2003, p. 208, and Luijf et al. 2009.
56. Commission of the European Communities 2004, p. 4.
57. Bekkers and Thaens 2005, p. 37.
58. For a compelling critique, see Davies 1996.

Part I
Connecting a Continent

Introduction

The chapters in Part I inquire about the emergence and governance of critical European infrastructure vulnerabilities from a transcontinental perspective. We focus in particular on natural gas (Chapter 2) and electricity (Chapter 3) networks. These have been at the heart of Europe's energy issues throughout the postwar era, and they still play a dominant role in its present-day energy supply. For instance, of the 1848 million tons of oil equivalent in the 2007 energy balance of the European Union (EU-27), 74 per cent was transported as natural gas or electricity to agricultural, industrial, service, and household users.[1] Although this figure includes huge transport and conversion losses, electricity and natural gas supplies surely have become "critical" to Europe's economic and social life.

As we shall see, natural gas and electricity system-builders have identified a number of criticalities and vulnerabilities since the very inception of these transcontinental networks. Accordingly, they have taken technical or organizational measures to reduce the vulnerability of their systems to internal and external threats (system vulnerability), and the vulnerability of users to system malfunctions (user vulnerability). The chapters in Part I investigate how and why these huge energy systems emerged, which and whose vulnerabilities were identified, prioritized, anticipated, and/or ignored in this process, and how actors chose to respond to them. In both gas and electricity, as we shall see, the overall logic of transnational system-building was closely linked with and affected by actors' vulnerability concerns. In the early days of both systems, the main challenge was to balance long-term supply and demand in such a way that the growth of the systems could proceed smoothly, and the corresponding vulnerability had to do with a fear of structural energy shortage. Later on there was a shift to a new type of vulnerability in terms of a fear of temporary disruptions, instabilities, and harmful environmental effects.

In Chapter 2, Per Högselius, Anna Åberg, and Arne Kaijser analyze the development of transnational natural gas systems in Europe during the Cold War and the remarkable growth of gas flows from the Soviet Union to Western Europe. They show that the construction of transnational gas pipelines in Europe was initially driven by importers' desires to access high-quality fuel that was not (sufficiently) available domestically. Importing gas from a single supplier was perceived as a risk, however, and as soon as an importer had built a pipeline to a foreign supplier, it felt highly motivated to follow this up with links to additional foreign sources. This strategy paved the way for Soviet natural gas to become an important source of energy in Western Europe – seemingly against the military and political logic of the Cold War: access to Soviet gas was seen as an efficient way to prevent Dutch, Norwegian, or Algerian natural gas from attaining a national or regional monopoly, while also stimulating relations with these alternative suppliers.

In electricity supply, as Vincent Lagendijk and Erik van der Vleuten show in Chapter 3, the possibility of power imports and exports likewise sparked visions of a transcontinental grid even back in the interwar years. Importers saw opportunities to reduce their domestic energy shortages, and visionaries argued that electricity made it possible to evenly distribute Europe's dispersed energy resources to all. However, not all stakeholders subscribed to this motive – with due implications for transnational system-building and vulnerability outcomes. In most parts of Europe, long-distance transmission of vast amounts of electric power remained marginal: on the eve of the neoliberal era in the 1980s, an average of only 5–6 per cent of all electricity crossed a border as an import or export commodity, which means that nearly 95 per cent did not. Although transmission lines connected the Atlantic to Siberia and the Arctic Circle to northern Africa, power companies and national governments used electricity trade only as a supplement to domestic energy autonomy. If cross-border connections nevertheless proliferated, it was to stabilize domestic systems rather than to trade in energy: connections across borders would facilitate emergency support in case of supply disruptions. Moreover, following the laws of physics, a synchronously coupled system would instantaneously counteract and correct any failures in the common voltage and frequency. The larger the system, the more counteraction to local disturbances. The fact that cross-border interconnections also introduced new vulnerabilities, such as cascading blackouts, only made vulnerability concerns more prominent in the building of electricity systems.

A central question discussed in Part I is how Europe's transnational energy infrastructure was governed. Electricity and natural gas systems both consisted of complex networks of trunklines transmitting energy over vast distances, combined with very dense local distribution networks that connected almost every household and every industry in Europe to the system. The systems were based on grids or pipes, which demanded sophisticated coordination. The complex governance structures of both systems functioned simultaneously at several levels, relying on the activities of commercial actors, multilateral organizations, local and regional actors, and national governments. Many of these played important roles in emergent transnational governance modes, taking the form of bilateral relations in the case of natural gas, and bilateral and multilateral relations in the case of electricity. Also, state governments were often more involved in shaping the European natural gas regime than in shaping electricity collaborations. The respective chapters explain why these differences emerged.

Intriguingly, both natural gas and electricity connections were built and used across the Iron Curtain. By the mid-1970s, natural gas networks in Austria, Italy,

West Germany, and France had already been connected to the Soviet and Eastern European natural gas system, and indirectly several other countries were linked to the communist gas infrastructure. These links across the Iron Curtain were often far more important than links between individual Western European countries. East–West integration was not at all as pervasive in electricity as it was in natural gas. The two chapters explain why Western European actors chose to make themselves dependent on steady gas flows from beyond the Iron Curtain, whereas in electricity they were less willing to take such steps. As we will see, again vulnerability considerations took center stage.

The chapters in Part I also analyze the nature, causes, and effects of "critical events." Contrary to common perception, the main types of critical event that actually occurred in natural gas infrastructure were of a similar type as in electricity, taking the form of unintended technical failures and logistic breakdowns. During the Cold War the Soviet Union never used natural gas as an "energy weapon" analogous to the OPEC "oil weapon." The only intentional critical events recorded in the natural gas sector during this period took the form of a series of strikes on Norwegian offshore oil and gas installations in the early 1980s. In the case of electricity, Eastern and Western European collaborations developed in parallel, with different vulnerability management priorities and corresponding implications for blackouts: Western stakeholders tended to focus on minimizing user vulnerabilities in their attempts to build high-reliability organizations, while Eastern stakeholders prioritized the system reliability of the main grid. As a result the Council for Mutual Economic Assistance (COMECON) power backbone rarely went down either, but user reliability was routinely sacrificed, as apartment blocks were intermittently shut off to secure the integrity of the overall system. This is why citizens in several Eastern European states remember the Cold War as the "disco era" – with lights constantly flashing on and off. Finally, major external critical events such as the oil crises of 1973/4 and 1979 had strong impacts on both gas and electricity, forcing system-builders and other actors to adapt to new realities.

What did Europe look like from the perspective of natural gas and electricity interdependencies and vulnerabilities? The two chapters approach this question by showing how the use of pan-European gas and electricity infrastructure, and in particular critical events such as blackouts and gas crises, revealed geographies of Europe that only partially corresponded to familiar political borders. In the gas case a "hidden regionalization" of Europe divided the Continent into three major regions unseen on "normal" European maps. In the electricity case, a similar hidden regionalization took the form of major synchronized blocks that did defy some political borders, such as the tight interconnection between continental Europe and northern Africa, yet reproduced others, such as the Iron Curtain. After 1989, Europe's "Electric Curtain" again followed political events; rather than vanishing, it moved eastward roughly to the borders of the former Soviet Union. The so-called European blackout of November 4, 2006 neatly displayed Europe's resulting electricity geography, cascading from northern Germany southward to the Mediterranean and crossing via the Iberian peninsula into northern Africa;

yet it affected neither the nearby Scandinavian peninsula in the north nor regions to the east of the new Electric Curtain.

Note

1. Commission of the European Communities 2010, p. 41. After conversion and transport losses, users receive just 289 million tons of oil equivalent (Mtoe) of electricity and 269 Mtoe of natural gas.

2
Natural Gas in Cold War Europe: The Making of a Critical Infrastructure

Per Högselius, Anna Åberg, and Arne Kaijser

Introduction

On January 1, 2006, Russian gas company Gazprom hastily decided to interrupt its delivery of natural gas to neighboring Ukraine. During a few dramatic days the Russian move raised concerns in large parts of Europe, since the interruption to Ukraine also had a direct effect on the gas supply to countries located further downstream the same pipeline. On January 2, gas companies in Hungary, Slovakia, and Austria reported a drastic drop in pressure – at a time of peak winter demand for natural gas. The crisis threatened the steady supply of electricity and heat to a vast number of industrial enterprises, power plants, hospitals, schools, households, and other gas users.

The immediate reason for the crisis was the failure to reach an agreement about a renewal of the Russian-Ukrainian gas export and transit contract. This problem, however, was in turn related to the general strain in relations between the two countries following the recent "Orange Revolution," after which Ukraine had embarked on more Western and less Russian-oriented political development. The acute problem of delivery was later solved through negotiations and the conclusion of a new Russian-Ukrainian gas contract, but the crisis gave rise to dismay and perplexity in Europe. Within the European Union (EU), demands for sanctions against Russia were raised. From a German perspective, the incident seemed to confirm the need for a new direct natural gas connection between Germany and Russia through the Baltic Sea – the Nord Stream pipeline – as an alternative to the apparently risky and unreliable transit through Ukraine, Slovakia, the Czech Republic, and Poland. However, in Central European media the proposed Nord Stream pipeline was interpreted as a threat. Poland's foreign minister, Radosław Sikorski, even dubbed the project "the Molotov-Ribbentrop Pipeline," since in his view – and many others' – it was unpleasantly reminiscent of the infamous Soviet-German pact of 1939.[1]

Similar "gas crises" became a more or less regular phenomenon in Europe during the following years, culminating in the much-publicized crisis of January 2009, which affected nearly all European countries in one way or another. From 2010, following the election of a new, more Russian-friendly Ukrainian president, gas relations in Europe seemed to be normalizing again – for the time being.

The supply crises of 2006 and 2009 are indicative of the explosive politics that have come to accompany natural gas in Europe, and of the deep embeddedness of the transnational pipeline infrastructure in European political and economic history. The existence of a European natural gas grid – with a complexity that has increased dramatically during only a few volatile decades – is intriguing: it provides a case of a truly pan-European infrastructure that today includes large-scale transnational flows not only within more narrowly defined European regions (such as the EU or Scandinavia) but also across the former Iron Curtain as well as between Europe and the Arab world. Many European countries have thereby developed strong dependencies on gas supplies from countries that in other contexts were typically regarded as untrustworthy or even as enemies. Excluding the major exporting nations, the average gas import dependence in Europe is now more than 90 per cent.[2] In this sense, natural gas has contributed to a far-reaching "hidden integration" between different European nations and regions as well as between Europe and the world beyond.

This chapter sets out to explain how this remarkable development has been possible and why so many countries have found it acceptable to engage in infrastructure relations that have made them vulnerable. In particular, it aims to explain how system builders were able to transcend political, ideological, and military divides like the Iron Curtain. Who supported and who resisted the emergence of the links, and why? How has the rise of new interdependencies and vulnerabilities stemming from the far-reaching transnationalization of natural gas been anticipated, perceived, and interpreted? To what extent has vulnerability shifted over time, and what have actors done to shape and respond to the perceived risks? And, last but not least, what kind of Europe can be discerned when we look at it through the lens of natural gas?

Previous research on these issues has been surprisingly rare. Whereas natural gas has been widely studied from economic, political, legal, and other social and natural science perspectives,[3] it has been a much neglected object of historical study, particularly when it comes to international aspects. Lacking, in particular, are deeper studies of the very emergence of transnational gas relations and the associated vulnerabilities that these have generated. While a few promising studies of this kind exist,[4] the present text is the first attempt to inquire into the history of natural gas on a European level.

To do this, the chapter uses primary research findings from Dutch (West and East), German, Austrian, Ukrainian, Russian, Danish, and Swedish archives. Our method is based on the principle of symmetry, taking into account documents from "both sides" in a given transnational gas relation. Our ambition has been to document Europe's natural gas history from the perspective of those people and organizations who have been – or tried to be – central in envisioning, negotiating, planning, building, operating, and regulating this infrastructure.[5]

The chapter is structured into several layers, each of which scrutinizes Europe's natural gas history from a different perspective. The first layer consists of a "horizontal" analysis of the European gas grid, centering on gas fields and the material pipeline infrastructure. It focuses on the sequence in which different links were envisaged, built, and used – as well as on links that were not built – and on the opportunities seen by system builders and other actors in the growing intertwinement of regions that were initially isolated from each other. The second layer shifts the emphasis to the governance of the European natural gas system. By analyzing transnational gas contracts, we try to discern how perceptions of vulnerability influenced the institutional setup of the European gas system. The third layer consists of an analysis of how actors, when engaging in transnational gas relations, have coped with vulnerability in practice. We analyze the concrete measures taken to reduce risks and prepare for action in times of crisis. The fourth, and last, layer focuses on the ways in which "Europe" can be understood through the lens of natural gas.

The emergence of a transcontinental network

The late rise of natural gas

Natural gas – or "earth gas" as it is called in many European languages[6] – has been known in Europe for centuries, but its large-scale use is a recent phenomenon. It was usually regarded as an annoying by-product of oil and the challenge was to get rid of it. Hence most natural gas was flared.

The history of natural gas in Europe began in earnest in the interwar years, when entrepreneurs in a few European regions started experimenting with it for various purposes, mainly as a substitute for manufactured gas.[7] The Second World War provided a further major impetus to the development, as natural gas was recognized, in those countries where gas finds were known, as a safe domestic energy source that could counteract the increasingly problematic reliance on imported coal and oil.

The first major natural gas networks to take form in Europe were built in countries (or regions within countries) where coal was not available regionally, particularly in Romania, Austria, northern Italy, southwestern France, and the eastern regions of interwar Poland. In Britain, Germany, Belgium, western Czechoslovakia, southern Poland, northern France, the Soviet Donbass industrial area, and other coal-rich regions, the coal-based gas infrastructure continued to predominate over the emerging natural gas networks up to the 1960s.

Up until this time, European gas infrastructure consisted of a number of isolated national or regional networks that seemed far from interconnected. In the years around 1960, however, vast new natural gas resources were discovered in several regions in Europe and beyond. The northern part of the Netherlands, the Sahara, eastern Ukraine, Central Asia, and northwestern Siberia emerged as particularly promising new gas regions. The estimated reserves of these fields were so large that, for the first time, substantial exports of natural gas seemed to become economically feasible.

In the absence of opportunities to import natural gas from far away, it would have remained a negligible source of energy in most European countries. As a result of international trade, however, the share of natural gas in the overall primary energy balance of Europe grew from around 3 per cent in 1960 to 25 per cent in 2008 – an increase that was even more dramatic in terms of the absolute gas volumes consumed. The growth in gas use and gas imports radically increased Europe's vulnerability to supply cutoffs, as an ever greater number of users and activities became dependent on reliable flows of gas. Imported natural gas began finding application as a fuel in a variety of industrial sectors (for the production of metals, cement, glass, etc.) as well as in households (for cooking and heating) and the energy sector itself (heat and power plants). There were also efforts to introduce natural gas for transportation purposes, though this has only recently seen some success. Apart from its use as a direct energy source, natural gas was used as a crucial feedstock in the chemical industry, particularly for the production of fertilizers (where the energy content of the gas was used indirectly to boost agricultural production).

The formative phase of transnational links

Europe's first major transnational pipelines were built to bring natural gas from the vast Groningen field in the Netherlands (discovered in 1959) and from a number of smaller gas fields in western Ukraine (which had been known since before the war) to major consumption centers in Western and Eastern Europe, respectively. Groningen gas was transported to neighboring Belgium and Germany from 1966, and to France via Belgium from 1967 (Figure 2.1).[8] In the east, Ukrainian gas was brought to Poland from 1944 and to Czechoslovakia from 1967. In the Polish case, the pipeline in question had been built as a domestic link before and during the war, but the postwar border shifts transformed it into a transnational link.

Within both the capitalist and communist parts of Europe, 1966–1967 thus marked a breakthrough in transnational system building. The pipelines from Groningen went to countries that, like the Netherlands, were members of NATO and the European Economic Community (EEC), and the lines seemed to fit neatly into a pattern of earlier transnational energy cooperation, notably in the form of the European Coal and Steel Community, the European Atomic Energy Community (EURATOM), and joint facilities for uranium enrichment and plutonium production. Soviet gas exports to Poland and Czechoslovakia similarly built on earlier experiences of cooperation within the Council for Mutual Economic Assistance (COMECON), whereby the much publicized "Druzhba" (Friendship) oil pipeline system played a special role.[9] The gas link to Czechoslovakia that was opened in 1967 was optimistically referred to as the "Bratstvo" (Brotherhood) pipeline.

In the next phase, however, the development took a different turn. The initial trend towards the formation of two regional natural gas networks – a "capitalist" and a "communist" network, separated from each other by an "Iron Curtain" – could not be sustained. The political logic in terms of the choice of partners gave way to a dominance of economic and geographical factors in transnational system-building.

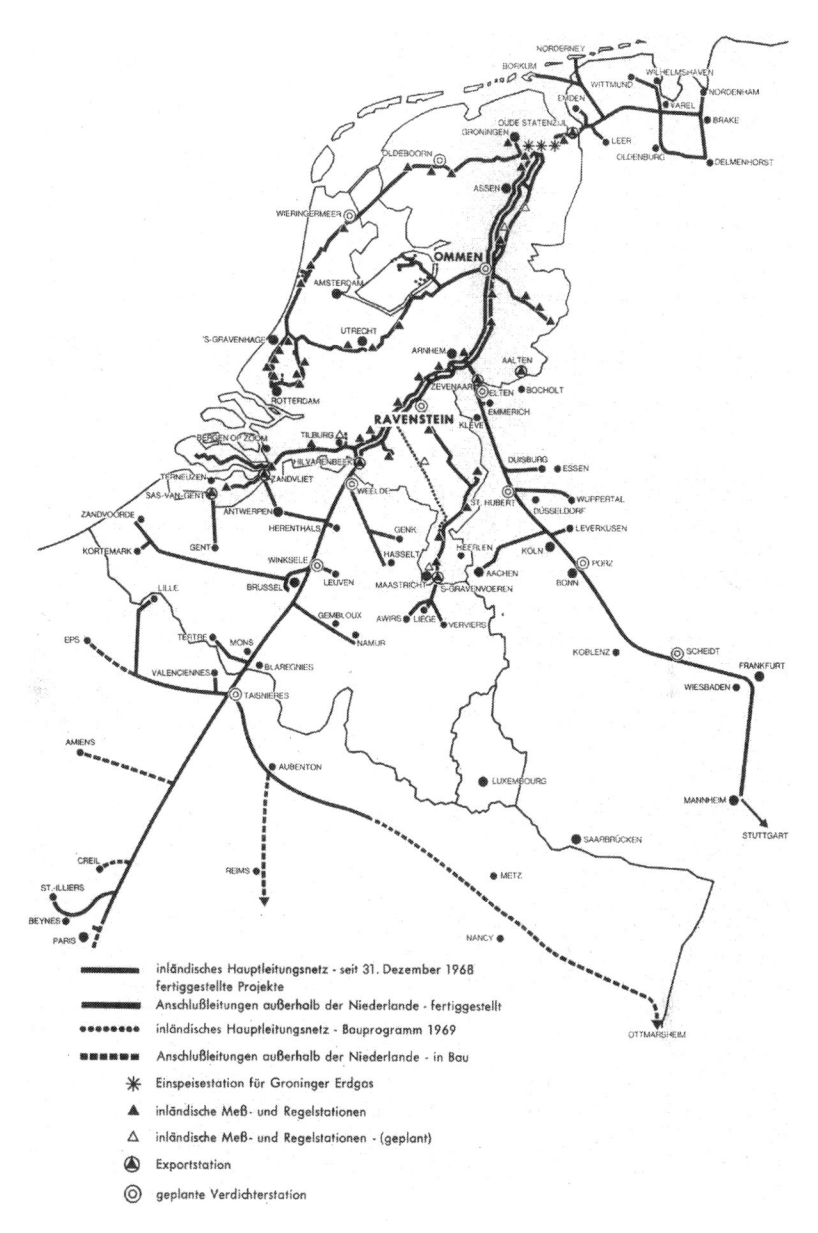

Figure 2.1 The natural gas export infrastructure with pipelines from the Netherlands to West Germany, Belgium, and France, as of 1968, with plans for 1969.
Source: *gwf*, November 21, 1969, p. 1302. Used by permission of DIV Deutscher Industrieverlag GmbH.

A Western European interest in Saharan gas had been obvious following the discovery of the large Algerian gas field Hassi R'Mel in the late 1950s, and possible trade regimes, most of which focused on shipments of liquefied natural gas (LNG), were discussed intensively, both nationally and within international organizations. The UN's Economic Committee for Europe (ECE) in 1956 launched an "Ad Hoc Working Party on Gas Problems," which became an important forum for discussing the prospects for Saharan gas with participants from both Western and Eastern Europe. The committee discussed the characteristics of gas systems in individual countries, legal aspects of transnational pipelines, and other fundamental aspects of a possible international gas trade.[10] Two pipelines were envisioned from Algeria to Britain, France, Spain, and Italy. In the end, the export of Saharan gas began not by pipeline, but in liquefied form (LNG). The task of laying a pipeline across the extreme depths of the Mediterranean was found to be too difficult, at least for the time being. A number of LNG agreements were concluded from the early 1960s with Libya and particularly with newly independent Algeria, for LNG imports to England, France, Italy, and Spain. These contracts were often negotiated in parallel with negotiations for Dutch gas, with the importing countries seeking to play different exporters off against each other and thus establish a competitive market.

However, although the hopes for Algerian gas to play a key role in Europe's gas supply were high, it turned out to be much more difficult than initially expected to turn the grand visions into reality. Uncertainties about the competitiveness of Saharan gas on European markets meant that many potential importers hesitated, whereas the Algerians suspiciously watched Western European attempts to exploit their former colony. Several prospective deals therefore failed and some contracts that had already been signed were cancelled before exports had begun. The agreements that *were* concluded in the 1960s and early 1970s were fairly small, particularly in comparison to the rapidly growing Dutch gas exports. The development took a more dynamic turn in 1976, when the Italian and Algerian state-owned energy companies ENI and Sonatrach, together with Tunisia as a transit country, concluded a major contract for Algerian gas exports to Italy and the construction of a submarine pipeline to Sicily and further on to the Italian mainland. The Trans-Mediterranean Pipeline, as it was called, was eventually inaugurated in 1983, and two decades later the volume of gas flowing through it corresponded to around a third of Italy's total gas demand. A similar pipeline project between Algeria and Spain, by way of transit through Morocco, initially failed but was finally realized in the late 1990s. In 2004, natural gas also started flowing through a pipeline from Libya to Italy.[11]

As a third potential supplier, in addition to Dutch and Saharan gas, the Soviet Union emerged. The export of Soviet natural gas across the Iron Curtain to Western Europe is one of the most intriguing aspects of Europe's natural gas history, and the motivations for Western European actors to import Soviet gas should therefore be discussed in somewhat greater depth here.

The capitalist country with the strongest motivation to import Soviet gas was Austria. Lacking domestic coal resources, it had been one of the forerunners in the

European natural gas industry. It possessed fairly large gas deposits in the region around Vienna, but the growing popularity of this fuel soon became a problem. Austrian state-owned oil and gas company ÖMV found itself struggling to meet the ever-growing gas demand, which in the long term could not be met through domestic production. When ÖMV heard about the Soviet-Czechoslovak Bratstvo project, it was extremely interested since the pipeline from the Soviet Union was to terminate just a few kilometers from the Austrian border – the distance separating the Bratstvo system from the Austrian gas network was only 5 km.[12] In addition, Austrian and Czechoslovak gas enterprises had already developed close cooperation to jointly exploit a large gas field situated right at the border between the two countries.[13] It seemed natural to extend this cooperation to involve cross-border pipeline construction as well. In 1965, ÖMV therefore inquired in Moscow whether there was any opportunity for Austria to become part of the Soviet-Czechoslovak gas brotherhood.[14] However, at this time the Soviet gas industry was struggling to meet even domestic demand, so the Soviets did not respond positively to ÖMV's request.

Soon afterwards, however, negotiations between the Soviet Union and Italy began, with the goal of exploiting the vast recent discoveries of natural gas in Siberia.[15] Italy's ENI was the Western European oil and gas company with the best relations with the Soviet Union and was already both a major oil importer and an exporter to the Communist Bloc of advanced equipment for the oil and gas industries.[16] Italy also had a strong Communist Party that sought ways to strengthen relations with the Soviet Union. ENI and the Soviets were discussing a pipeline that was referred to in both Western and Eastern media as the Trans-European Pipeline. It was to originate in the Siberian gas regions and to reach northeastern Italy by way of transit through Hungary and Yugoslavia.[17]

When ÖMV heard about this project in summer 1966, it initiated a new charm offensive vis-à-vis Moscow, while also approaching Rome in the hope of becoming part of the Soviet-Italian project. ÖMV managed to achieve this goal by cooperating with state-owned Austrian steel company VÖEST, which offered the Soviets large amounts of steel pipe, to be used for the gas pipeline, in return for gas imports. As a matter of fact, VÖEST did not have the capacity to produce these pipes but only the thick steel plates that served as an intermediary product. However, VÖEST's director, Rudolf Lukesch, agreed on a cooperative deal with the pipe-producing plants belonging to the large German steel companies Thyssen and Mannesmann. These operated Europe's most modern factories for the production of large-diameter steel pipe, which the Soviets were keen to obtain for the exploitation of its Siberian gas fields.[18] Hence Germany also became indirectly involved in the Trans-European project. For the Germans, this was seen as highly advantageous at a time when a NATO embargo on pipe exports to the Communist Bloc was in force. The embargo had been imposed on the initiative of the United States following the construction of the Berlin Wall in 1961 and the Cuban Missile Crisis in 1962.[19]

Bavaria's minister of economy, Otto Schedl, sought to extend the Soviet-Italian-Austrian-German arrangement by inquiring about possibilities for southern

Germany to import Soviet natural gas. Bavaria was an underdeveloped region in postwar Germany, and Schedl thought that the key to a modern, industrial Bavaria was access to competitively priced energy. Bavaria was traditionally dependent on north German coal, which had to be transported over long distances and therefore gave it a competitive disadvantage vis-à-vis northern Germany. In the early 1960s, Schedl had managed to arrange for the import of cheap oil from the Middle East by way of pipelines from Mediterranean harbors, and he identified access to Soviet gas as a further way to strengthen Bavaria's energy independence and industrial competitiveness. Schedl was a Christian Democrat but he believed in the benefits of cooperation with Soviet communists; to him, Russia and Germany belonged to the same cultural sphere and should therefore cooperate, whereas the real danger to European civilization was the threat from China, where Mao Zedong at this time had just proclaimed his "Cultural Revolution."[20]

The German government, however, which at this time was a grand right–left coalition headed by Chancellor Kurt Georg Kiesinger of the Christian Democrats, found it much too risky and uncertain to import natural gas from the East. The government's energy experts at the Federal Ministry of Economy thought that the Soviet Union might, on the one hand, use the threat of interrupted gas supplies for political blackmail; on the other hand, they might seek to flood the German market with cheap natural gas, deliberately aiming to disturb the politically sensitive coal industry in the Ruhr, which was already facing severe difficulties due to inefficiencies and competition from abroad. Moreover, Ruhrgas – owned by the coal industry but also partly by Shell and Esso, companies that had come to dominate the natural gas industry in both Germany and the Netherlands following the Groningen discovery – argued that southern Germany could be supplied more efficiently through domestic German gas and imports from the Netherlands. Otto Schedl's attempts to obtain Soviet gas thus failed – for the time being.[21]

Austria was luckier. ÖMV managed to persuade the Soviets and the Italians to build the "Trans-European" pipeline through Czechoslovakia and Austria rather than through Hungary and Yugoslavia. ÖMV saw a chance of establishing itself as a future hub in the envisaged East–West gas trade, in which France was also expected to become involved. The Soviets were willing to support this plan not least because at the time Austria was actively seeking closer association with the EEC. For the Soviets, the incorporation of Austria into the Eastern European energy system became a way to counteract this political trend. Natural gas was thus becoming a pawn in the geopolitical European game (Figure 2.2).[22]

ÖMV concluded a deal with the Soviet Union in June 1968, and the gas started flowing on September 1 the same year – only ten days after Warsaw Pact forces invaded Czechoslovakia.[23] Italy's ENI failed to reach a corresponding agreement with the Soviets and the negotiations broke down in late 1967. In 1969, however, the Soviet-Italian talks were reinitiated. At the same time, new trends in the West German government's Eastern Policy allowed for a re-evaluation of the opportunity to import gas from beyond the Iron Curtain. The key persons behind the new policy were the minister of foreign affairs, Willy Brandt, his close advisor, Egon Bahr at the Foreign Office, and the state secretary Klaus von Dohnanyi at the Ministry of Economy – all of them Social Democrats who had previously worked

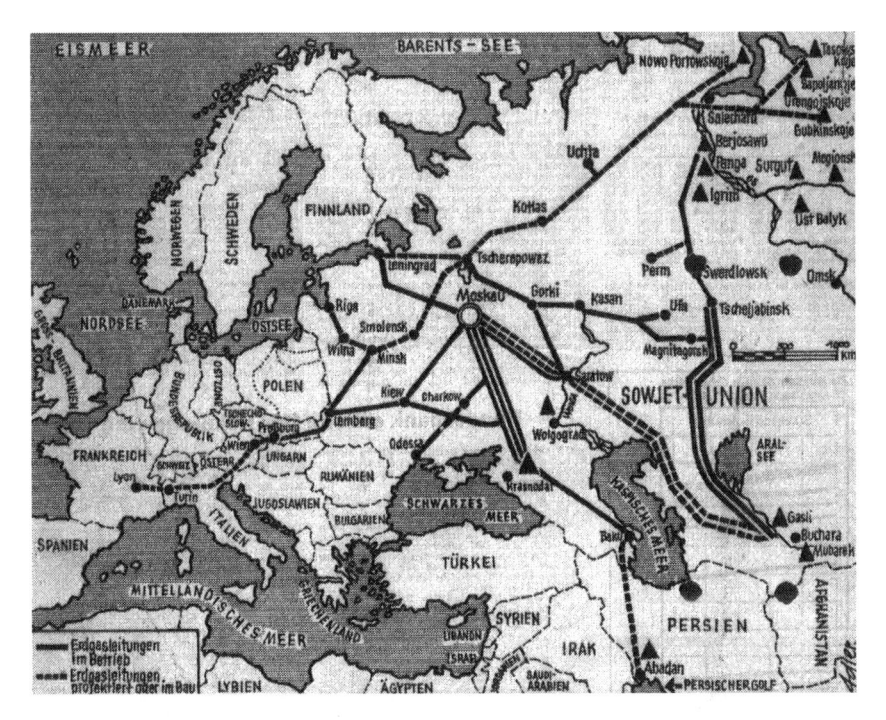

Figure 2.2 In 1967 it still appeared unlikely that West Germany would have access to Soviet natural gas. In media reports, the Soviet exports were generally expected to take a more southern route from Vienna to Italy and from there to France. Here a sketch published in *Süddeutsche Zeitung*, April 22, 1967. Note the unclear status of East Germany and Poland on this map. The GDR was recognized by West Germany only in 1972; up to then it was referred to as "The Eastern Zone."
Source: *Süddeutsche Zeitung*, April 22, 1967. Used by permission.

together in the senate of West Berlin. Brandt, who advanced to become chancellor in 1969, integrated the East–West gas pipeline scheme into his new "Ostpolitik." The Bavarian minister of economy, Otto Schedl, the Christian Democrat who had initiated the West German overtures to purchase Soviet gas back in 1966, was largely bypassed in the process, which was a matter of concern to the regional Bavarian gas industry as well.[24]

Agreements about Soviet gas exports were reached with both Italy and West Germany around the beginning of 1970. Interestingly, the United States did not object to the German deal. Esso, which regarded Soviet gas as a competitor to its own German activities, lobbied Washington in an effort to prevent the deal, though to no avail. The German government inquired at the US embassy whether it would object but was given the green light from the Nixon administration.[25]

Finland also negotiated a gas deal with the Soviet Union. Although relations between Finland and the Soviet Union were peaceful, guided as they were by the Agreement of Friendship, Cooperation and Mutual Assistance of 1948, the discussions surrounding the natural gas pipeline were lively.[26] The official discussion mainly concerned operational reliability, but there was an underlying fear regarding the consequences of being totally dependent on the Soviet Union, which might be tempted to use the pipeline to exert political pressure.[27]

Despite these discussions, a deal with the Soviet Union was signed in 1971, and in terms of price the Soviets were "surprisingly accommodating," as one Finnish negotiator described the situation.[28] The Finns thus seem to have expected tougher price negotiations. Despite a clear upward trend in the market price for natural gas in Europe, the final cost was roughly equal to that negotiated by the Austrians three years earlier.[29] This seems to indicate that the Soviet Union was eager to capture the Finnish market. The Soviets were probably motivated not only by the prospects of increased technology imports from Finland but also by the possibility of using Finland as a point of entry into the larger Scandinavian market.[30] For Finland, the project was more than a way of importing energy. As in the Austrian case, it was an important means to balance bilateral trade with the Soviet Union, as well as a demonstration of friendly relations between the two countries.[31]

The first cubic meters of Soviet gas flowed into West Germany in late 1973 and into Italy in spring 1974. Finland also began importing Soviet gas in 1974, and France followed in 1976. Negotiations with Sweden were held but ultimately failed. Except for Finland, all importing countries received their gas through the same export pipeline, which went through Ukraine and Czechoslovakia. Within the Eastern Bloc the German Democratic Republic (GDR) was linked to this system in 1973, using a different pipeline but the same route. In 1974, Bulgaria started importing Soviet gas, through a more southern pipeline through Moldova and Romania. Hungary, which was already importing Romanian natural gas, followed suit in 1975. In 1978, Yugoslavia also became an importer of Soviet gas, whereby its southern provinces received its shipments through Hungary, and the northern regions by way of transit through Czechoslovakia and Austria.

North Sea gas and the Nordic failure to create an integrated gas infrastructure

The oil crisis in 1973/4 further boosted the popularity of natural gas in many European countries. Natural gas was seen as a suitable way of diversifying away from oil and in particular from reliance on the Organization of the Petroleum Exporting Countries (OPEC). An advantage was seen in the fact that the oil exporters coincided only to a limited extent with the major gas exporters. The only notable gas exporter that was also a major OPEC oil exporter was Algeria, but OPEC coordinated only oil exports, not gas. Vulnerability considerations regarding oil hence contributed strongly to an expanded European natural gas trade. In addition, the high hopes for cheap nuclear power met with disappointment in the form of technical problems and environmental criticism. Coal was also subject to environmental concerns. This served to further boost the popularity of natural gas as an alternative, ecofriendly fuel.[32] The resulting increase in natural gas demand, however, also meant that the level of import dependence increased further and that ever more remote gas fields had to be linked to the major consumption centers.

In the Netherlands the oil crisis, in combination with growing concern about the (un)safety of nuclear power, triggered an intense debate about energy policy in the mid-1970s. The result was a major policy change concerning gas exports. Since cheap nuclear power was no longer seen as a probable future option and OPEC had demonstrated the vulnerability of oil imports, the Netherlands decided to save as much as possible of its gas resources for the future rather than export it. Dutch gas company Gasunie was obligated to fulfill the gas contracts that it had already signed but was instructed not to sign any additional export contracts.

From this perspective, the discovery of vast oil and gas fields in the North Sea starting in the late 1960s was highly welcome for those regions that had become heavily dependent on imports from the Netherlands. Gas was discovered in the British, Danish, Dutch, and Norwegian sectors of the North Sea. On this basis, the creation of a new, submarine pipeline network was initiated.

For Northern Europe, which, with the exception of Finland's link to the Soviet Union, had so far been isolated from the rest of European system-building, the North Sea presented a major opportunity. In the first half of the 1980s, exports of Danish natural gas to southern Sweden and Germany started. There were also attempts to bring about an integration of North Sea and Soviet gas, by way of pipelines through Sweden and across the Baltic Sea (Figure 2.3). Finland was interested in this possibility since it seemed to offer a way of decreasing the country's total dependence on Soviet gas. The Swedes were also interested in the idea, which prompted them to approach the Finns during their negotiations with the Soviet Union in 1970–1.[33] However, the vision ultimately did not materialize, mainly as a consequence of the small amounts of gas that the Soviet Union offered the Swedes.[34]

The pipeline through which Danish gas was exported to Sweden, the Sydgas (South Gas) pipeline, was constructed between 1980 and 1985. It is still the only transnational pipeline connecting two Nordic countries. Despite many grandiose plans and visions aimed at the creation of a Nordic gas system, no further pipelines have been built.[35] In light of pipelines being constructed across the Iron Curtain and the Mediterranean, it is remarkable that neighbors such as Denmark, Finland, Norway, and Sweden – with friendly relations and a similar culture – have not been able to connect to each other in the field of natural gas. There have been various reasons for this failure. Overall, the projects have been strongly influenced by the debate about nuclear power, renewable energy interests, and competition with other pipe-bound energy systems, such as electricity and district heating. But there have also been economic and market arguments working against the grand Nordic visions.[36]

The gas industry facing political turbulence

With large reserves of Dutch and North Sea gas available, it appeared acceptable, from a security of supply perspective, to expand the import arrangements from non-European sources. Contracts with Soviet and Algerian suppliers were extended. From the mid-1970s the quest for Middle Eastern and in particular Iranian gas also intensified, which was either to be transported to Europe by pipeline or shipped in liquefied form in LNG tankers. In this, Western European gas companies faced competition from both the United States and resource-poor Japan. The former was also interested in imports of Algerian and, possibly,

38 *Natural Gas in Cold War Europe*

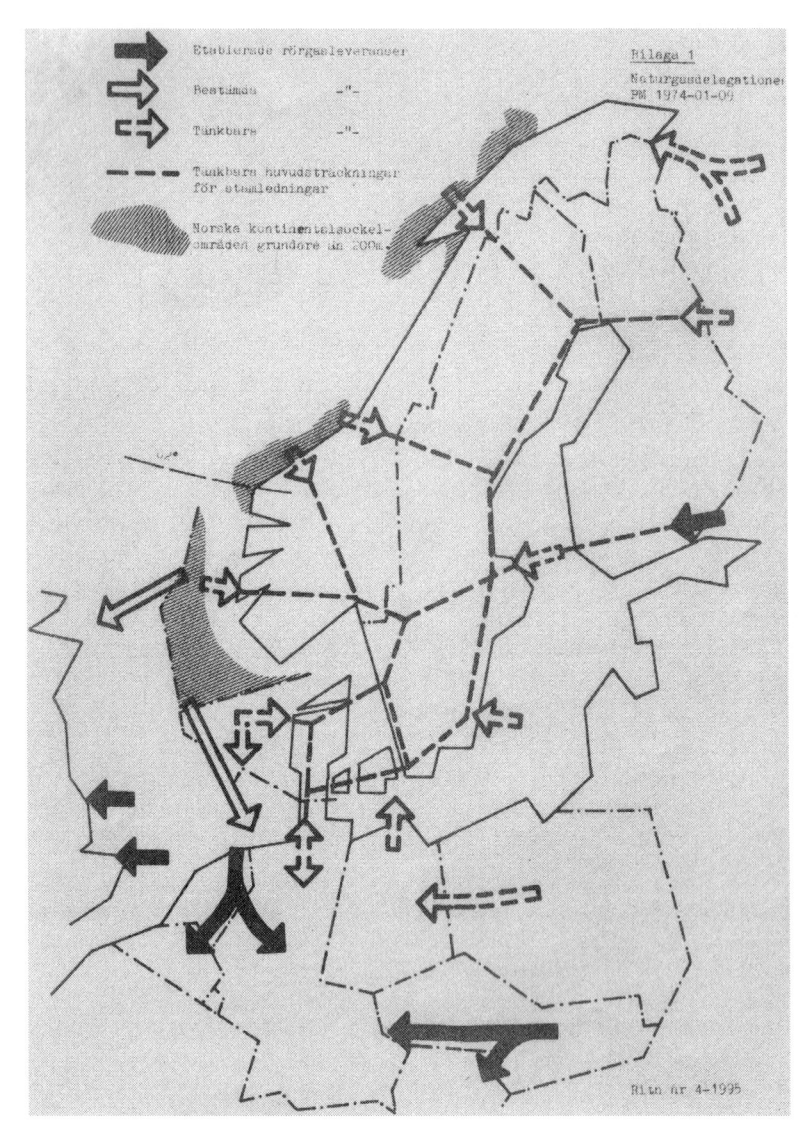

Figure 2.3 Possible natural gas flows through the Nordic countries, as envisaged by the Swedish Natural Gas Committee in 1974.
Source: Swedish National Archives, Stockholm. Used by permission.

Soviet LNG. Natural gas seemed to be on its way to becoming an increasingly globalized fuel.

Two competing pipeline projects were being negotiated for the import of Iranian gas to Western Europe. The Soviet Union, which had imported natural gas from Iran for the supply of its trans-Caucasian republics (Georgia, Armenia, and Azerbaijan) since 1970, sought to exploit this experience by offering itself as a transit country for Iranian gas to the West. The main alternative was transit through Turkey, from where the gas could either be piped through southeastern Europe or brought in liquefied form to Mediterranean harbors in Italy or France. The large German gas company Ruhrgas was the main proponent of the first alternative, whereas Italy's ENI supported the Turkish project.

In the end the Soviet alternative seemed to have the best chances of being realized, so that a contract was signed in November 1975 in which Ruhrgas, ÖMV, and Gaz de France were to import 11 billion cubic meters (bcm) of Iranian gas annually.[37] Construction began on the requisite pipeline in Iran and plans for natural gas system building in the importing countries were adapted accordingly. In 1978–9, however, Iran was shaken by revolution. The Shah was forced to abdicate and political power was seized by Ayatollah Khomeini. Iranian deliveries of gas to the Soviet Union along the already existing pipeline were interrupted and construction of the new one stopped completely. All contracts were declared invalid by the new Iranian regime, which from now on intended to use its natural gas for domestic purposes only.[38]

At about the same time, the East–West political climate started to worsen, particularly after the Soviet invasion of Afghanistan in December 1979 and the election of Ronald Reagan to the US presidency a year later. The United States had so far taken a more or less passive stance on Western Europe's imports of "communist" gas. The Carter administration had "reckoned that increased exports of Soviet energy would ease supply–demand pinches worldwide and lead to the moderation of energy prices."[39] Reagan had a totally different interpretation of the flourishing East–West gas trade and thus tried to halt its further growth. This was due in part to his own preferences, but it was partly also a result of anti-Soviet tendencies in US politics more generally. Opposition to the East–West gas trade peaked after martial law was declared in Poland in December 1981, at which time the largest East–West gas deal ever – for the construction of the "Yamal" pipeline – was just about to be finalized. The Polish crisis gave rise to new fears of military confrontations in Europe. Shortly afterwards, the CIA reported to Reagan that an

> increased dependence on Soviet gas will almost certainly influence European decision-making, despite likely efforts to provide a cushion against supply cutoffs. The Soviets conceivably could exacerbate European differences with the US over future economic sanctions against the USSR or even over more sensitive issues such as NATO force modernization.[40]

Fearing political dependence of Western Europe on the communist world and a divergence in terms of loyalties, Washington launched a major campaign to prevent construction of the Yamal pipeline. However, the Americans failed to convince their European NATO partners, who perceived the project as a logical extension of the cooperation established already, and did not view the vulnerabilities as all that alarming.[41] The much publicized export pipeline began operation in 1984 and, between 1985 and 1990, Soviet gas exports nearly doubled.[42]

The fall of the Berlin Wall in 1989 and the collapse of the Soviet Union in 1991 put the European gas system under stress. Given the political and economic chaos

Figure 2.4 A Soviet stamp from 1983, issued to commemorate the completion of the major export pipeline – the "Transcontinental Export Pipeline" as it is called here – from Siberia to Western Europe.

in Central and Eastern Europe, and the appearance of new transit countries on former Soviet territory, Western European gas companies and users feared that the transnational pipeline infrastructure would cease to function. But in the end the infrastructure proved robust and the end of the Cold War did not result in any major discontinuity in Europe's natural gas history. Supply patterns continued to evolve along the same lines as before. Russia, Algeria, Norway, and the Netherlands remained the main suppliers.

All in all, the emergence of the European natural gas grid presents a stark contrast to the case of electricity (see Chapter 3). The natural gas system became a truly pan-European system, stretching from Siberia to Ireland, with large-scale transnational flows of fuel. Tables 2.1 and 2.2 illustrate the general trends in the changing gas import and export structure in Europe. Europe as a whole has become much more dependent on imported natural gas than before. Moreover, whereas trade within Western Europe still accounted for 67 per cent of all imports in 1982,

Table 2.1 Western European gas imports in 1982, by exporting country (bcm).

	Western Europe	Soviet Union	Algeria	Libya
Austria	0.1	3.0		
Belgium	9.2		0.4	
Finland		0.7		
France	10.0	3.7	6.7	
West Germany	32.6	10.9		
Italy	4.9	8.6		0.1
Netherlands	3.6			
Spain			1.3	1.0
Switzerland	1.2			
Britain	10.7		16.0	
Total	72.1	26.8	8.4	1.0

Source: IEA 1984.

Table 2.2 Western European gas imports in 2005, by exporting country (bcm).

	Western Europe	Russia	Algeria	Nigeria	Qatar
Austria	1.8	4.9			
Belgium	10.9	0.9	3.4		
Finland		4.4			
France	20.4	10.2	8.3	3.9	
Germany	49.2	38.3			
Italy	14.1	23.9	28.1		
Netherlands	17.2	3.5			
Portugal			3.0	1.9	
Spain	2.4		16.4	5.9	5.3
Sweden	1.0				
Switzerland	2.8	0.3			
Britain	12.4		0.4		
Total	132.2	86.4	59.6	11.7	5.3

Source: UN 2006.

this figure had fallen to 45 per cent in 2005 – despite the surge in North Sea gas production and Norwegian gas exports. Of the non-Western exporters, Russia was the most important one, with a share of 29 per cent.

Transnational governance

How was it possible for such an extensive, nearly pan-European natural gas system to be created, given the far-reaching interdependencies and vulnerabilities with which the international gas trade was linked? One important prerequisite was that actors in different countries were able to build mutual trust and understanding, and to develop appropriate forms of transnational governance. In parallel, those involved in the gas trade created mechanisms and strategies that they could pursue on their own to cope with infrastructure vulnerabilities. In the next section we discuss institutional mechanisms for governing the transnational gas infrastructure. In the subsequent section we then turn to technical arrangements for coping with vulnerability and uncertainty.

Developing reliable partnerships

Given the uncertainty in trying out something completely new with a variety of different partners, both importers and exporters of natural gas perceived a high degree of "system vulnerability" (see Chapter 1) when embarking on the first transnational pipeline projects. As in other radical technological projects, no one could be certain that the new system would work as envisioned. An advantage, however, was that the "user vulnerabilities" involved were still fairly low. This was a consequence of the fact that natural gas was still of negligible importance to Europe's overall energy supply. In other words, even in case of a major gas disruption, the actual effects on economy and society were bound to be limited. The low level of user vulnerability made it easier for system builders to experiment with transnational arrangements, particularly in cases where the cooperative projects did not demand any major investments. As the volumes of gas being consumed and traded grew, however, so did the vulnerabilities and the need to respond to them.

The enormous investments necessary for the construction of transmission pipelines, compressor stations, and other components in the gas infrastructure demanded an atmosphere of trust between the involved partners. The pipelines would have no alternative use if gas exports should cease for any reason, and this would have a severe economic impact on both exporters and importers. Reducing vulnerability was therefore first of all a matter of developing reliable partnerships. Exporters had to persuade importers of their intentions to actually deliver the gas, and of their technical and organizational capacity to do so. Conversely, importers had to persuade exporters that they actually intended to receive and pay for the gas and that they were technically able to do so.

It was easier to establish trust when the partners had experience of prior cooperation in areas other than the gas trade. Much of Western Europe's imports of Soviet natural gas, for example, emerged as an extension of an already well-established oil trade. ENI in Italy and ÖMV in Austria were examples of Western European importers of Soviet gas that had long years of experience in dealing with the Soviets for the purpose of oil imports.

Conversely, it was more difficult to build trust when such earlier experience was lacking. In some cases, vulnerability considerations of a political nature thereby prevented transnational pipeline projects from being realized. For example, the German Ministry of Economy was initially suspicious about Bavaria's desire to import Soviet natural gas:

> It may be expected that the Soviet price bid, for political reasons, will be manipulated to be sufficiently low, if there is a serious intention to deliver natural gas to the FRG. In the case of a far-reaching dependence in the gas supply of the Federal Republic or in parts of it upon Soviet deliveries, it must be feared that different political considerations from the Soviet side could lead to an increase in price or to a curbing or suspension of deliveries.[43]

This was the main reason for Bonn's disapproval of Bavaria's efforts to conclude a contract with the Soviet Union in 1966–7.

In the case of Western European imports from Algeria, it was even more difficult to establish trust than in the case of imports from the Soviet Union. Algeria had been a French colony until 1962, and the country's independence was followed by political turmoil and economic chaos. The new political leaders often changed their minds regarding what was to be regarded as a "fair" gas price. As a result, several "agreements in principle" and even a few final contracts with Algerian state oil and gas company Sonatrach did not ultimately materialize. In Germany an LNG terminal under construction at Wilhelmshaven for receiving Algerian gas was never completed. Several Western European gas companies with an interest in Algerian gas eventually judged that cooperation with Algeria was too risky.[44] Although far-reaching visions for a key Algerian role in Europe's gas supply had been promoted since the 1950s, it was not until the 1980s that exports from Algeria actually experienced a breakthrough.

In the longer term, hesitant would-be importers were often convinced of the trustworthiness of a certain exporter by its performance in terms of exports to other countries. Pilot projects, in which the economic stakes were not that high, were thus of a certain significance. Transnational gas relations often began with agreements between countries (or regions within countries) whose existing gas infrastructures offered convenient interconnection possibilities. In such cases the necessary investments in cross-border pipelines were not very large and the perceived economic risks were low. For example, when Austria began importing natural gas from the Soviet Union by way of Czechoslovakia, only 5 km of new pipelines had to be built to interconnect the already existing national infrastructures of Austria and Czechoslovakia. The trade, which started in 1968, was reported to function satisfactorily, and this became an argument for the German Ministry of Economy to change its mind regarding the Soviet Union as a gas exporter. The perceived trustworthiness of this country then increased gradually. By 1969, an import from the East corresponding to up to 10 per cent of total German demand was considered acceptable from a security perspective.[45] Three years later the perceived vulnerability had decreased, so that a level of 14 per cent was not regarded as problematic.[46] By the early twenty-first century, Russian gas covered around 35 per cent of total German gas demand.[47]

The foreign partners also included participants in third countries that were needed for transiting the gas. Some transit routes were seen as too risky. West Germany, for example, did not wish to import Soviet gas by way of the GDR at a time when West Germany had not even recognized the existence of its neighbor as a sovereign state. Importing gas by way of Czechoslovakia, in contrast, was politically acceptable. Similarly, the GDR and the Soviet Union favored Czechoslovakia as a transit country over the politically less reliable Poland. The plan for the GDR's gas supply was originally to transit the gas through Poland, but after the violent strikes in northern Poland in 1969, the Soviets changed the plan to supply the GDR instead through the much longer, but arguably safer, Czechoslovak route.[48] Relations with transit countries thus strongly influenced the geography of the European network in natural gas.

Contractual arrangements

Carefully designed long-term bilateral gas contracts became the core feature governing the emerging transnational infrastructure. The first of these contracts, which were typically signed for a period of 20–25 years, were negotiated between the Dutch gas company NAM and importing companies in Germany, Belgium, and France. As Esso and Shell were dominant shareholders in both NAM and several of the importing companies, they were able to exert a strong influence on both the general design and the specific contents of the contracts.

The Dutch export contracts became a model for governing transnational gas relations in Europe. Soviet exports, in particular, were largely modelled on the Dutch experience. When the Soviet Union started to consider gas exports to Western Europe, the first thing Moscow did was to arrange a meeting with representatives from NAM and Thyssengas (the first large German importer of Dutch gas), seeking to understand how Dutch exports had come about and how they were organized.[49] When Ruhrgas in West Germany negotiated the price of gas with Soviet foreign trade organizations, the Dutch export price was also a central point of reference to which all other issues had to refer.[50] The Dutch contractual model also spread to countries that were not directly involved in the original deal with NAM, such as Sweden and Denmark.

The contracts contained extensive paragraphs defining technical aspects such as gas quality and how it was to be measured, but the key features concerned the gas price. The governments in importing countries took an active role in assuring a "harmonious" entry of Dutch gas onto their fuel markets. Hence the gas price would have to be competitive but not too low. To reduce the risk that the gas would outcompete, or be outcompeted by, other energy sources, it became important to adapt the gas price to the price of other fuels, of which the most important was fuel oil.

This was fairly unproblematic as long as oil prices remained stable. The first contracts for exports of Dutch and Soviet gas that were signed in the mid-1960s did therefore not explicitly link the gas price to the price of fuel oil, although they allowed for a revision of the gas prices "in the event that economic circumstances beyond the control of the parties" occurred.[51] However, with growing price volatility on world fuel markets from around 1970, almost all transnational gas contracts that were signed included a paragraph that linked the gas price to the price of fuel oil, and gave the parties the right to renegotiate the gas price if the oil price changed substantially – upwards or downwards. European gas markets thereby became linked to world market prices for oil.[52]

The gas contracts also regulated potential critical events that might take place in the transnational gas trade. Detailed paragraphs identified formulas for penalties to be paid by the exporter in case of non-delivery or failure to deliver the right gas quality, while listing events in which penalties would not have to be paid. Other ways of dealing with interruptions in gas supply was through clauses regarding mutual assistance in cases of emergency. This included the mutual use of gas supply, but also the will to start negotiations in case of severe economic "hardship."[53]

As a rule, the gas contracts were placed under the jurisdiction of the exporting country. Thus, in the case of the Swedish-Danish gas deal, all juridical actions would take place in a Danish court, according to Danish law.[54] However, in the contracts signed between Western European countries and the Soviet Union, this was never the case. Instead, the contracts would answer to the laws of a third country.[55] Thus the Soviet-Austrian contract of 1968, as well as the Soviet-Finnish contract of 1971, stipulated that in case of a conflict that could not be resolved in a friendly way, an independent court in Sweden would resolve the matter.[56]

The role of the state

The state's role in negotiating the contracts and governing the transnational gas trade varied from country to country. The Dutch state played a reticent role behind the scenes when the first exports of Groningen gas were negotiated. Up to 1975 it was not formally part of the contractual arrangements. Instead it was NAM – that is, Shell and Esso – that carried out the negotiations with prospective foreign customers. De facto control over the export contracts, however, was exercised by another company, Gasunie, which was owned 50:50 by NAM and the Dutch state. This meant that the Ministry of Economic Affairs had to agree to all contracts signed. NAM was a façade used by the Dutch government to avoid the impression that it was actively involved in hydrocarbon exports, a sensitive issue at a time when many OPEC countries were pushing for stronger state participation in their national oil industries – at the expense of foreign companies, such as Shell. In 1975, however, when OPEC governments had seized control over most production plants, the NAM façade was abolished and Gasunie became the formal signatory to contracts.[57]

The importing countries did not face the same problems. Clearly, the governments in all importing countries wanted to take part in shaping transnational governance. The ways in which they did so varied, however. In West Germany, which did not have any national oil or gas company, the state sought to play a "facilitating" role when the contracts were negotiated. The German Ministry of Economy sent its gas expert, Norbert Plesser, as an "observer" to the German-Soviet negotiations in 1969, and the government made clear to all involved parties what it would and would not accept.[58] Formally, however, all import contracts had the status of private business deals. The government's most immediate role was to act as a guarantor for export credits that were agreed upon in connection with the countertrade arrangements with the Soviet Union.

In Italy, relations between ENI and the government were problematic and conflict-ridden – despite the fact that ENI was state-owned – as ENI acted much more independently than the government thought reasonable.[59] In the Austrian case, the government was not formally involved, although it was very active in shaping the countertrade arrangements with the Soviet Union.[60] However, since Austrian gas company ÖMV was a fully state-owned company and its relations with the government were less strained than in the Italian case, the government had a certain indirect control over the negotiations and the contracts. This was also the case when the Sydgas pipeline between Denmark and Sweden was negotiated, as both the exporting and importing companies were state-owned. In this case, however, the respective governments were more prominent in the negotiations, with the initial talks leading to the contract being organized by state officials from the Swedish Ministry of Industry.[61] The final negotiation about the gas price was even held in private between the Swedish and Danish ministers of energy, Carl-Axel Petri and Poul Nielsen.[62]

Bilateral vs. multilateral governance

The natural gas story provides an example of transnational governance in which international organizations have had very limited influence. Instead, a patchwork of bilateral relations between major gas companies formed the backbone of the European natural gas regime. The EEC and other international organizations formed arenas for discussions but did not reach much beyond that, and they appear not to have decisively influenced any major natural gas deals. This also concerns sector-specific organizations such as the ECE's Working Party on Gas Problems, mentioned above, and the International Gas Union, each of which had both Western and Eastern European members.[63] Gaz de France in the late 1950s and early 1960s actively sought to use the ECE to promote its vision of Algeria, at that time still a French colony, as the key to the future regarding Europe's oil and gas supply. With the extent of Dutch and North Sea gas reserves still unknown, it sought an alliance with the Soviet Union, seeing the future of Europe in "gas from the sands and gas from the steppes." But the French did not find any support for this idea in the international community – not even from the Soviet Union.[64]

Among Western European importers, who shared a common interest in enabling secure inflows of gas, there were attempts to develop closer cooperation on the regional level. The cooperation initially took the form of customer consortia that sought to increase their bargaining power vis-à-vis exporters by acting collectively. These attempts started with the joint Western European negotiations for Algerian gas in the early 1970s and also characterized later negotiations concerning Norwegian, Iranian, and Soviet gas.[65] In some cases, however, it was not only a shared feeling of resource scarcity that allowed such consortia to emerge but also a feeling of cultural community. Hence the Netherlands participated in some of the importing consortia despite the immense natural gas reserves at its own disposal. But the Dutch were also very active in raising the gas issue at the level of the EEC when their own exports seemed threatened by competition from Algeria and the Soviet Union, arguing that natural gas from outside the EEC should not be given the same rights as Dutch gas on the internal market.[66]

It is an interesting question as to whether it would have been possible to establish a multilaterally rather than bilaterally focused form of governance in the European natural gas industry. The 1960s, which was the formative decade for the transnational European gas infrastructure, was a period when the need for a "common energy policy" was hotly debated within the EEC. However, the widely diverging interests of individual EEC member countries – particularly between exporters and importers – made it difficult to agree on a common natural gas policy. Moreover, it appears that once the first major bilateral contracts had been signed, anticipating a very long-term cooperation, the fundamentals of the European supply structure had already been defined for the foreseeable future, reducing the prospects for establishing an alternative, multilateral regime. Considerable momentum in terms of styles of governance was thus quickly established. It was only from the late 1980s, when "deregulation" of electricity and gas began to be discussed within the EEC, and the TEN-E program for support to transnational system building was launched, that the EEC managed to take a more active part in shaping the European natural gas system.

Coping with vulnerability in practice

From the perspective of gas importers, the worst-case scenario was a long-term, total interruption of gas imports from a major foreign supplier. One way to reduce this risk was, as we have seen in the previous section, to choose foreign partners that were deemed both trustworthy and competent, and to formulate detailed contractual obligations for the exporter (and, in some cases, transiteers). As a complement to these arrangements, importers developed methods and strategies on their own to prevent gas disruption, should it actually occur, from (seriously) affecting users.

The quest for diversification

One strategy pursued in this context was the diversification of supply sources. Diversification became particularly important as the role of natural gas and the absolute volumes consumed increased. It was considered risky to depend on a single foreign supplier. Therefore, as soon as agreement had been reached for imports from one country, importers had strong incentives to negotiate additional gas imports from other countries, seeking to reduce dependence on the first supplier. As long as different suppliers balanced each other, it was considered acceptable to increase the overall level of import dependence, in both relative and absolute terms. For example, in 1972 the German Ministry of Economy argued that

> to the extent that the German gas industry can obtain additional natural gas volumes from other areas, imports from the Soviet Union could also be increased without further ado.[67]

This logic became an important driving force in the further transnationalization of natural gas in Europe. In other words, the quest for diversification tended to accelerate the pace of transnational integration. The diversification logic became particularly evident in large countries, such as Germany, France, and Italy, which were situated in reasonable proximity to various potential suppliers and which, as a result, came to build the most transnationalized gas systems in Europe. Germany saw it as advantageous to complement imports from the Netherlands with imports from the Soviet Union and later from Norway, whereas Italy's initial imports of LNG from Libya were quickly complemented by imports of Soviet, Dutch, and eventually also Algerian and Norwegian gas.

An interesting issue in the context of diversification was the fact that gas from different sources was not necessarily interchangeable. In particular, Dutch gas from Groningen and north German gas had a lower heat value than Soviet and Norwegian gas. Groningen gas was referred to as L-gas, as opposed to the H-gas delivered from the Soviet Union, North Africa, and the North Sea. In the 1960s, when Dutch gas started to be exported, it seemed that L-gas would become the standard in those countries that imported Dutch gas, largely coinciding with the EEC area. Similarly, H-gas appeared to be the natural standard gas in Eastern and Southern Europe.

However, when Soviet gas began to be imported into Germany, where Dutch gas was already used to a great extent, a major question was how to deal with the two gas qualities. One possible strategy was to build separate pipeline networks for Dutch and Soviet gas. Another was to transform H-gas into L-gas by adding nitrogen. Ruhrgas concluded that separate gas networks might increase vulnerability, since gas of one type would then not be able to come to the rescue in case of an interrupted gas supply from another source. For this reason, Ruhrgas initially aimed to invest in expensive facilities to harmonize the heat value of Soviet gas with that of Dutch and north German gas, so that gas from different sources could replace each other. The need for such investments, which made little economic sense, was used as an argument for negotiating a lower gas price with the Soviet Union in 1969.[68]

Before deliveries of Soviet gas had commenced, however, Ruhrgas changed its strategy, deciding to enable the use of Soviet gas in Germany – initially in Bavaria – without transforming it. This was because in the meantime agreements had been signed with Norway for the import of North Sea gas and with Algeria for the import of Saharan gas (the latter project failed). Both concerned H-gas. A further contract was also under negotiation with the Soviets. It therefore increasingly appeared that Dutch gas, after all, would not become the dominant gas type in Germany. It also meant that Norwegian and Algerian gas would be able to help the Bavarians out in case of a supply crisis, without any costly transformation. Norwegian gas, in particular, was expected to play a crucial role as emergency fuel in the case of interrupted supplies from the Soviet Union – and vice versa. The arrangement was codified through contractual agreements between Ruhrgas and Bayerngas for the eventuality of shortages.[69]

The only problem was that Norwegian gas was not available at the time when Soviet exports to Bavaria started in 1973. The first deliveries from Norway were expected in 1976. The regional Bavarian government came under pressure from opposition parties to deal with possible supply disruptions in the interim. Eventually it was decided to build an interconnecting pipeline between Bavaria and Baden-Württemberg, which was supplied from the Netherlands, and an expensive conditioning facility for transforming Dutch L-gas into Soviet H-gas quality. Since the investments were uneconomic – no one knew whether the arrangement would ever be needed – the facilities ended up being financed by Munich and Bonn.[70]

Soviet H-gas thus became the de facto standard in most parts of Germany, although a separate L-gas network continued to exist in areas close to the Dutch border. With the decrease in Dutch gas exports to Germany, this area gradually shrank, thus changing the geography of vulnerability. This process had consequences for users. In northern Bavaria, complaints were loud among the population when users in 1971–2 first had to switch from coal gas to Dutch natural gas, and then in 1974–5 once more from Dutch natural gas to Soviet natural gas, whereby each transition required the purchase of new gas burners or the costly modification of old ones.[71]

It was thus not an easy thing, in practice, to use diversification of supply as a method for countering vulnerability. Moreover, not all countries succeeded in diversifying their supplies. As we have seen, both Finland and Sweden, which were totally dependent on the Soviet Union and Denmark, respectively, failed to access additional supplies.

The need for diversification was felt not only in Western Europe but also in the communist countries of Central and Eastern Europe. The desire to diversify inspired a number of attempts to link up with non-Soviet gas sources, though without much success. East Germany tried in vain to get access to Dutch natural gas as a complement to Soviet imports, and Czechoslovakia similarly failed to link up with LNG supplies from Algeria by way of Adriatic harbors. In the 1980s, Poland discussed a possible pipeline from Norway but also failed. Likely reasons were opposition from Moscow and failure to raise sufficient funding, but it may also be that the communist would-be importers did not manage to persuade Algeria and the Western exporters of their ability and willingness to actually receive and pay for the gas.

The lack of freedom in influencing the emerging natural gas infrastructure was particularly obvious for those countries and regions that had ceased to exist after the war, having been incorporated into the Soviet Union – namely, the Baltics and the eastern regions of what had been interwar Poland. The Baltic republican governments, particularly the Latvian one, were suspicious about the single pipeline through which the Baltic region was connected to the gas fields in Ukraine. The republican governments tried, through numerous petitions to the Gas Ministry and Gosplan in Moscow, to gain support for a more diversified supply structure.[72] It took until 1972 before this wish was fulfilled through the construction of a new pipeline from northern Russia to Riga. By that time the Baltic republics and Belarus had already come to experience the hardships that followed from numerous gas outages, particularly during winter.

Countering vulnerability through domestic reserves and underground storage

As an alternative to diversification, importing countries and regions could use their local gas fields as a backup gas source. This strategy became particularly important in countries that were able to import gas from only one other country. The existence of domestic gas reserves was an important condition for Austria's willingness to conclude a first contract with the Soviet Union in 1968. The contract was of enormous importance for Austrian gas company ÖMV, which foresaw a rapid increase in Soviet deliveries; within only a few years, Soviet gas was to contribute more than half of national consumption. Such a development would have been unthinkable had the country lacked access to sizeable domestic gas fields as a backup source. The gas geography was particularly convenient in this context: the main Austrian fields were located near the Slovak border, where the Soviet gas was to arrive, and ÖMV could thus treat the imports logistically as just another large gas field. This state of affairs made it possible for the Austrians to be very patient when the Soviets initially had great difficulty in living up to their export promises specified in the 1968 contract. During ÖMV's first years of gas imports, partial or total interruptions in supply were the rule rather than the exception.[73]

The reasons for these interruptions were linked to the general weakness of the Soviet economy and the central planning system. The Soviet gas minister, Alexei Kortunov, argued that the problems in supplying Austria stemmed not from within the gas ministry itself but from the failure of other ministries to deliver steel pipes, compressor stations, and other equipment necessary to guarantee the new export regime. Sometimes the equipment had been produced, but logistical problems hindered delivery, so that pipeline construction was delayed. In autumn 1968, for example, the Western steel pipes that the Soviets had bought in return for gas, and which were to be used for constructing the export infrastructure, were lying idle in Leningrad's harbor because the necessary railway cars were not available.[74]

The Austrians, unaware of the disastrous logistics of the centrally planned Soviet economy, regarded the delays as "teething problems" with the new arrangement. They could afford to adopt such a perspective because it was easy for ÖMV to quickly balance any shortages in Soviet gas by way of accelerated domestic production.[75] Hence Austrian gas users never noticed these early problems, which were not publicly known. Moreover, ÖMV's optimistic interpretation became an argument for Italy, France, and Germany to go ahead with their plans to negotiate very large imports of gas from the Communist Bloc.

Apart from domestic gas reserves, importing countries also constructed artificial reserves in the form of underground storage facilities. These came to play a key role in combating supply interruptions from abroad, although their main purpose was to enable "peak shaving" and, more generally, load factor management, thereby seeking optimal utilization of the pipeline infrastructure. The dual purpose of gas-storage facilities – for load factor management and for security of supply – was seen as an advantage since importing countries did not want to openly let their foreign partners know that they felt skeptical about their intentions and abilities to fulfill their export obligations.

The most radical European case of a gas-storage facility built for security reasons was in West Berlin. This involved the creation of a huge underground reservoir, with the help of which the isolated city would be able to secure its gas supply for a year even in the case of a total interruption of supplies, without users noticing it. The gas was to come from Soviet sources, and a further political uncertainty involved transiting this gas through the GDR. Before the storage facility had been completed, however, its main rationale suddenly disappeared, as a consequence of the fall of the Berlin Wall.[76]

The quest for a gas-storage facility in Sweden provides a further interesting case. Sweden was particularly vulnerable to potential import disruptions since it received all of its gas by way of a single pipeline (from Denmark) and did not possess any domestic gas fields that could be used as a backup. Construction of a storage facility was therefore considered to be an important investment, despite the high cost and technological challenges that it entailed. There was major interest from various actors to invest in the project, which was a recurring theme throughout the 1980s and part of the 1990s, despite the relatively small amounts of natural gas that were being imported. However, geological factors prevented the facility from being built and, in the end, Sweden handled its security of supply in other ways – for example, by negotiating with Denmark about the use of Danish storage facilities.[77]

The strategies of diversification, the use of domestic gas reserves, and the construction of gas-storage facilities to counter vulnerability were all dependent on the existence of an efficient domestic pipeline infrastructure. Without a good infrastructure, it was impossible to redirect gas flows quickly in case of an unexpected interruption in supply from abroad. Security considerations in this sense stimulated the construction, further development, and effective maintenance of national gas grids. They also stimulated the development of efficient organizational arrangements, and of advanced information and control systems to keep track of the gas and rapidly calculate the most efficient gas flows in the case of an interruption from abroad. Of course, these arrangements could also be used in the case of unexpected interruptions within national systems.

Interruptible customers

An additional fundamental strategy for countering vulnerability was to supply natural gas only to those customers that had access to alternative fuels and could switch to these quickly and easily in case of delivery problems. Such arrangements were common particularly in the early phase of development in the European natural gas industry. The strategy was particularly important in countries and regions that were located at the remote end of a single pipeline and lacked domestic gas resources or strategic storage facilities. Sweden, for example, not being able to construct a domestic underground storage facility, used interruptible contracts as a method to improve the efficiency and security of its gas system. Finland, being totally dependent on the Soviet Union, was another case in point.

Stockpiling coal and oil were the most common forms, and the users in question were either large industrial companies or electric power plants. The same method was also used for load-management purposes, so that natural gas was used in a power plant, for example, only during certain hours of the day or certain months of the year, when gas demand among other consumers was low. An interesting case is Leningrad's gas supply. Situated on the far end of a pipeline that stretched from the northern Caucasus over Moscow to the former imperial capital, in the 1960s the Soviets developed a regime according to which Leningrad's pattern of use was to be the reverse of Moscow's. Hence Leningrad's gas supply decreased in winter (when demand was high in Moscow) and increased in summer. Leningrad simply received the volume of gas that Moscow did not need.

Europe through the lens of natural gas

The history of natural gas in Cold War Europe, in which East and West were crucially interlinked, contradicts much of mainstream European historiography, in which an "Iron Curtain" is thought to have divided Europe into two antagonistic camps, cooperation between which is usually argued to have taken place only on the margins. The huge gas pipelines that penetrated both the Iron Curtain and the Mediterranean enable us to discern a different Europe. This section discusses Europe through the lens of natural gas, first, by looking at the European "vulnerability geography" that resulted from the gas trade and, second, by analyzing the "hidden regionalization" of Europe that emerged from the late 1960s and onwards.

The vulnerability geography of the European gas system

Which parts of Europe have been most vulnerable to critical events in natural gas, such as interruptions in gas supply? Western Europe's fear of falling victim to interruptions in supplies from the East was not realized during the Cold War. The Soviets had difficulty in supplying gas for exports during a few initial years, but from 1974 the impression in the West was that the Soviet Union had turned out to be a trustworthy partner, since the contracts were "fulfilled precisely."[78] Nevertheless, worries among the population in importing countries and pressure on the political level forced gas companies to implement security mechanisms of the types discussed above. Expensive security-related projects were thereby often partly or even fully financed by the state. When Washington opposed a scaled-up Western European gas import from the Soviet Union in the early 1980s, the Reagan administration used this circumstance to argue against the Yamal pipeline:

> If Europeans factored in all the costs of importing Soviet natural gas – including the expensive measures necessary to insure against politically induced or technically caused supply interruptions, military spending to offset Soviet technology acquisitions made possible by gas sales, and interest rate subsidies for pipeline construction – the pipeline project would prove financially unattractive.[79]

But Western European fears of supply disruptions from non-Western exporters also stimulated and accelerated the process of strengthening the links between the gas networks of EEC member countries. Several Western European politicians, notably in West Germany, argued that close Western European cooperation in the construction of a unified gas transmission infrastructure was necessary in order to make it possible, in the case of supply interruptions from the Soviet Union or Algeria, for those regions affected to access gas sources quickly from elsewhere in the larger system – that is, from other member states. Fritz Burgbacher, a prominent German member of the European Parliament and deputy chairman of the Economic Committee of the NATO Parliament, argued as early as 1967 that gas produced in EEC member states should be regarded as "domestic." A central task for strengthening security of supply would thus have to be the construction – with NATO contributing financially – of a unified EEC gas grid, "so that the balancing of our energy supply with neighbor states and allies, which is especially necessary in the case of crisis, can be carried out."[80] Vulnerability to interruptions in gas supply from outside Western Europe could be reduced, they believed, through deeper integration among the Western European countries themselves. Western European gas integration was thus boosted by the perceived vulnerability to interruptions from the outside. In other words, in natural gas the idea of "Western Europe" was largely a product of perceived vulnerability.

While importers in Western Europe, accordingly, built effective protection mechanisms for countering disturbances both nationally and at the EEC level, the most vulnerable gas users in the context of the East–West gas trade were, paradoxically, gas users in the Soviet Union. This was because domestic Soviet users were supplied from the same gas fields as were the foreign importers. When gas production was insufficient to meet the total demand, the Soviet gas ministry faced the delicate choice of either failing to honor its export commitments or sacrificing the gas supply to industries and households in the Soviet Union. The Soviets chose the second option.

The result was devastating for industries as well as for the local population, not only in western Ukraine but also in Belarus, Lithuania, and Latvia, all of which competed with Czechoslovakia and Western Europe for access to scarce west Ukrainian gas. Families found themselves living in ice-cold houses without any way to cook. Many industries were forced to a standstill as logistical failures prevented them from accessing sufficient volumes of reserve fuel in the form of oil or coal. The population used their local Communist Party organizations to vent their anger, and desperate letters were sent to Moscow, begging the country's leaders to force the gas ministry to improve the situation.[81]

Gas users in the Soviet Union faced an increasingly problematic situation due to the anarchic chaos that followed in times when supply failed to match demand. In cooperation with the gas ministry, Gosplan, the powerful Soviet planning organization, worked out detailed lists that prescribed how much gas a certain factory was allowed to use in case of a critical event. But these instructions were rarely followed and users located on the far end of pipelines became defenseless victims, despite repeated attempts by Moscow to prevent upstream users from using more gas than they were entitled to.

The completion of large pipelines from Siberia improved the situation, but it remained far from harmonious. In the stagnating Soviet economy, investments in and maintenance of the export pipelines and compressor stations were often neglected. The result was frequent accidents, explosions, and temporary interruptions of a "technical" nature. In the mid-1980s, Soviet gas transport along the largest transmission lines thus experienced about 28 stoppages per year, each averaging more than four days.[82] Western European countries, which had always been suspicious of Soviet gas deliveries, though from a political rather than from a technical point of view, were well protected against these temporary breakdowns. In Central Europe, however, where gas-storage facilities and other emergency arrangements were often lacking, industries and households were affected directly. In the post-Soviet era, this historical legacy continued to play a major role in shaping Europe's gas vulnerability geography, and thus it was the post-communist countries of Central and Eastern Europe that were most severely affected by the recurring Russian-Ukrainian gas crises cited in the introduction to this chapter.

Europe's vulnerability geography, as outlined above, can be taken as evidence of an East–West divide in Europe's natural gas history. Obviously, vulnerability was influenced not only by the behavior of suppliers but also by the overall functioning of economy and society in the user regions, along with the extent to which a region or a country possessed mechanisms for countering gas disruptions. Against this background it is not surprising to find a strong correlation between the ways in which countries were affected by supply problems, and their political and economic systems. From this perspective the history of natural gas can arguably be said to fit with the view of Cold War Europe as a continent radically divided between East and West: on the one side the communist, centrally planned economies that were members of the COMECON and the Warsaw Pact; on the other side the capitalist, market-centered economies, most of which were members of the EEC and NATO.

The gas regions of Europe

The making of the European natural gas system has been characterized by close and long-term transnational cooperation that was not divided by the "Iron Curtain." Austria, Italy, and Bavaria became part of the Soviet-based natural gas system before they linked up with the Dutch and North Sea-based infrastructures, and they became more dependent on Soviet than on Dutch and North Sea gas. Greece, a NATO and EEC member, came to rely on Soviet gas only, as did Finland.

At the same time, we find an interesting relationship between the "hidden integration" of Europe, in the form of the construction and use of a transnational gas infrastructure, and the politically and economically more visible integration that proceeded in parallel. For the Soviets, for example, the incorporation of Austria into the Eastern European natural gas system – at a time when the country was pushing for an association with the EEC – became a way of counteracting the overall political trend of Western European meso-integration. In the same way, Egon Bahr and Willy Brandt in Germany deliberately sought to place the German-Soviet natural gas negotiations into the framework of developing a new German "Eastern policy" from 1967 on. Natural gas was thus an integral part of broader political developments in Europe.

The latter aspect also applies to developments within the Soviet Union and the COMECON region. Here, transnational gas networks were of a certain significance both politically and symbolically. After the Second World War the emerging all-Soviet natural gas system became a very concrete and material way of integrating the newly annexed Soviet territories with the rest of the country, tying Galicia (in what had been eastern Poland) and the three Baltic countries (which had become three Soviet republics) to Belarus and central Ukraine. Natural gas became a tool of Sovietization and an instrument in erasing the perceived Europeanness of Galicia and the Baltics. Stimulating a system-building process that made the new Soviet republics highly dependent on each other for their gas supply, the Kremlin also fostered a perceived vulnerability of these republics to attempts at breaking out of this emerging system. From the early 1960s, Moscow proceeded by seeking to extend the emerging transrepublican system of gas dependencies to the communist satellite states in Central Europe. The Soviets complemented this strategy by preventing the Central European countries from signing import contracts with alternative suppliers in Western Europe or North Africa. Moreover, the

Soviets succeeded in turning the construction of transnational gas networks into a showcase of successful cooperation within the COMECON.

The Cold War era was also one of decolonization in Africa, with bitter wars of independence and strained relations between the new African states and the former colonial powers, particularly France, for whom the loss of Algeria became a traumatic experience. Independent Algeria embarked on a new, radical development that made its political and economic situation radically different from that of France, and the Algerians partly hostile to the French. Nevertheless, France started importing Algerian natural gas on a large scale. This seems to indicate that the close links between France and Algeria in the past, Algeria's familiarity with the French language and culture, the existence of cross-border epistemic communities, and Algeria's geographical proximity (and thus the economic feasibility of transporting gas) were stronger than political tensions between the two countries.

If a divided Europe is to be constructed analytically on the basis of transnational dependencies, it is a Europe consisting of three major mesoregions, defined by their major source of natural gas. These regions can be labelled Eurasia, Eurafrica, and North Sea Europe. The boundaries between them have evolved and continue to evolve over time, a process that might be thought of as a "hidden regionalization." The boundaries between the gas regions rarely coincide with national borders; the major division lines cut through the hearts of Germany, France, and Italy. The main gas companies in these three countries – Ruhrgas, Gaz de France, and ENI, respectively – have consistently pursued a diversification strategy, making them part of at least two regions (see Tables 2.1 and 2.2). All other countries clearly belong to one particular region only (Figure 2.5).

Some parts of Europe, however, were not part of any region at all; they remained white spots on Europe's natural gas map. To these belonged, throughout the Cold War era, most of Poland, which for a long time was not integrated to any significant extent with any one of the major regions. Today the main remaining white spot on the map is formed by a large part of Scandinavia where natural gas is not used at all. Despite a long tradition of cooperation and a sense of cultural community, the Nordic countries have failed, with the exception of the Danish-Swedish pipeline, to create a transnational gas infrastructure. This must be regarded as highly intriguing, especially when contrasted with the deep integration across the East–West political divide.

All in all, it thus proved possible to build pipelines between countries that one perhaps would not have expected to cooperate. Conversely, possible pipelines between countries that were close to each other in a political and cultural sense often were not built. It would have been perfectly possible to create a transnational European pipeline infrastructure coinciding precisely with the established political, military, or economic "blocs" in Cold War Europe, but this did not happen. The reasons for actors' willingness to transcend the major borders between different blocs – notably the Iron Curtain – have been scrutinized in the previous sections. Overall, the impression is clearly that the "blocs" – both in natural gas and from a more general political perspective – were in reality not as sharply defined as European historians have wanted them to be. They were often much

Figure 2.5 European mesoregions for natural gas, by dominant source of supply.

weaker constructs, and the barriers within blocs were in some cases higher than between them. The construction of a transnational infrastructure for natural gas was used as an instrument for lowering barriers between blocs even further.

It is only through such an adjusted, more complex political map of Cold War Europe that we can understand the overwhelming extent to which factors such as geographical and cultural proximity, epistemic communities of trust, and economic rationalities could be transformed by actors into convincing arguments for building dependency-generating pipelines that transgressed the Iron Curtain, the Mediterranean, and other allegedly major barriers in Cold War Europe.

To a certain extent the major gas pipelines in and to Europe can be said to follow routes that lead rationally and logically from major gas-producing regions to major gas-consuming regions. It is worth repeating, however, that Europe's natural gas geography could have looked different. For example, Dutch, Algerian, and North Sea gas could clearly have supplied all of Germany, France, Austria, and Italy, making these countries independent of deliveries from the Soviet Union. There were no objective pressures or arguments that compelled these countries to engage in natural gas imports from certain regions rather than others, as long as it could be economically justified. The importers' choice of partners was deliberate, and might have been different if other system-builders had been involved, with different agendas, visions, and worldviews.

The choice to import large volumes of Soviet gas seemed controversial in challenging the postwar ideological divide, but it seems less surprising if seen in a longer historical perspective. Natural gas was but the latest among the raw materials and agricultural products that had been imported from the East for centuries, and which had formed part of the backbone of Europe's economic geography. In particular, Western Europe's import of Soviet natural gas built on a long tradition, stretching back to the mid-nineteenth century, of importing Russian coal and oil.

The attempts of the United States to prevent Western Europe from trading with the Communist Bloc became a major hallmark of the Cold War period. Western European countries themselves were less inclined to give up their Eastern trade relations, with their deep historical roots, for the sake of ideological and military considerations. From this perspective, the struggle over imports of Soviet natural gas was also a struggle relating to what Europe should be. Washington preferred a divided Europe and sought, instead, to favor a tightly integrated mini-Europe in the West, which was to be closely linked to North America. In contrast, most European countries, judging from our material, regarded a much more open Europe, with large-scale flows of energy and technology between East and West, as the natural and historically justified path. As we have seen, key actors in the West – such as Rudolf Lukesch in Austria and Otto Schedl in Bavaria – sought to use their historical traditions of trading with the East when attempting to link up with the Soviet natural gas infrastructure. They were not only pragmatic but also historically aware and culturally sensitive personalities whose worldviews – and views of Europe – differed markedly from those of Dwight Eisenhower or Ronald Reagan.

Not all US presidents opposed the European natural gas trade. At the formative moment, when Western Europe was negotiating its first contracts with the Soviet Union, Washington opted to take a passive stance. This seemed to fit with the more relaxed relations between the two superpowers at the time. Washington's stance changed under Reagan in the early 1980s, when it sought to deliberately prevent natural gas imports to European NATO members from the Soviet Union. By then, however, the momentum of the natural gas infrastructure had already reached a high level and the attempts to prevent the East–West gas trade from scaling up failed.

The natural gas story thus demonstrates how during the Cold War, which also coincided with the transition from colonial to post-colonial relations, Europe was able to build on economic and cultural experiences from the past to overcome political and military divides.

Notes

1. For a good discussion of the 2006 gas crisis, see Fredholm 2006, pp. 4ff.
2. See IEA 2004, pp. 317–319.

3. The focus has been on, for example, the functioning mechanisms of the natural gas market, on effects of liberalization, on policy aspects of building the natural gas network, on game-theoretical analyses of the interdependent relationships between East and West, on comparisons between the organization of the natural gas industry in different countries, on natural gas from an energy economy perspective, on the competitiveness of natural gas in relation to other energy sources, on problems in price formation within the gas industry, and so on.
4. The promising exceptions include Hayes's study of the Transmed and Mahgreb pipeline projects that linked Algeria with Italy and Spain, respectively (Hayes 2006), as well as Kaijser's studies of the emergence of the Dutch natural gas regime and of gas and electricity from a comparative Nordic perspective (Kaijser 1997, 1999). There is also an interesting literature on the history of gasworks and gas production, although the focus there is usually less on gas as an infrastructure or system, not on transnational aspects. A good collection of case studies of early gas histories from various parts of Europe is provided in Paquier and Williot 2005.
5. With the exception of West Germany, where we were not granted access to original documentation from Ruhrgas, we studied documents from both governments and gas companies. As for government materials, we sought access to the archives of both economy and foreign ministries, although in the Russian and Ukrainian cases we were not granted access to the foreign policy collections.
6. There is a – perhaps important – "linguistic divide" in Europe regarding methane terminology. The term "natural gas" is used in English, Greek, Turkish, and most Romanic, Nordic, and East Slavic languages (gaz naturel, naturgas, природный газ, φυσικό αέριο, etc.), whereas Dutch, German, Finnish, Estonian, and most West and South Slavic languages use the notion of "earth gas" (Erdgas, aardgas, maagaas, etc.). It is notable that this divide goes straight through the former Yugoslavia, with Bosnian, Serbian, and Slovenian using "earth gas" (zemni plin, zemni gas, zemeljski plin) but Croatian "natural gas" (prirodni plin).
7. By adding nitrogen to natural gas, the latter could also be used as a gaseous fuel with the quality of coal gas.
8. An important motivation for the Netherlands to export part of its gas (instead of saving it for future domestic use) was the anticipation that nuclear power would fairly soon become a low-price competitor to gas. The export of gas therefore seemed an option to earn as much as possible from the available gas reserves (Kaijser 1999, p. 46).
9. The Druzhba system was an immense transnational infrastructure of oil pipelines through which Soviet oil was pumped to Central Europe. It consisted of two major pipeline routes, one going through Poland and continuing into East Germany and the other taking a more southern route, heading for Czechoslovakia and Hungary. It became operational in 1962.
10. ECE, Ad Hoc Working Party on Gas Problems, "International Gas Pipelines: Notes by the Secretariat," January 24, 1957, UNOG Archives, G.X 11/13/11, file 25459.
11. Hayes 2006.
12. "10 Jahre Erdgasimport," *ÖMV-Zeitschrift*, no. 3/1978, p. 1.
13. Rambousek 1977, p. 79f.
14. "Soviets have big pipeline plans for '65," *Oil and Gas Journal*, December 28, 1964; *gwf*, May 20, 1966, p. 531.
15. The first major Siberian gas fields were confirmed in 1962 and 1964. In 1965 the first supergiant field, Zapolyarnoe, was discovered, which boosted both Soviet and foreign interest in bringing Siberian gas westwards.
16. ENI subsidiary SNAM Progetti exported several large gas refineries to the Soviet Union in the early 1960s. See Semichastnov (deputy minister of foreign trade) and Sidorenko (deputy minister of gas industry) to the Soviet Council of Ministers, August 11, 1966, Russian State Archive of the Economy (RGAE), fond 458, series 1, file 103.

17. The decision is quoted extensively by the gas minister, Kortunov, in a letter to the Soviet Council of Ministers, October 5, 1966, RGAE, fond 458, series 1, file 109.
18. "Niederschrift über die 3. ordentliche Aufsichtsratssitzung/Geschäftsjahr 1966 der VÖEST AG, abgehalten am 24. Oktober 1966," Österreichisches Staatsarchiv (OeStA), AdR, ÖIAG-Archiv, Box 325; Willy H. Schlieker to the German Foreign Office, December 12, 1966, Bundesarchiv (BArch), B102–152193.
19. For a detailed analysis of the pipe embargo, see Rudolph 2004.
20. Otto Schedl, "Warum Osthandel? Ansprache vor dem Union International Club e.V. am 20. September 1967 in Frankfurt/Main," Bayerisches Hauptstaatsarchiv (BayHStA), NL Schedl, Box 188.
21. Grimm, January 23, 1967 and Plesser, January 27, 1967, BArch, B102–152193.
22. This interpretation was made by, for example, Willy H. Schlieker, a semiretired high-level German insider in the steel industry and an advisor to the federal government. See Schlieker to the Foreign Office, December 12, 1966, BArch, B102–152193.
23. For a detailed discussion of the relationship between the start-up of gas exports to Western Europe and the Soviet-led invasion of Czechoslovakia, see Högselius 2013, Chapters 6–7.
24. After an intense struggle, it was Ruhrgas rather than Schedl and regional gas company Bayerngas that took control over the negotiations with the Soviet side. Similarly, Ruhrgas insisted that it and not Bayerngas would build and operate the new transnational pipeline infrastructure, despite the fact that this was to be built on Bavarian territory, where Ruhrgas had no operations, and with the gas destined mainly for Bavarian users.
25. Plesser to Lantzke, August 19, 1969, BArch, B102–152193.
26. Mikkonen 2008, pp. 2 and 8.
27. Ibid. p. 8
28. Swedish Embassy in Helsinki to the Swedish Foreign Ministry, May 5, 1971, Riksarkivet (RA), H53:40 Er, Dossier 7.
29. Ibid.
30. This was used as an argument by the Finns during the negotiations. See "Memorandum o peregovorakh po prodazhe sovetskogo prirodnogo gaza v Finlandiyu," March 15–17, 1971, RGAE, Minvneshtorg SSSR, 31:2, 4447, Zanitsi besed rukovodstva minvneshtorga SSSR s predstavitelyami Finlandii, February 3–December 7, 1971.
31. Haaland Matlary 1988, p. 13.
32. Cf. Damian 1992, pp. 596–607.
33. "Sovjetrapporten", Arbetsutkast, Swedegas AB, p. 5, Swedegas Archives; Stern 1980.
34. Ibid. This insecurity regarding the supply status at the time was due to both production shortfalls and a gas contract concluded in the mean time between the Soviet Union and West Germany.
35. Examples of Swedish pipeline projects that were not completed include (among many others) the PGT pipeline from northern Norway through Sweden to Denmark, and one connecting Sweden to the Finnish grid through a pipeline across the Baltic Sea.
36. For example, it has been argued that Sweden, through which any such pipeline would have to pass, is too sparsely populated for this to be profitable, and that Sweden would not be able to pay as high a price as customers on the Continent, due to energy market structures. See, for example, Swedish Government Bill 1978/79:115, Bilaga 1, Industridepartementet, p. 154; "Rapport från Erik Grafström till Statsrådet Rune Johansson," October 18, 1973, Naturgasdelegationen, 1:1:3.
37. For details of the contract, see ÖMV, November 20, 1975, OeStA, AdR, ÖIAG-Archiv, Box 138.
38. Ritzel to the Foreign Office, June 8, 1979, BArch, B102–313208.
39. "Soviets Push for Construction of Strategic Yamal Gas Line," *Oil and Gas Journal*, November 2, 1981.

40. "Overview of the Siberia-to-Europe Natural Gas Pipeline", CIA, Office of Soviet Analysis, February 9, 1982, p. 1.
41. See e.g. "End to Yamal Pipeline Embargo Eyed," *Oil and Gas Journal*, November 8, 1982.
42. Victor and Victor 2006, p. 134.
43. Plesser, January 27, 1967, BArch, B102–152193.
44. Swedish Government Bill 1978/79: 115. pp. 154–156.
45. Plesser, "Besprechung bei Herrn Staatssekretär Dr. von Dohnanyi am 3.7.1969," July 10, 1969; Schöllhorn to Franz, May 12, 1970, BArch B102–152194.
46. Lumpe, July 3, 1972, BArch B102–152196.
47. IEA 2004, p. 318.
48. The construction of an additional East–West pipeline through Belarus and Poland continued to be debated throughout the 1970s and 1980s, but it was only after the end of the Cold War that it could be realized.
49. As reported by VGW, "Aktennotiz über ein Gespräch am 26.4.1967 in Frankfurt/Main betreffend Erdgaslieferung aus der UdSSR in die Bundesrepublik," BArch, B102–152193. VGW was represented by Dr F. Gläser and Mr H. Utzerath.
50. As discussed, for example, by Wedekind, "Einfuhr sowjetischen Erdgases," May 16, 1969, BArch, B102–152194.
51. ÖMV and Soyuznefteexport, "Vertrag über die Lieferung von Erdgas," Vienna, June 1, 1968, OeStA, AdR, ÖIAG-Archiv.
52. See Davis 1984 and Estrada et al. 1988. See also "Protocol Concerning the Delivery of Natural Gas from the Soviet Union to Finland," May 21, 1971, RA, H53:40 Er, Dossier 7 (here the term used is "competing fuels").
53. For the Soviet-Austrian case, see ÖMV and Soyuznefteexport, "Vertrag über die Lieferung von Erdgas," Vienna, June 1, 1968, OeStA, AdR, ÖIAG-Archiv. In the Swedish-Danish case, the section on force majeure, enumerating the different critical events that could be envisioned, stated that none of these events was to be considered a legitimate reason for breaking the contract. Instead, the contract could be suspended, or the problems solved through mutual aid.
54. Gas-transportation contract between Dansk olie og Naturgas and Swedegas AB of February 1980, Swedegas Archive.
55. Torkel Ösgård, author's interview, May 30, 2008; Michael Schultz, author's interview, November 11, 2008; "Protocol concerning the delivery of natural gas from the Soviet Union to Finland," May 21, 1971, RA, H53:40 Er, Dossier 7.
56. ÖMV and Soyuznefteexport, "Vertrag über die Lieferung von Erdgas," Vienna, June 1, 1968, OeStA, AdR, ÖIAG-Archiv; "Protocol concerning the delivery of natural gas from the Soviet Union to Finland," May 21, 1971, RA, H53:40 Er, Dossier 7.
57. Kaijser 1999.
58. Plesser, "Deutsch-sowjetische Erdgasverhandlungen über die Lieferung sowjetischen Erdgases am 31.7. und 1.8.1969 in Moskau," August 7, 1969, BArch B102-152194.
59. This was observed by, for example, VÖEST in Austria. See "Niederschrift über die 2. ordentliche Aufsichtsratssitzung/Geschäftsjahr 1967 der VÖEST AG, abgehalten am 29. Juni 1967," OeStA, AdR, ÖIAG-Archiv, Box 325.
60. See e.g. "Protokoll über das Gespräch mit Ministerpräsident Kossygin am 19. März 1967," OeStA AdR, II-Pol, UdSSR 1967.
61. Minutes from Danish-Swedish deliberations concerning a natural gas cooperation, from June 1979 onwards, Swedegas Archive.
62. Thyberg to the Swedish Foreign Ministry, February 12, 1979, Swedish Foreign Ministry Archive; Poul Nielson, author's interview, April 1, 2008, Carl-Axel Petri, author's interview, May 29, 2008.
63. Davis 1984, p. 152, claims that the ECE "has a largely toothless informational role and confines its activities to the issues of periodic statistical and technical reports."
64. See, for example, William R. Connole, "Will Europe Come to Depend on Russian Natural Gas?," *Oil and Gas Journal*, August 28, 1961, p. 60, and Alois Bauer, "Die Gaswirtschaft

der Europäischen Wirtschaftsgemeinschaft am Jahresanfang 1967," *gwf*, May 1967, p. 594.
65. Dahl 1998.
66. See, for example, Hankel to Schiller, January 21, 1970, BArch, B102–152194.
67. Lumpe, July 3, 1972, BArch, B102–152196.
68. Plesser, "Deutsch-sowjetische Erdgasverhandlungen über die Lieferung sowjetischen Erdgases am 31.7. und 1.8.1969 in Moskau," August 7, 1969, BArch, B102-152193.
69. Staatsminister Anton Jaumann to Rudolf Hanauer, Präsident des Bayerischen Landtages, December 21, 1972, BayHStA, MWi, Box 27208.
70. Rummer to Lantzke, June 18, 1971, BArch, B102–277764.
71. This issue was debated in the regional Bavarian parliament in 1974 and 1975. See Bayrischer Landtag, January 23, 1974, February 20, 1974 and February 26, 1975. BayHStA, StK, Box 18790. The Netherlands, in contrast with Germany, invested in an expensive transformation station at Ommen, where nitrogen was mixed with high-calorific gas from the North Sea to make it compatible with other Dutch gas.
72. Ruben (chairman of the Latvian Council of Ministers to the Soviet Council of Ministers), "O stroitelstve vtorogo gazoprovoda v Latviiskuyu SSR," April 11, 1967, RGAE 458–1–495.
73. "Niederschrift über die Sitzung des Arbeitsausschusses des Aufsichtsrates der ÖMV AG am 17. September 1969"; ÖMV, "Bericht über das 1. Quartal 1971," OeStA, AdR, ÖIAG-Archiv.
74. "Spravka o perspektive gazosnabzheniya g. Kieva v zimniy period 1968–1969 gg.," September 3, 1968, Central State Archive of the Highest Organs of Government and Administration of Ukraine, Kiev (TsDAVO Ukrainy), 337–3–20.
75. ÖMV ARS-Sitzung, September 21, 1970; ÖMV, "Bericht über das 1. Quartal 1971." OeStA, AdR, ÖIAG.
76. Landesarchiv Berlin (LAB), various files.
77. "Agreement on the Storage of Natural Gas between Dangas and Swedegas of October 1, 1989," Swedegas Archive.
78. Jaumann to Hanauer, November 17, 1978, BayHStA, StK 18790. The Soviets also concluded that in contrast with previous years, 1974 was the first year when the country's export obligations could actually be met. See Orudzhev and Shcherbin to the Soviet Council of Ministers, "Ob itogakh vypolneniya plana postavok sovetskogo prirodnogo gaza na eksport v 1974 godu," January 22, 1975, RGAE, 458–1–3963.
79. "US Mulls Effect of Ban of Exports to USSR," *Oil and Gas Journal*, February 22, 1982.
80. Fritz Burgbacher, "Energieprobleme in Europa – Möglichkeiten ihrer Lösung", *gwf*, December 22, 1967, p. 1462.
81. For example, "To the General Secretary of the Central Committee of the Communist Party, comrade L.I. Brezhnev," December 11, 1973, signed by 50 residents, TsDAVO Ukrainy, 337–15–440.
82. Gustafson 1989, p. 159.

3

Inventing Electrical Europe: Interdependencies, Borders, Vulnerabilities

Vincent Lagendijk and Erik van der Vleuten

Prologue: Contours of a critical event

November 4, 2006, late Saturday evening. German electric power transmission system operator E.ON Netz disconnects an extra-high voltage line over the Ems River at the request of a northern German shipyard. This should allow the large cruise ship **Norwegian Pearl** to pass safely from the yard to the North Sea. Other power lines are supposed to take over the duties of the disconnected line as usual in this routine operation.

This evening is different, however. When E.ON Netz switches off the line the burden on other lines in the network increases, as expected. Several of these are now operating near their maximum capacity. Further fluctuations of electric currents cause one line to overload and automatically shut down at 22:10:13. The following sequence of events is astounding. Within a mere 14 seconds, a cascade of overloads and power lines tripping spreads throughout Germany from northwest to southeast, each tripped line increasing the burden on the rest of the system. In the next five seconds the failure cascades as far as Romania to the east, Italy to the south, and Portugal to the southwest. The incident affects electricity supply in about 20 countries, and supply is cut selectively to some 15 million households. Via the Spain–Morocco submarine cable the disturbance reaches Morocco, Algeria and Tunisia, where lines trip and consumers are left in the dark.[1]

The failure is soon repaired but the event continues to live on in the press and in political debate.[2] Newspaper reports, such as the Associated Press article "German-Triggered Blackout Exposes Fragile European Power Network," quote pro-European Union (EU) politicians, who argue that the blackout reveals an intolerable fragility of Europe's electric power grid.[3] They blame this fragility on decentralized, insufficiently "European," power-grid governance by transmission companies and their international associations. Romano Prodi – Italian prime minister and former European Commission president – sees "a contradiction between having European [electric power] links and not having one European [electric power] authority... We depend on each other but without being able to help each other, without a central authority."[4] The EU Energy commissioner, Andris Piebalgs, stresses that "these blackouts... are unacceptable" and "confirm the need for a proper European energy policy."[5] "European" here denotes EU intervention. Next follow debates about an EU-level regulator, formally binding EU legislation, and an EU priority interconnection plan.[6]

Power sector representatives, however, completely disagree. If the events of 4/11 (European notation) prove anything, it is that existing security measures and governance structures work well, for the failure was quickly contained and repaired. The inconvenience to consumers remained minimal. The lights stayed on for the overwhelming majority of European households and businesses. Where supply was cut, it was cut by protection gear in a way that facilitated quick restoration, generally within 30 minutes (and completely within two hours). President André Merlin of the French transmission system operator RTE and others argue that "Europe's power network had worked smoothly."[7] The international transmission system operator organization UCTE (currently ENTSO-E) confirms that "a Europe-wide blackout could be avoided. The decentralized responsibilities of TSOs have demonstrated their efficiency."[8] A secure electrical Europe exists, successfully built and operated by the power sector, not the EU: the UCTE system connects some 450 million people "from Portugal to Poland and from Belgium to Romania" at an "electrical heartbeat" of 50 Hz.[9]

Introduction

For a brief moment in November 2006, Europe's electric power network became highly visible to the broader public by virtue of its disturbance. The event exposed a magnificent technological collaboration that spans and transcends the subcontinent. Since their inception in the 1880s, electricity networks had proliferated, and by 1970 Europe had been linked up electrically from Lisbon to Moscow and from Trondheim to Naples. This vast technological system, normally performing silently in the background, came to bind Europe's households, industries, and nations in electrical interdependency. Proponents were delighted: this transnational system allowed electric utilities to supply cheaper and more reliable electricity to consumers – for, contrary to Romano Prodi's complaint cited above, electric utilities have helped each other for about a century, providing mutual security and system stability across national borders. Yet the events of 4/11 also suggested that transnational electric power collaboration came with new transnational vulnerabilities, in which an incident in northern Germany can make the lights go out in Portugal and Tunisia.[10] Such events are all the more serious since during the twentieth century, households, industries, administrations, and other institutions developed an addiction to cheap and steady electricity supply.[11] Either way, for good or ill, electric power networks tie together economies and societies in a much more mundane way than EU politicians and institutions. Electrical interdependency constitutes a major site for Europe's "hidden" integration, occurring largely outside the spotlight of popular media, history books, and the formal European integration process represented by the EU and its direct forerunners – at least until very recently.[12]

The 4/11 failure teaches us yet another important lesson. In particular, its subsequent discursive career in EU policy-making shows that notions of "vulnerability" and "European" were, and are, interpreted and contested among stakeholders. Analysts should not take these terms for granted in any essentialist or predefined way: the very same events could mean proof of fragility and non-Europe to some, but reliability and successful European cooperation to others. Taking one of these interpretations as our a priori definition would imply taking sides and lending voice to one particular stakeholder while silencing others. Instead, our narrative and analysis must capture how electrical Europe and its associated interdependencies, reliabilities, and vulnerabilities were negotiated, shaped, and interpreted as part and parcel of one and the same historical process.[13]

This, then, is the aim of this chapter. We set out to inquire how, by whom, and why Europe's electrical interdependencies were built; how they were interpreted in terms of reliability and vulnerability, and how different forms of vulnerability were anticipated and reconciled in the process; and what was "European" about all of this, in terms of electrical integration and fragmentation, inclusions in and exclusions from institutional collaborations, and discursive claims to the notion of "Europe."[14] To achieve this we synthesize older, nation-centered electricity historiography with recent work on electric internationalism into a transnational history that appreciates and inquires about the complex, multilayered shaping of electrical Europe.[15] We have consulted the archives of relevant international organizations as well as contemporary government and engineering publications to investigate the role of perceived vulnerabilities herein. As we shall see, changing notions of interdependency and vulnerability were heavily implicated in the electrical wiring of Europe.

Inventing electric (inter)nationalism

Winter, 1921. An extraordinary drought in northern Italy reduces the production of Italian hydroelectricity and threatens the industries of the country's economic heartland, the Po Valley. Local governments ration the available power to industry, while foreign power companies come to the rescue. Coal power stations in Nancy and Vincey, France, produce electricity for Zürich and Geneva, Switzerland. This move frees production capacity of the Alpine hydropower plants at Brusio and Thusis, Switzerland, for emergency power supply into neighboring Italy. These emergency measures are possible thanks to recent interconnections of the power systems of the French, Swiss, and Italian utilities involved, and successfully prevent the shutdown of Po Valley industry.

After the crisis, this event, too, continues its career in politics. In March 1922, Paolo Bignami, engineer and member of the Italian chamber of deputies, reports to a League of Nations committee that the way in which northern Italy's problem was solved is "perhaps a first step towards the solution of wider and more interesting problems."[16] Why should the collaboration between utilities providing mutual backup stop at national frontiers?

So began international electric interdependency and vulnerability debates in the League of Nations (established 1919, succeeded by the United Nations in 1946), which would become an important setting for debating Europe's electrical integration in the interwar years. The 1921 event illustrates that, by then, several electrical cross-border collaborations had been established, but these constituted only "a first step" and much work remained to be done. In addition, the incident underscores the relevance of economic and electrical vulnerability perceptions as a *leitmotif* for this endeavor. Yet this time, cross-border collaboration by the electric power sector counted as a reliability-improving measure countering the threat of blackout, not as a source of vulnerability and blackout as in the EU view of 4/11. Also note that Bignami spoke of international and national power networks; the notion of a "European power grid" had not yet been born.

The birth of cross-border collaboration

Such talk of international and national electricity infrastructure itself was rather recent. Prior to the First World War, the national–international distinction had been much less an issue. High-voltage, alternating current transmission had developed rapidly since the 1890s but was rarely interpreted in national, let alone international, contexts. Early electric power systems instead served local or (micro)regional purposes. Since national borders were not yet key obstacles, and state governments not yet important players, such local or microregional systems were established within, as well as across, political borders.

Early cross-border systems took many forms. For instance, between 1894 and 1898 a dam and hydropower plant were constructed on the Upper Rhine at Rheinfelden, a binational town on the border between the Swiss canton of Aargau and the grand duchy of Baden, Germany. The formal border was the so-called *Thalweg*, the deepest continuous line along the Rhine watercourse. A bilateral agreement entitled each side to half of the electricity generated. The power system was co-funded by electrochemical firm Elektrochemische Werke, which built a plant near the hydroelectric station and became a major consumer. The system grew to supply nearby villages on both sides of the border, and from 1906 it extended to Guebwiller in Alsace, France, by means of a 40 kV line. The Rheinfelden system was now microregional, connecting consumers in three countries.[17]

Other models of early cross-border power systems include the Alpine hydropower station of Brusio in the Swiss Canton of Ticino, erected in 1907 for the purpose of commercial electricity export to northern Italian factories. The Silesian city of Chorzow became implicated in cross-border electricity exchange because of a border change: the city became Polish in 1922 but stayed electrically connected to the system of Zarborze, then still a German town (it later became Polish too). In still other cases, existing utilities connected across borders for mutual benefit. From 1915, collaborating municipal utilities in southern Sweden exported surplus hydropower to the thermal power-based rural district system of NESA, north of Copenhagen, Denmark, using a submarine power cable under the Øresund strait. The cable had been paid for by the receiving power company.[18]

In the continued absence of national power grids, microregional cross-border initiatives ensued after the First World War. Czechoslovakian utilities with access to large coal reserves near the German border engaged in cooperation with utilities in the German states of Bavaria, Silesia, and Saxony. In Hungary the first stretch of cross-border transmission line followed the electrification of the Budapest–Vienna railroad completed in 1932. The Rheinisch-Westfälisches Elektrizitätswerk (RWE) in Western Germany, still one of Europe's largest power companies today, connected its coal-fired system based in the Rhine-Ruhr area to the hydropowered system in the Austrian province of Voralberg by means of a 600 km transmission line. This well-advertised engineering feat was completed in 1930 and promised a bright future for long-distance power transmission.[19] By then, despite the lack of a single integrated network, individual utilities in Austria, Czechoslovakia, Denmark, France, Germany, Finland, Hungary, Italy, Lithuania, Luxembourg, the Netherlands, Norway, Poland, the Saar region (a League of Nations protectorate between 1920 and 1935), Sweden, and Switzerland engaged in some form of cross-border exchanges.[20]

These developments coincided with the propagation of electrical collaboration, though not necessarily across political borders, as state-of-the-art electrotechnical science. The argument had already been pushed in the 1910s – for instance, by a prominent international authority, professor at the Berlin Institute of Technology, and director of the large electrotechnical manufacturer Allgemeine Elektrizitäts Gesellschaft, Georg Klingenberg.[21] At that time, power stations were usually located near consumers, mostly in cities, and operated in isolation from other power plants. New electrotechnical science developments, however, enabled the interconnection of different power plants by high-voltage, alternating current power lines. The trick was to run such interconnected power plants synchronously, meaning that all electromagnetic machines operated in tune at one frequency. Once this was accomplished, existing power stations could run in parallel and form one power pool, in which multiple power stations jointly supplied much larger areas. Moreover, distant power stations sited near mine mouths or hydropower resources could be integrated into such pools. Adversaries pointed out that such power pools came with huge investments in high-voltage, alternating current power grids.[22] International authorities like Klingenberg, however, stressed their vast economic advantages. Electricity could be produced wherever it was cheapest in the pool at any given moment, thus exploiting the complementary characteristics of large (achieving economies of scale) and small (avoiding overproduction when demand was low) power stations, and of hard coal, lignite, diesel, and hydropower plants. Importantly, cooperation also reduced the necessary investment in local backup units for emergencies or maintenance: instead of guaranteeing full backup capacity for each and every power plant, this could be drawn instantly from the pool and thus be shared among partners. Electrical collaboration in power pools was therefore accompanied by a different way of providing backup and reliability management. In the following decades, secure and undisturbed operation thanks to mutual backup and system stabilization (the larger the pool, the more power stations instantaneously counteract any disturbance of the shared frequency) became a key motive for setting up ever larger synchronized power pools.[23]

Nationalization and internationalization

Along with the notion of power pools, however, came new actors and new categories for electrification. While existing electric utility owners – large and small commercial companies, municipalities and other lower governments, rural cooperatives – jostled for position in the booming electricity sector, national governments became an important new player. During the First World War, many state governments not only introduced obligatory border passport requirements that stayed in place after the war, but also increasingly committed to economic nationalism. After November 1918, as hostilities still loomed and coal markets remained distorted, governments often tried to interfere in electricity supply. Where successful, they tightened their grip on prices, hydropower resource development, electricity exports, electricity as a national service, and national power grid planning.[24] The nation-state, in short, became a potent additional category for electrification.

Indeed, Klingenberg's early call for synchronized power pools had already appealed to state governments as carriers of this development. While the suggestion was very controversial on Klingenberg's home ground – Germany – such schemes were actively discussed in the state governments of Saxony, Baden, Bayern, Prussia, and Württemberg before the end of the First World War.[25] By then these discussions had been picked up elsewhere in Europe.[26]

To the West, the British government was alerted to the coal-saving advantages of power pools during the war. By 1919, power pools were officially identified as a potential source of national industrial strength in an Electricity Act that pushed utilities towards voluntary national collaboration. By 1926, Britain's loss of national prestige and power was blamed explicitly on its continued backwardness in electrical technology compared with Germany and the United States. In response, a new Electricity Act set up the Central Electricity Board to build a synchronized national power grid (power stations remained private until after the Second World War), by and large operational ten years later.[27]

To the East, Russian electricity generation and distribution were completely nationalized and forged into the largest electric power collaboration to date. As in Britain, the decision process was rife with fears of electrical backwardness and its implications for the national economy. Vladimir Lenin famously argued that

> *Communism is Soviet power plus the electrification of the whole country.* Otherwise the country will remain a small peasant country... Only when the country has been electrified, and industry, agriculture and transport have been placed on the technical basis of modern large-scale industry, only then shall we be fully victorious [original emphasis].[28]

Lenin then initiated and supervised the State Commission for Electrification of Russia (GOELRO) in 1920, producing the *План электрификации Р.С.Ф.С.Р*, an electrification plan for the entire Russian socialist republic including some 30 high-priority large power stations and extensive transmission networks. Less centralized electrification schemes, privileging smaller urban and rural systems, existed but were bypassed. This national electrification scheme, too, was largely realized within a decade.[29]

In between East and West, governments interfered in various ways to varying degrees of success. For instance, in the 1920s the Belgian, French, Luxembourg,

Portuguese, and Swedish governments adopted national power grid schemes.[30] In countries where direct national government interference was ultimately rejected, such as Denmark or the Netherlands, they still influenced the pattern of electrification.[31] Overall, the strategic importance of energy and electrification for the national economy was increasingly emphasized. In countries possessing vast black, brown, or white coal energy resources, these were relabeled as "national resources," which demanded national government control. In France and Switzerland, the debate was about nationalizing hydropower resources.[32] Berlin pushed hard to hold on to Upper Silesia and its coal resources, without which "Germany will fall apart completely."[33] Austria lost most of its coal assets after the war, and its government embarked on a national electrification scheme to utilize hydropower.[34] Governments of energy-importing countries often pushed national electrification schemes in order to optimize the use of resources on a national level – electrification should benefit the national economy rather than urban or microregional ones.[35] Electric nationalism came with notions of national economy, autarky, and what we today would call national energy security. Thus, in our interpretation, state-initiated electrification often aimed to counter perceived national economic vulnerabilities. As a result, the national element in electrification progressed steadily and constituted an additional layer to microregional and local electrification patterns.

This development of electric nationalism, however, was contested not only by existing players such as private, municipal, or cooperative utilities, resisting state interference with varying degrees of success; often national grids only emerged after the Second World War. Electric nationalism was also countered by a new electric internationalism.[36] Much of this movement was initiated and carried by industry. Electrotechnical manufacturers and financial institutions joined forces in multinational holding companies in order to construct power systems worldwide. Examples include Elektrobank (Allgemeine Elektrizitäts Gesellschaft combined with German and Swiss banks) and Motor (Brown Boveri with Swiss finance).[37] Moreover, the electrotechnical industry and electric utilities used international organizations to create larger markets for equipment and to liberalize cross-border electricity flows. The emerging debates at the League of Nations were an example from the highest political stage of the attempt to shift the dominant concern for national energy security towards the promises of mutual cooperation. In addition, international cooperation was strengthened by setting up a series of new international organizations addressing the electricity domain. The standard-setting International Electrotechnical Commission had already existed since 1906; in the 1920s followed the International Council on Large Electric Systems, established in 1921 to provide a platform for the exchange of information about electricity generation and high-voltage transmission in large systems; the World Power Conference (currently World Energy Council), conceived in 1923 to restore a shattered European electricity industry, although the agenda immediately expanded to cover all forms of energy; and the International Union of Producers and Distributors of Electrical Power (currently Eurelectric), established in 1925 on the initiative of Italian, French, and Belgian utilities.[38]

The rise of the national as a category for electrification thus came with new structures for international collaboration. Notably, while the new international organizations gladly used terms like "international" and "world" in their names, their membership was overwhelmingly, and sometimes exclusively, European. Still, "Europe" as such had not yet been claimed as a lead category for electrification. So far when we have spoken of "Europe" we have implicitly projected a broad yet imprecise geographical notion of the term – thus made sure to include Russia, unlike many present-day political identifications of "Europe" with the EU. When "Europe" became an explicit actor category for electrification around 1930, it did indeed have a pan-European scope (Figure 3.1).

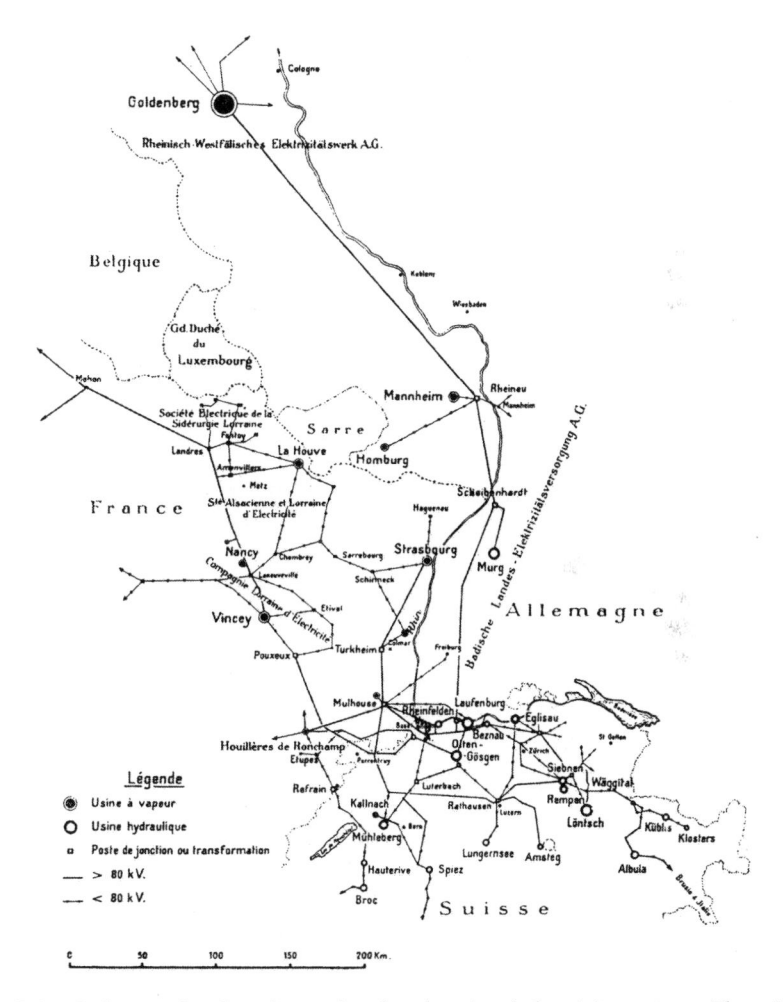

Figure 3.1a Early cross-border microregional and national electricity systems. The microregional system around Rheinfelden extending into three countries around 1926.
Source: Niesz 1926, p. 1026. Used by permission of the World Energy Council.

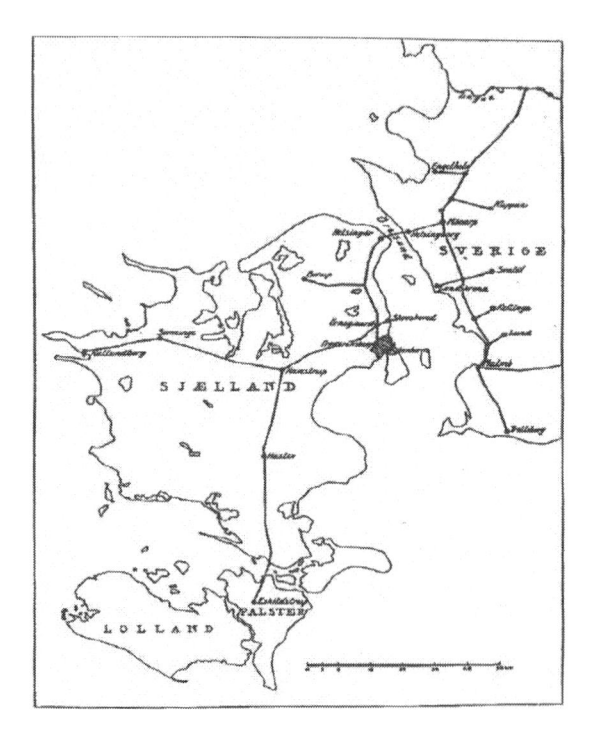

Figure 3.1b The southern Swedish–eastern Danish power pool across the Øresund straits by the mid-1920s.
Source: NESA 1927. Used by permission of DONG Energy.

Imagining electrical Pan-Europe

November 20, 1932. The Dortmunder Rundschau *newspaper enthusiastically reports from an exhibition by architect Hermann Sörgel.*[39] *The exhibition displays a plan of unprecedented imagination, ingenuity, and magnitude. For one, it displays technical drawings and scale models of a dam closing the Mediterranean at the Straits of Gibraltar. Since more Mediterranean water evaporates than rivers contribute, this would cause the water level to decrease and create new space for human settlement. Next, hydroelectric power plants situated at the Gibraltar dam and several secondary dams would produce more electricity than all Europe's existing power plants combined. This electricity would be distributed all over the continent by means of a pan-European high-voltage grid. Uniting Europe's states in electrical interdependency, the scheme would provide unity, prosperity, and peace for a war-prone continent: "the integration of Europe by power lines is a better peace warranty than treaties on paper; because in destroying these power lines, each nation would destroy itself."*[40]

The plan is originally presented as the "Panropa project" to express support for the flourishing Pan-European movement. Its new 1932 name of "Atlantropa project" further denotes that Europe and Africa will be forged into a new continent able to withstand the rising powers of Asia and America. After the Nazi takeover the plan is adjusted

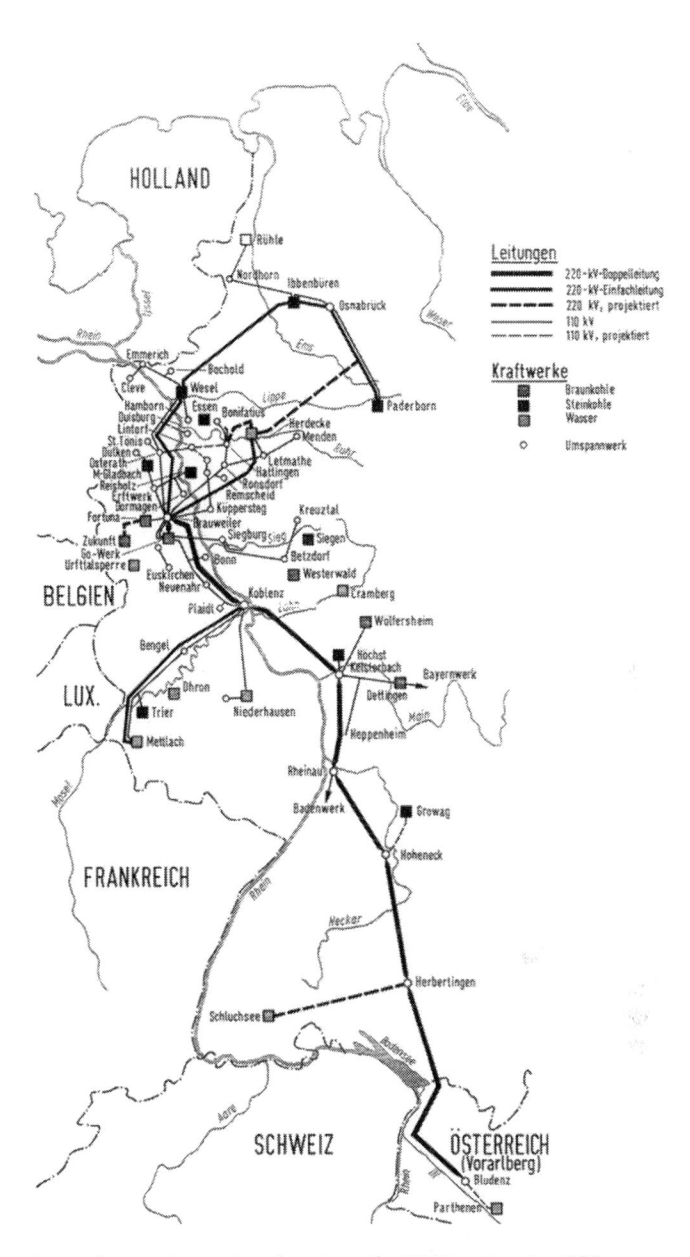

Figure 3.1c A very large microregional system: the RWE system in 1928.
Source: Boll 1969, p. 45. Used by permission of BDEW Bundesverband der Energie – und Wasserwirtschaft e.V.

72 *Inventing Electrical Europe*

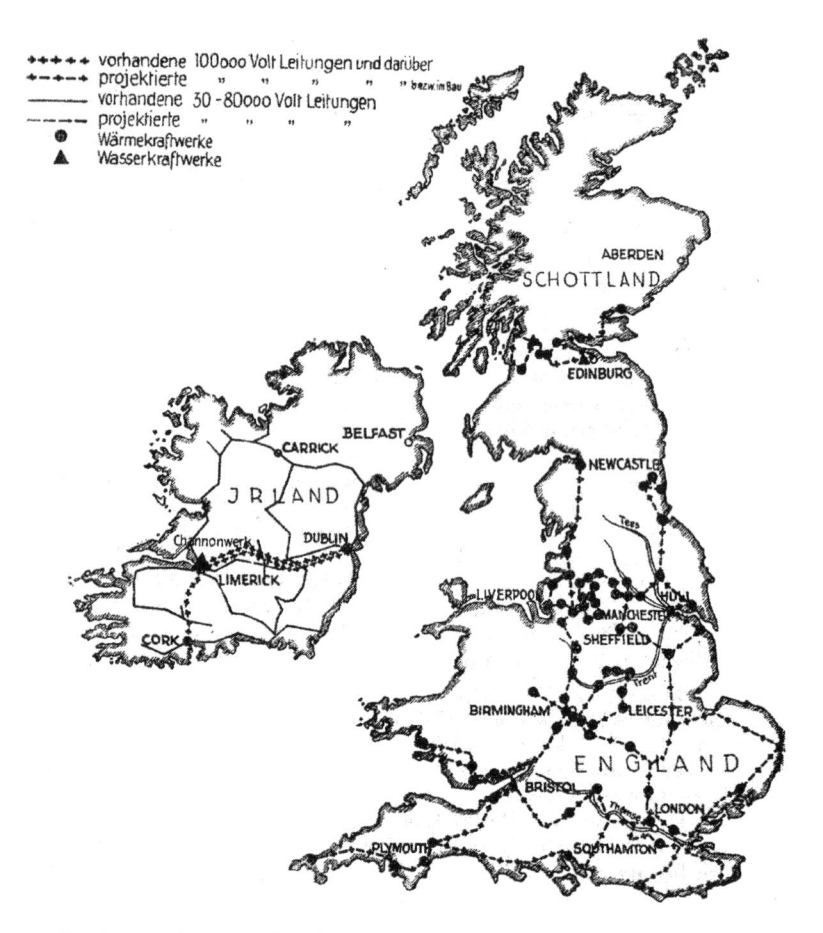

Figure 3.1d The British national grid.
Source: Legge, 1931, p. 123.

to Nazi ideology, posing Greater Germany and the Italian Empire as Atlantropa's pillars. In 1949 the Atlantropa Institute advertising the project has about 700 members and eight branches in different German cities. In 1952, Sörgel dies, nuclear power takes over hydropower's role in [the] political and public imagination, and Atlantropa fades to the background. The Atlantropa Institute is closed in 1960. Decades later, Sörgel's Gibraltar Dam re-emerges in public discourse as an example of technocratic megalomania and ecological nightmare. By contrast, his transmission grid design implicitly echoes in present-day sustainable energy visions of a "supergrid," a single high-capacity power grid integrating off-shore wind parks in the North Sea, Baltic Sea, and Mediteranean; Nordic and Alpine hydropower plants; and Sahara and Arabian desert solar power plants, thus joining Europe, the Middle East and north Africa in energy and environmental unity.[41]

Thus reads an abridged biography of the most imaginative of interwar visions of electrical Europe. It was conceived around 1930 along with several other schemes of what we today would call a "supergrid."[42] In 1930, George Viel, president of the southeastern section of the French Association of Electricians, proposed a power pool including 3000 km of ultra-high-voltage power grids stretching from Trondheim, Norway, in the north to Naples, Italy, in the south, and from Lisbon, Portugal, in the west to Russia in the east. It would integrate Europe's massive, yet scattered, hydropower resources into one energy economy. At the World Power Conference held in Berlin that same year, Oskar Oliven, director of the Gesellschaft für elektrische Unternehmungen, presented a European electric power program involving 9750 km of power lines. This power pool had a similar geographic reach and was fed by large hydropower stations, mostly in Scandinavia and the Alps, and thermal power plants near Europe's major coal deposits. In 1930, Ernst Schönholzer published another, fourth, vision of a "European power grid" in the leading Swiss engineering journal. His scheme involved no less than 15,000 km of power lines from Dublin and Lisbon to Istanbul and Moscow.[43]

Three aspects of this sudden boom in European electrification schemes are important to our analysis. First, "Europe" became vigorously promoted as a category for electrification. While Viel presented his design as an add-on to a French national power grid, the others foregrounded "Europe" as the preferred unit for electrification. Their designs differed in detail but all clearly interpreted Europe on a macroregional scale, embracing or even transcending the Continent.[44] Schönholzer, like Sörgel, prioritized the promise for Europe's future that electrification held and did not eschew technological challenge. His design, accordingly, included power lines reaching Moscow in the East and Dublin, Glasgow, and Manchester in the West; the latter came with a dam across the English Channel.[45] Sörgel's scheme, as we saw, did not even accept the Mediterranean as Europe's southern border: the sea should be connector rather than border, as it had been in ancient times, and his power grid extended well into northern Africa and the Middle East – in a clearly colonialist mindset. Oliven and Viel, by contrast, were more concerned with the technical and financial feasibility of their schemes. They discussed state-of-the-art electrotechnical science and construction possibilities, and provided cost estimates. This led Oliven, deterred by the technical challenges of crossing the English Channel and the vast distances of Russia, to exclude Britain, nearly all of Russia, and the Baltic states. Still, his design had quite a pan-European scope, stretching from Lisbon and Calais in the West to the Donets River basin (a River Don tributary) across the Ukrainian-Russian border in the East. After all, he added, if freight transport, telecommunications, and radio networks crossed the Continent, why not electric power systems? Certainly electricity grids were much less difficult to establish than railroad lines, which by then traversed the Continent from Lisbon to Vladivostok.[46]

Second, in our reading, pan-European electric integration was articulated as a response to several perceived political and economic vulnerabilities. In this respect they resembled nation-based electrification schemes aimed at countering national economic and political problems. Yet they differed in spotlighting nationalism itself as the main problem. The authors here drew on increasingly popular ideas of European political unification and the European movement, which experienced an apogee in this period.[47] Oliven connected to this tendency superficially by framing electric power as a challenge "for all peoples of Europe" and emphasized how "the idea of peaceful cooperation between all people... is steadily gaining currency."[48] Sörgel and Schönholzer explicitly announced support for the pan-European movement and imported some of the fears of this movement into the electricity domain. Pan-European movement spokespersons sought to unite Europe politically as a counter move to, on the one hand, the intrinsic capacity of Europe's states to prioritize national self-interest at the expense of economic fragmentation, military expenses, and the permanent threat of war and self-destruction; and, on the other hand, the rising powers of the United States, the Soviet Union, and Asia.[49] Schönholzer and Sörgel explicitly reproduced these concerns: In Schönholzer's words, "what if we, usually so 'clever' Europeans,... set aside our 'political tensions' once and for all, and created *international power highways* as a symbol of a basic cultural community, which will not bring military expenses and war to individual states but profits for the economy? [original emphasis]"[50] For Sörgel, electrical interdependency was a better peace guarantee than paper treaties: "Europe is a large cage with singular cells [the individual countries]. Those who dare open their cage for the sake of a beautiful idea [Europe's political unification] become prey of the others. Only a common, simultaneous interlinking in a high-voltage network creates a European Union."[51] This unification was all the more urgent since Europe was increasingly squeezed between the rising powers of Asia and the Americas, and Sörgel envisaged a world of three great powers: the three As – Asia, America, and Atlantropa.

Third, it is important to note that none of these schemes was realized. Though for the most part technologically and financially possible, they did not gain sufficient support. Only for a brief period of time did ideas of a pan-European power pool gain strong momentum. Political support came particularly from the League of Nations and the International Labour Office.[52] The League added European electricity system planning to the agenda of its Commission of Enquiry for European Union at the suggestion of Belgian representatives.[53] The International Labour Office promoted a European electricity grid to diminish international political tensions and provide employment during the Great Depression. However, their envisioned model of top-down construction of a European power grid, backed by political will and international financing, became a road not taken. In a context of economic depression and increasing national strategic interests, international financing plans for a European power grid were torpedoed. Domestic pressures, not least coal-mining interests, caused even the Belgian initiators to shift sides.[54] Many engineers now favored a gradual and decentralized approach to European interconnection, based on national electrification schemes that could subsequently be connected. The Europeanists became isolated, and the push to build a supranational electricity system ended. System-building activity was left to power companies and national governments. The concept of a European power grid countering economic and political vulnerabilities, however, was there to stay. It was now seen as a patchwork of gradually emerging and collaborating national networks, rather than a supranational system to be built from scratch.

At the end of this period, another unrealized pan-European electrification scheme added yet another aspect to the theme of electricity and vulnerability. The role of electricity supply in war had been acknowledged since the First World War. English and German governments found power pools attractive instruments to economically power their war industries in times of fuel shortages; the French government pushed transmission lines to cater to the Maginot Line, the fortifications on the Franco-German border; and the entire Dutch-Belgian border was sealed by a 1.80 m high, 2,000 Volt electric fence to electrocute war refugees, deserters, volunteers to the Allied forces, spies, and smugglers.[55] In response, electricity supply system elements themselves became important military targets.[56] Worse, in the age of aviation, bomber planes could follow power lines to key centers of consumption, including strategic war industries. When designing a European power grid to integrate an envisioned Neuropa from the Atlantic to the Ural Mountains, Nazi engineers therefore opted for an underground system. Fritz Todt, general inspector for water and energy and a civil engineer, argued that underground cables were safe from atmospheric disturbances, air attacks, and sabotage. Besides, they did not disfigure the landscape and did not interfere with electric communications.[57] This underground system was not realized either, as its implementation was delayed and started only a few months before the final defeat of the Nazi regime. The military vulnerability of overhead power systems, however, was widely recognized after the war, especially when interrogated Nazi military leaders stated their surprise that Allied bombers had neglected this major vulnerability of their war economy (Figure 3.2).[58]

Wiring and securing mesoregional Europe

Bretagne, western France, January 12, 1987. Very cold weather and massive use of electrical heaters by consumers trigger the failure of three out of four active units of the Cordemais thermal power station. Nine thermal and nuclear units in neighboring power stations fail in turn. Lights go out in Paris and Le Havre, and the disturbance threatens the integrity of the French system as well as the synchronized power pool of the Union for the Coordination of Production and Transport of Electricity (UCPTE), which by now covers most of Western Europe.[59] Network operators of Électricité de France massively disconnect consumers in order to rebalance production and demand. In addition they draw power from Spanish, German, and Belgian partners. Belgian operators, in turn, import power from German and Dutch plants. Belgian network operators prevent further electricity export as their own system threatens to break down; Swiss operators start two hydropower units to counter a domestic frequency dip; and Italian dispatchers start additional power units to stabilize their frequency. Yet consumers in these countries, and in eastern France, do not notice the stress on their power grids at all; the failure is successfully contained and repaired. The incident inspires a sharpening of French security measures but is widely cited as an example of effective international collaboration to contain and counter power system disruptions.[60]

The events of January 12, 1987, exposed a large-scale increase in power pools in the postwar era. Power supply in western France was now embedded in the

76 *Inventing Electrical Europe*

Abb. 40. Das Raubtier „Mensch". Europa ist ein großer Käfig mit Einzelzellen.

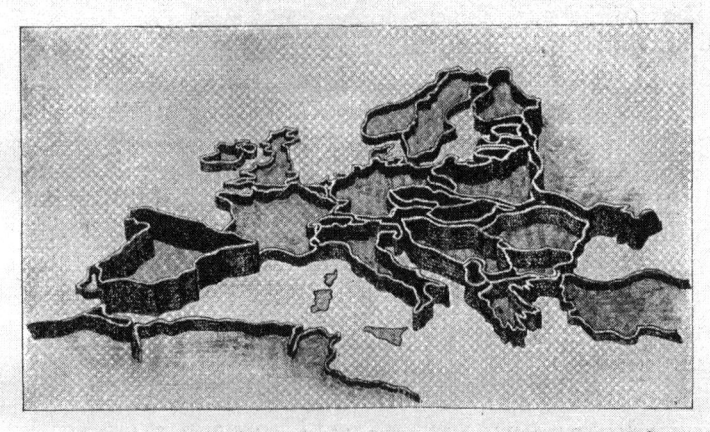

Wer es einer bloßen schönen Idee zuliebe wagen würde, seinen Käfig zu öffnen, wäre die Beute der anderen.

Abb. 41. Statt trennender Mauern: bindende Leitungen!

Nur eine gemeinsame, gleichzeitige Verkettung durch ein Groß≠Kraftnetz schafft eine Europa≠Union.

Figure 3.2a Early proposals for a European "supergrid." Sörgel's Atlantropa Plan.
Source: Sörgel 1938, p. 91.

transnational synchronized power pool of the UCPTE, which we characterize as a mesoregional collaboration (as opposed to subnational microregional power pools and imaginary interwar macroregional, pan-European pools).[61] By 1987 it included power companies from many countries in Western and Continental Europe, but excluded Scandinavia, Britain, and so-called Eastern Europe. These areas possessed synchronous transnational power pools of their own. Cooperation

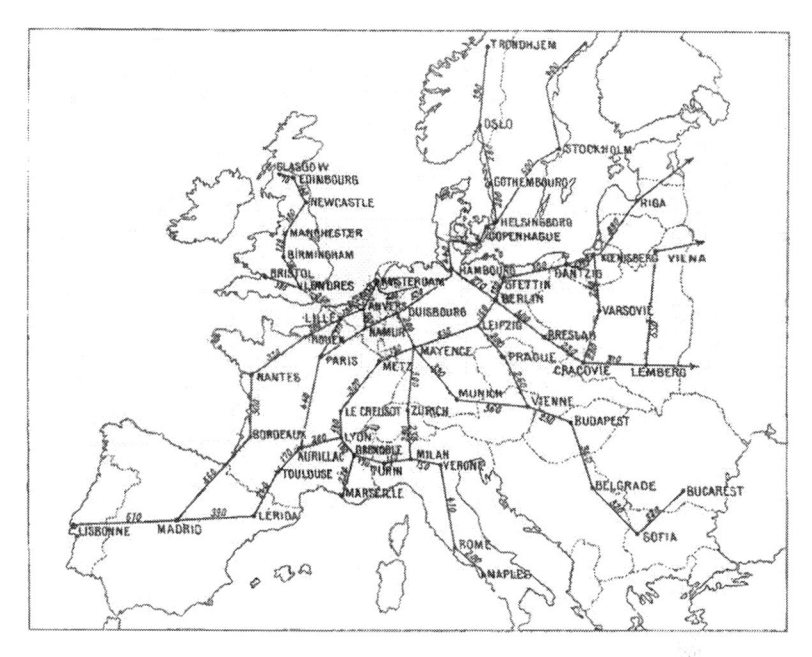

Figure 3.2b The European plan of George Viel.
Source: Viel, 1930.

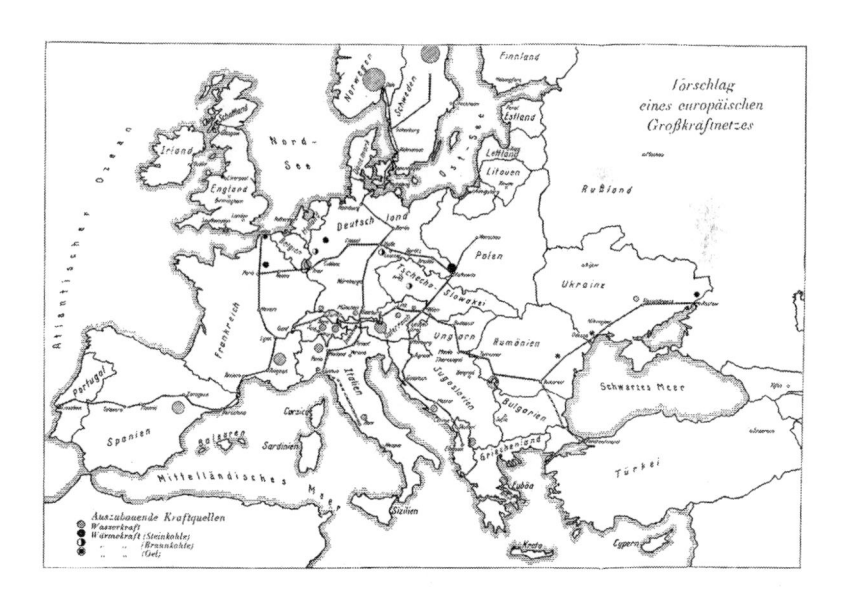

Figure 3.2c Oskar Oliven's plan for a European system.
Source: Oliven 1930. Used by permission of the World Energy Council.

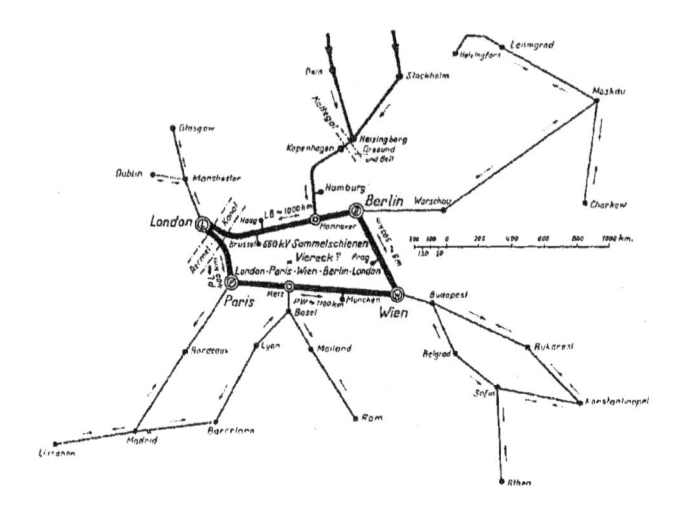

Figure 3.2d A European grid plan by Schönholzer.
Source: Schönholzer 1930. Used by permission of the Schweizerische Technische Zeitschrift.

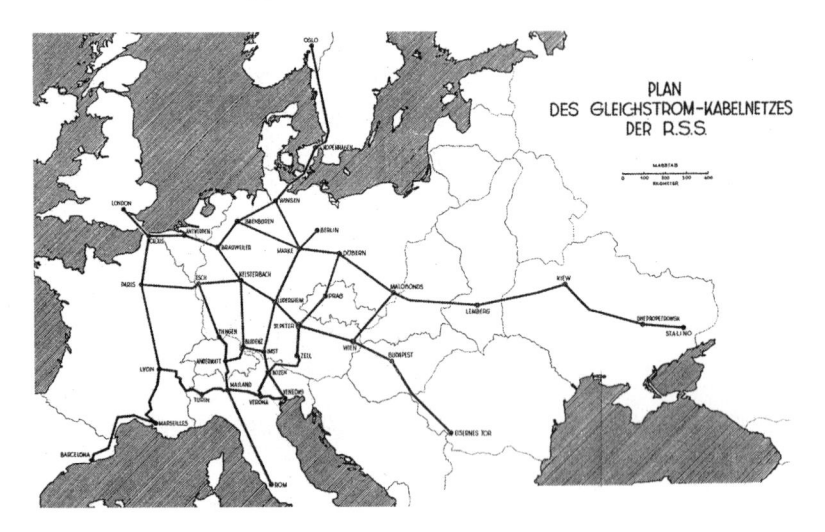

Figure 3.2e The Nazi proposal for a European network.
Source: Maier 2006, p. 131. Used by permission of Helmut Maier.

between such mesoregional power pools existed but took an asynchronous, and therefore less immediate and tightly coupled, form.

The 1987 events also reveal an important change in vulnerability perceptions and priorities. Synchronous power pools originally served to reduce socioeconomic, political, and military vulnerabilities. Once in place, European economies and societies increasingly depended on their undisturbed functioning, and securing a power supply itself became a major concern. The 1987 events confirmed to many observers that transnational power pools did this job well. By the late

1980s, some organizational sociologists even called electric power pools "high reliability organizations": in an age of increasingly complex technological systems, they provided a remarkably high degree of service reliability.[62] The 1987 events also demonstrate a particular form of coping with potential failures: emergency response in Western Europe was decentralized in the hands of individual power companies, not their international organizations or government institutions. We now turn to the historical processes shaping this particular material, institutional and discursive makeup of postwar electrical Europe.

Two models for mesoregional collaboration

Postwar mesoregional collaborations came in two different models of electrical interdependency, each with its own implications for vulnerability. A first and very influential model developed in the continental part of Western Europe. After several years of debating and preparations, power company representatives from Belgium, the Federal Republic of Germany, France, Italy, Luxembourg, the Netherlands, Austria, and Switzerland founded the non-governmental UCPTE in 1951. Its chief aim was to coordinate a transnational power pool. Seven years later, synchronized operation in the UCPTE supply area was operational.[63]

Two observations on this collaboration are particularly important to us here. First, UCPTE spokespersons regularly claimed to work for "Europe." In 1955, Heinrich Freiberger of the Vereinigung Deutscher Elektrizitätswerke hoped that the UCPTE "shall be allowed to continue to work silently and effectively for Europe and therefore for the greater good of humanity and of peace."[64] On the occasion of its 20th anniversary, Italian UCPTE president P. Facconi emphasized the organization's "historic importance for its remarkable contribution to the ideal of a 'United Europe'."[65] Most of the time, however, European integration ideals were absent. The 1954 statutes do not speak of "Europe" at all but foreground internal power sector advantages. Importantly, these had an economic and a reliability component. As for economics, a transnational power pool enabled an economic mix of power stations, and should in particular help to eliminate losses of excess hydropower in postwar Europe. In a synchronous power pool, all available water could be led through the turbines and fed into the power pool, instantaneously enabling a fuel cost reduction in thermal power stations elsewhere in the system. Hydropower wastes had largely been eliminated in the UCPTE system by 1970.[66] As for reliability, the key motive was that in a synchronous collaboration, any power-station failure would be counteracted in a matter of seconds by other generators in the pool. In this way, "all production units in the synchronous system jointly counterbalance the disturbance of one power station, regardless if this power station is located in Lisbon, Palermo or Hamburg, Le Havre or Vienna."[67]

Second, and contrary to the next model of transnational collaboration we discuss below, these concerns for economic and reliability advantages translated into a decentralized model of transnational organization. This choice had been in the making for several years. After a devastating Second World War, Western European policy-makers and utility representatives looked to the United States for examples. US Marshall Plan (1947–1951) negotiators pushed supranationally owned and financed European power plants in a centrally planned and controlled power pool.[68] Such a system would boost the Western European economy and thereby provide a barrier to the spread of communism. Accordingly, the Marshall Plan's International Power Program should finance "projects [...] selected without regard to national frontiers."[69] As in the 1930s, however, electric utility representatives preferred a looser collaboration. Visiting the United States on a Marshall Plan Technical Assistance Mission, they were impressed by the centralized, state-of-the-art Pennsylvania–New Jersey Interconnection, which used a single control center to manage electricity production and load management of the entire collaboration in an effort to optimize the overall system economy. However, they found this system unfit for Europe. The South Atlantic & Central Areas Group example would serve better: this huge interconnected system connecting the Great Lakes to the Gulf of Mexico and was organized in a decentralized way as a voluntary association of over 80 partner companies. Each partner managed power supply in its own supply area.

Just as the operators of the South Atlantic & Central Areas Group found that by far the larger part of the economic advantages of interconnected operation could be gained within the relatively small systems of single companies, so it has been found in Europe that the major advantages are to be gained within national frontiers.[70] Back in Europe, these power company representatives accordingly managed to divert funds from the International Power Program to distinctly national projects. Additional cross-border power exchanges were to be left to free negotiations between partners, and the UCPTE was established to coordinate this effort.[71]

In this scheme the UCPTE was intentionally set up as a non-governmental, coordinating body of power company and power authority representatives who participated on the basis of personal membership and voluntary adherence to UCPTE recommendations.[72] Power companies in the UCPTE pool remained fully in charge of their control centers, network-building, and supply in their own supply areas. They also decided on, financed, built, owned, and operated cross-border connections. The UCPTE merely provided coordination and facilitation.[73] Importantly, UCPTE spokespersons stressed that "decentralization is indispensable for economy, security, and continuity of supply on the regional level," for individual power companies knew the particulars of their situation best. Thus "a European centralized control centre...does not exist and could not function properly, because it would not be able to see the needs of the separate regional networks."[74] The events of January 12, 1987 illustrate UCPTE procedures in which power-grid disturbances were not countered centrally by the UCPTE but by individual power company operators restoring supply in their respective supply areas.

This decentralized organizational form was reflected in power-grid construction and electricity flows. In some areas, utilities were internationally minded and developed power grids and exchanges accordingly, most notably in the case of Austrian and Swiss power collaborations with neighboring power companies. In other parts of the UCPTE zone, cross-border grids and exchanges remained minor, and microregional or national power circulation was clearly dominant. To the dismay of the European Commission, by 2000, countries such as Germany and France had a poor "interconnection capacity" (the import capacity relative to domestic generating capacity) of less than 10 per cent. Italy, Greece, Spain, and Portugal, which had joined the UCPTE later, did not even reach 5 per cent.[75]

A second model of mesoregional collaboration developed in the Soviet Union in the late 1950s, when the Russian United Power System embraced other Soviet republics into one huge transnational synchronized system. Incidental links had preceded this initiative, such as a 1955 link to Estonia. Yet in 1959, Nikita Khrushchev unrolled a formidable electrification scheme in his Seven Year Plan, which was accepted by the 21st Party Congress. The plan envisaged a set of mutually interconnected power pools including a Center Pool (around Moscow) interconnected to a Middle Volga pool and a Ural pool, a Southern Pool (Ukraine and Moldova), and a Northwestern Pool (the Baltic region and Belorussia).[76] Not unlike the Russian GOELRO plan of the 1920s, the rationale was to boost industrial growth by pooling power resources scattered throughout the Soviet Union, thus allowing efficient deployment of available power stations, avoiding load peaks by combining consumers in six different time zones, and sharing backup capacity "to maintain the reliability of a power supply."[77] In 1965, experts calculated that a power pool in the European part of the Soviet Union could save more than 1,000 MW of installed capacity and another 600 MW by reducing peak loads. By the late 1980s the United Power Systems consisted of no fewer than nine interconnected power pools, extended into Central Europe, Siberia, and the Trans-Caucasus, and covered some 10 million sq. km – equalling the size of conventional geographical Europe from the Atlantic to the Urals (Figure 3.3).[78]

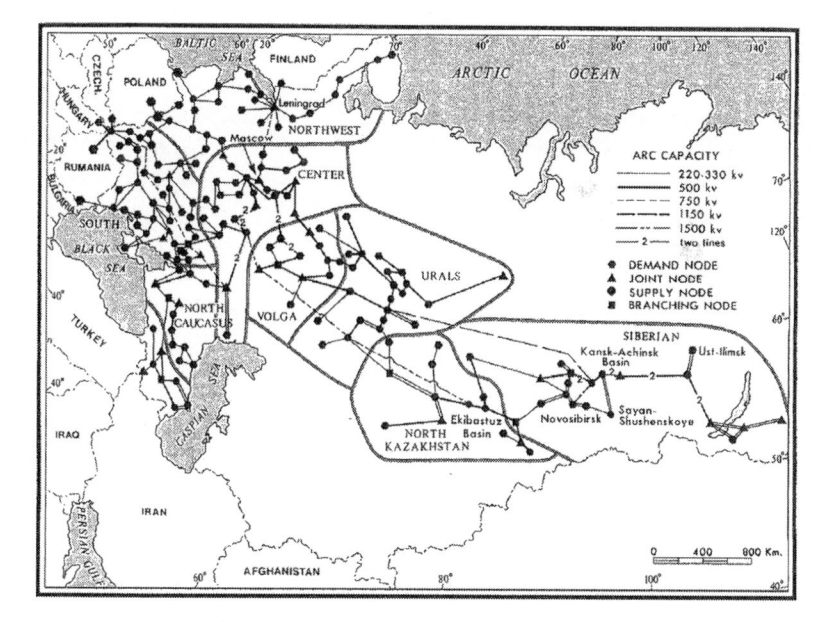

Figure 3.3 The United Power Systems by the early 1980s pierces the Urals as a potential border of Electrical Europe.
Source: Sagers and Green, 1982, p. 292. Reproduced by permission of *The Geographical Review* and the American Geographical Society.

This impressive collaboration differed in several ways from the UCTPE collaboration discussed above, with due implications for vulnerability issues. In line with prevailing paradigms of centrally planned economies, the Soviet Union's transnational system was centrally planned, managed, and controlled. Since the inauguration of the Moscow Central Dispatch Center in 1926, additional control centers had been established for the Southern pool (Ukraine/Moldova) in 1940, for the Urals in 1942, for Siberia in 1959, and for the Middle Volga in 1960. Yet to manage the new mesoregional collaboration, in 1967 a renewed Central Dispatch Center was set up in Moscow to serve the entire Soviet Union. This monitored the other integrated systems, controlled the trunk lines interconnecting them, and administered power exchanges between collaborating pools.[79]

This leads us to a second and related difference. While both collaborations aimed at industrial and economic growth, the UCPTE partners focused on exchanges and projects within national borders and set up the UCPTE itself as sort of add-on. The Soviet scheme, by contrast, was designed to transport huge amounts of energy across the borders of participating republics. This was particularly urgent as 90 per cent of the Soviet Union's energy resources lay outside the urbanized areas in the "center zone." Thus the Center Pool around Moscow massively imported power from the Northwestern, Southern, and Volga systems.[80] Notably, transporting large capacities across vast distances required a "backbone supergrid system" for high-capacity exchanges, which in Western Europe had been envisaged in the 1930s but never got off the ground.[81] This, in turn, demanded massive investment in ultra-high-voltage transmission technology. By the end of the Cold War the Soviet system operated interconnections up to 750 kV, transported capacities between participating power pools up to 5 GW, and was preparing for 1,150 kV transmission. By comparison, UCPTE partners used transmission voltages of up to 380 kV.[82]

Finally, we observe that the UCPTE claimed to work for "Europe" even though, as critics would have it, it included only a string of states on the western side of the peninsula.[83] By contrast, the Soviet system covered much larger parts of geographical Europe and beyond, but eschewed any reference to the term "Europe." This absence partly reflects the fact that Europe's Ural border was erased by electric power networks. In addition, it follows a broader discursive change. During the revolutionary period, Trotskyist authors had interpreted "Europe" as an economic term and included Russia in an economically modernizing Europe, discursively opposed to "Asian" tsarist autocracy and traditions. From the Second World War, however, "Europe" was increasingly perceived as an area divided between a "true" socialist half and a "false," US-dominated capitalist half.[84] In Khrushchev's famous words in the journal *Foreign Affairs*, the main category for economic development now became the "community of socialist countries," increasing their economic power and consolidating world peace, since "the material might and moral influence of the peace-loving states will be so great that any bellicose militarist will have to think ten times before risking going to war."[85] Accordingly, transnational electrification schemes rhetorically bypassed the notion of "Europe" and foregrounded first the Union of Soviet Republics and then the Socialist Brotherhood, regardless of its geographical position, as its primary object. This discursive shift, by the way, did not prevent pragmatic explorations of electricity collaboration and interconnection to Western European partners by the mid-1960s, motivated not least by prospects of massive energy exports to Central and Western Europe.[86]

Electrical alliances on the move

These two models of mesoregional electrical collaboration inspired similar initiatives elsewhere on the subcontinent. Their nearly simultaneous establishment in the first half of the 1960s suggests a mutual influence. By 1970 these externally connected, internally synchronized transnational power pools linked up power stations and consumers from Lisbon to Moscow. This particular configuration of electrical Europe is illustrated in Figure 3.4.

The UCPTE model was more or less copied in Northern and Southern Europe, although forms of collaboration between mesoregional groupings might differ. In Southern Europe, Iberian UCPTE membership was complicated: the Spanish and Portuguese dictatorships were international *personas non grata* and sought political and economic isolation.[87] Spanish, Portuguese, and French power companies therefore set up their own Franco-Iberian Union for Coordination and Transport of Electricity (UFIPTE) in 1963. Its motives – hydropower pooling and mutual system stabilization – and statutes were similar to those of UCPTE. Through France, UFIPTE operated synchronously with the UCPTE pool from 1964:

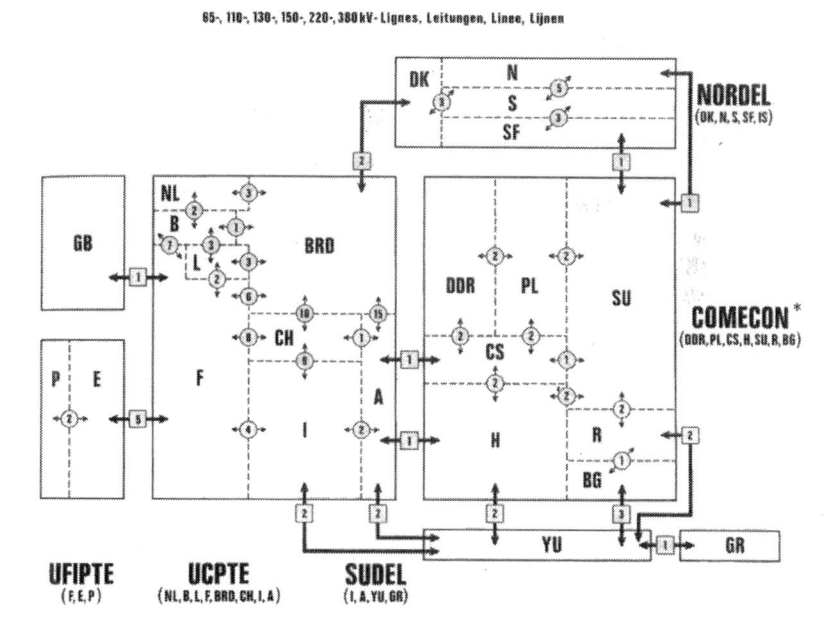

Figure 3.4a Electrical Europe by 1976 is represented by mutually connected mesoregional power pools. Numbers represent power lines.
Source: UCPTE 1976, p. 199. Reproduced by permission of ENTSO-E.

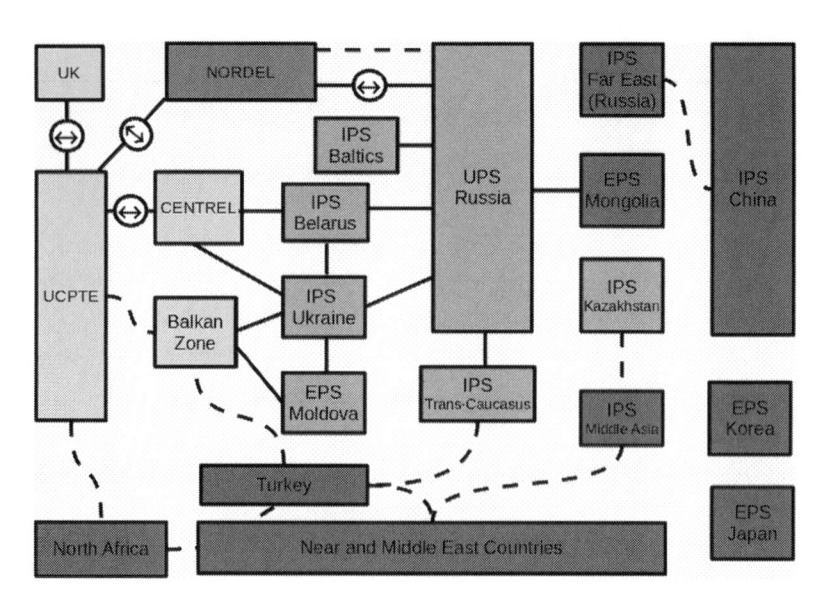

Figure 3.4b Electrical Europe in the early 1990s. The symbol ↔ represent asynchronous connection. Dotted lines represent planned projects.
Source: Based on Hammons et al. 1998.

institutional fragmentation masked physical integration, until the Iberian partners became full UCPTE members in 1987.[88] Likewise, Austrian and Italian power companies, desiring a politically sensitive collaboration with hydropower-rich Yugoslavia, founded SUDEL in 1964, which again resembled the UCPTE. SUDEL and the UCPTE cooperated synchronically from 1975 to achieve greater reliability, particularly for the Yugoslavian (and the soon-to-participate Greek) system. SUDEL members also became full UCPTE members in 1987.[89] All parties agreed that expansion of the synchronous zone improved the stability and reliability of the joint system. Scandinavian power companies also mimicked the UCPTE model but adopted asynchronous collaboration with other groupings that still holds today. In postwar Northern Europe, a Nordic political and economic integration process initially was considered to be a valid alternative to Western European integration, which resulted in a Nordic Council (1952), a Nordic Passport Union (1954), and – at the suggestion of the Nordic Council – the Nordic power collaboration NORDEL (1963), coordinating a Nordic power pool. NORDEL, too, was set up along the decentralized and voluntary model of the UCPTE.[90] The two groupings collaborated on asynchronical high-voltage direct current submarine cables, which do not transmit frequency, do not require tuning of both systems, and accordingly lack the advantages of immediate mutual system stabilization and support. Plans for synchronous collaboration were discussed in the 1960s but rejected as expensive and risky; the necessary modifications to the existing system would not outweigh the gains.[91] The exception that proves the point was the NORDEL partner in continental western Denmark, which for similar economic reasons chose to maintain its traditional synchronous collaboration with northern German UCPTE partners. It collaborated by asynchronous direct current links with its NORDEL partners, including eastern Denmark.[92]

The vulnerability implications of direct current connection were foregrounded in the British choice. In the late 1950s a study committee recommended a synchronous alternating current connection to the UCPTE pool to benefit system stabilization, among other reasons. Yet the committee noted that asynchronous, direct-current connections had other reliability advantages, such as providing a barrier to cascading frequency disturbances that can only travel in systems with frequency synchronization. Intensive Swedish lobbying on behalf of Swedish direct-current cable manufacturer ASEA ultimately won over the French and British parties for a direct-current connection. The British national grid was connected to France by direct current in 1961 and remains so today.[93]

Finally, a major division in postwar electrical Europe followed the so-called Iron Curtain or, in this case, the "Electric Curtain" between East and West.[94] Central Eastern European utilities were inspired by the system in the Soviet Union. In the context of the Council for Mutual Economic Assistance (COMECON, 1949), members discussed the pooling of fuels and an international power grid by 1954. In 1956, COMECON discussed the construction of interconnections between the German Democratic Republic (GDR) and Poland, with a possible extension to Czechoslovakia. An internationally interconnected electricity system was seen as the next move.[95] Rules of cooperation were established in December 1957.[96] In a first phase the GDR, Poland, Hungary, and Czechoslovakia were connected through 220 kV lines between 1957 and 1960. Western Ukraine followed in 1962, with Romania and Bulgaria in 1963–4. Their Interconnected Power Systems, also known as the Mir or Peace Grid, now involved seven socialist states. In terms of governance the new power pool partly followed the Soviet model: bilateral negotiations between national power authorities continued to dominate in practice as in the West, but a common, centralized control center established in Prague in 1962–3, the Central Dispatch Organization, was allowed to implement electricity exchange schemes between member states on a day-to-day basis.[97]

Externally the Central Eastern European pool was synchronized with the Soviet system in 1962.[98] In the 1960s, both pools formed a bipolar system with key control centers in Prague and Moscow. In the 1970s, however, the collaboration increasingly functioned as a single centralized power pool, as the Moscow control center took charge of frequency regulation as well as the exchange programs of individual countries.[99] Conversely, collaboration across the Electric Curtain was difficult and marginal – certainly when compared with the successful establishment of East–West trade in natural gas (Chapter 2, this volume). In electricity supply, however, energy trade was not the sole driver of transnational collaboration, as we saw above. System stabilization was an equally important, if not more important, concern, and in this respect the UCPTE was hesitant to pursue synchronous collaboration with Central and Eastern European systems that did not comply with UCPTE security norms.[100] This is not to say that visions of large-scale power trade were absent; yet they were less dominant than in the case of natural gas pipelines, and more easily thwarted. Thus a promising 1963 plan to export Polish coal-based power via Czechoslovakia to Bavaria in West Germany was successfully blocked by NATO.[101] Only a few asynchronous connections between East and West materialized, including the link via Yugoslavia (supported by NATO to lure Yugoslavia further away from the socialist block), a link between Czechoslovakia and neutral Austria, and Finnish-Russian and Bulgarian-Greek links.[102]

The end of the Cold War did not eliminate the Electric Curtain but pushed it eastward. Polish, Czech, Slovak, and Hungarian power companies now set up yet another organization, CENTREL (established in 1992; terminated 2006), halted synchronous collaboration with the former Soviet system and began synchronized cooperation with the UCPTE in 1995.[103] Their motives included envisioned lucrative power exports to the West, besides the traditional arguments of pooling reserve capacity, emergency support, and frequency stabilization. Full UCPTE membership was obtained in 1999. Traditional partners, such as the western Ukrainian, Romanian, and Bulgarian power authorities, followed in 2002 and 2004.[104] In the northeast, Estonian, Latvian, and Lithuanian power companies developed similar plans, but economic interests in power exports to Russia prevented that move for the time being.[105] Europe's Electric Curtain now roughly followed the border of the late Soviet Union.

The UCPTE power pool, meanwhile, had grown considerably. After absorbing Southern, Central, and Eastern European members, it began synchronous collaboration with Moroccan, Algerian, and Tunisian power companies via the 1997 Spain–Morocco submarine cable. Again, reliability considerations weighed heavily: the cable was designed for an anticipated change to more economical direct-current operation, but this change was not implemented because synchronous connection greatly improved the stability of the Moroccan system.[106] The huge synchronized area became known as the Trans European Synchronously Interconnected System.

Thus "Electrical Europe" emerged as it (by and large) still looks today. Vulnerability considerations informed choices for either synchronous or asynchronous collaborations between distinct power pools. Synchronized power pools provide instantaneous backup and stabilization to participants; asynchronous links did not have these advantages but are able to halt cascading blackouts of the sort that threatened the synchronous UCPTE system on January 12, 1987. The "European blackout" of November 4, 2006 clearly exposed the present geography of electrical interdependency: the frequency disturbance traveled from northern Germany to the Iberian peninsula, Central Eastern and Southeastern Europe, and northern Africa. By contrast, it could not cross the asynchronic barriers to Scandinavia, Britain, the Commonwealth of Independent States, the Baltic Republics, and Turkey.[107] Notably, in 2010, Turkey joined the synchronous "European" power pool, illustrating that the attractions of synchronous collaboration may still outweigh its risks, and that the dynamics of European electrical integration differ from those of political integration – negotiations on Turkish entry into the EU remain cumbersome.

The making of high-reliability organizations

How, then, did transnational collaborations deal with this trade-off between the pros of automatic system stabilization and emergency support, which increased with the size of synchronous power pools, and the cons of potential cascading blackouts? How did they produce such high levels of reliability that caused organizational sociologists to view electricity supply as high-reliability organizations? In the exemplary case of the UCPTE we have already seen that system stabilization was a major argument for the establishment and subsequent expansion of the power pool. Yet from its beginning the organization acknowledged that synchronized collaboration also introduced the possibility of cross-border cascading failure, where frequency disturbances are transported throughout the network. Indeed, it quickly decided to make system reliability a cornerstone of its activity.[108] The UCPTE developed a double strategy: working for the expansion of synchronized collaboration was accompanied by measures to prevent or contain this new form of failure. By 1965, when large-scale rolling blackouts in the United States prompted a renewed sense of urgency, the UCPTE had identified a number of potential hazards and associated countermeasures that its members should implement.[109] The overall strategy was that its power pool should consist of interconnected, yet separately managed, networks, and that decentralized network managers were responsible for reliability in their own supply areas. Decentralized organization and vulnerability management thus went hand in hand. Another crucial principle was that allowing short time disruptions was "more acceptable than the effects of a comprehensive network disturbance with an unavoidable interruption of supply for a long time."[110]

These principles inspired a set of precautionary measures. A number of design principles were intended to reduce the chance of disturbance in all member areas. If disturbances should occur nevertheless, it was important to prevent long-lasting damage. Therefore all system elements should have protective equipment, to automatically disconnect the element whenever system parameters fell below predefined thresholds, shutting it down before it burned out. Once the system parameters rose back above their thresholds, the element should be automatically reconnected. In the blackouts of 1987, 2006 and so on, it was such automatic protection gear that caused the line and generator trippings, and soon after brought the equipment back online.

To further contain and counter such failures, UCPTE members were to provide for sufficient backup capacity throughout the interconnected system. In the 1960s, members were to run extra generator units at all times, corresponding to some 3–5 per cent of the expected load or the largest power station in the pool. In addition, they were supposed to invest in emergency generators that could be started relatively quickly, and cross-border interconnections in particular were to have ample spare capacity to be used in case of disruptions. Later the general rule became that the entire system must always be operated with at least what was called "single backup capacity" (so-called N-1 backup), denoting that if one system element fails, the other elements are able to absorb the additional load.[111]

Should cascading failure happen despite these measures, cascading overloads would be countered by automatically tripping generators, while cascading underloads were to be contained by selectively disconnecting consumers. For this purpose, members were supposed to develop predetermined load-shedding programs – that is, emergency plans preparing the controlled disconnection of electricity users (households, industry, and pump storage plants) if the frequency dropped below a certain threshold. These should preferably be executed automatically by means of frequency relays. The blackouts of 1987 and 2006 were due not to malfunctioning equipment but to such deliberate and controlled load shedding, which temporarily sacrificed selected consumer areas in order to secure others. Next, to restore the system after failure, UCPTE members were responsible for improving system parameters in their own supply areas. To facilitate the coordination of such a decentralized response, telephone and telex connections were to be established between the control centers of neighboring members. A final measure proposed in the mid-1960s was the introduction of monitoring equipment to detect irregularities in the operation of power stations, load centers, and international tie lines. These latter grew into data-processing programs, such as Supervisory Control and Data Acquisition systems, and Energy Management Systems (compare Chapter 8, this volume).

These measures required considerable investments in the 1950s and 1960s but seemed to pay off: in the 1970s and 1980s the UCPTE system was considered to be highly reliable. Simulations suggested that local failures did not lead to cascading failure and did not compromise overall system security.[112] Incidents such as the 1987 failure confirmed this picture. On the eve of neoliberalization, the UCPTE concluded that although it could not provide absolute guarantees, its coordinated purposeful action produced "a very high degree of reliability of power supplies, without incurring costs which are out of all proportion."[113] As noted, the organization emphasized time and again that such reliability was best achieved in the informal and decentralized governance model of the UCPTE, for, as observed above, the partners knew the particularities of their own systems much better than any centralized organization could ever hope to.

In the centrally planned, managed and controlled power pools of the Soviet Union and COMECON, however, one may find similar discourses of high reliability. According to Vladimir Semenov, long-time employee of the Moscow control center and professor at the Moscow Power Institute, "centralized control disciplines and standard protection schemes, coupled with advances in technology, have continually improved the security and reliability of this transmission system."[114] Thanks to "this high standard of service," major Soviet system blackouts were few and far between, including a blackout in Moscow in December 1948 and one in Kazakhstan in 1975.[115]

A number of measures resembled those in Western Europe. For instance, the COMECON system was equipped with protective gear against short circuiting, telephone circuits for communication between control centers, and measuring devices. In terms of operation, the system would function at 50 Hz with a maximum deviation of 0.5 Hz; if such deviations lasted longer than 30 minutes, the load dispatcher was allowed to intervene directly in the planned electricity exchange scheme or shed part of the load.[116] And as in the West, the 1965 blackouts in the United States inspired renewed attention to the reliability of the Eastern European systems.[117]

Different from the West, however, was the hierarchical nature of balancing supply and demand, both in planning and in operation. In the Soviet system, hierarchical planning meant adjusting generating capacity and power line capacity on a 5- to 20-year basis. In addition, Soviet power authorities developed a three-tier hierarchical system of operational and emergency control, in which over 60 regional control centers were subordinated to the area control centers of the regional power pools, which in turn answered to the central Moscow dispatch center. Orders coming from higher levels were mandatory; lower dispatch control levels had the freedom to counter local problems only within these operational guidelines. Since the operating staff at the highest level was responsible for the security and economy of the overall system, preserving the overall system had institutionalized priority over subsystems. High reliability discourse in communist Europe thus applied to the integrity of the primary grid, rather than continuous supply to individual power consumers. When praising Soviet reliability management in the late 1980s, Moscow Power Institute engineers observed that the grid operated most reliably with average outages of up to merely six "system minutes" per annum, without any system collapses in the last decades. They did not provide any information on outages for consumers, which became the primary indicator of reliability in Western Europe.[118]

The technological and organizational means to achieve primary grid reliability, accordingly, included the central control of power station output and power flows in the grid. In addition, a comparative study found that "auto-regulation of consumption" played a large role compared with decentralized systems of the UCPTE or NORDEL.[119] In other words, "disconnecting some of the least essential consumers" was a key strategy for balancing supply and demand.[120] In large parts of the centrally controlled grids of communist Europe, coping with periodic power rationing was a daily routine for end users. In Bulgaria, communist-era power supply is still remembered as the "disco era" since the lights flashed on and off.[121] In Byelorussia, blackouts were usually quite short and selective – for example, alternating between large apartment blocks.[122] In this scheme of securing the primary grid first, the overall system could be kept up despite ensuing shortages. Construction delays persisted especially during the 1980s, supply shortages were common, particularly in winter, and operational reserve capacity of about 1 per cent was way below the planned level (and below the level in the decentralized UCPTE system).[123]

These different control regimes collided in 1991 when Central Eastern European power authorities announced their wish to disconnect their synchronous links with the Soviet system and connect to the UCPTE instead. The UCPTE demanded tighter frequency control and national defense plans, whereas load shedding had previously been arranged by the Prague Central Dispatch organization.[124] Interestingly, these and other changes were consistently phrased as "power quality improvements" rather than adaptation to a different system, reflecting the quality perceptions of the UCPTE collaboration. After four years of preparation, the new collaboration became operational.

The invention of vulnerability

September 28, 2003, 3:20 a.m. Sunday. A severe storm tips a tree over a power line carrying Swiss electricity exports to Italy, igniting overloads in Swiss, French, and Italian power systems. In marked contrast with the events over 80 years earlier at the Swiss-Italian border, French and Swiss power authorities now cut their connections to Italy to prevent blackout at home. Soon the entire Italian peninsula plunges into darkness. In Rome, where a million people are participating in the celebration of the Notte Bianca ("White Night") festival, subways and elevators come to a halt, trapping passengers inside. Traffic lights fail and cause massive traffic jams, while 110 trains carrying over 30,000 passengers come to a sudden standstill. Hundreds of people panic. Nationwide, hospitals report a surge of accidents involving elderly people.[125]

After reparation of the failure, Italian, French, and Swiss power authorities blame each other, but their conflict fades into the background when EU officials get involved. Earlier European Commission energy security debates had focused on fuel imports and bunkers. But a week after the "Italian blackout" the security of energy systems, in particular electrical power, tops the agenda for the upcoming EU energy ministers meeting.[126] Two months later the European Commission proposes its first directive for the security of electricity infrastructure. Further encouraged by the "European blackout" of November 4, 2006, a new EU Agency for the Cooperation of Energy Regulators is set up in Ljubljana, Slovenia. Furthermore, the power sector yields to EU pressure and terminates international associations like the UCPTE (now UCTE) and NORDEL, which had dominated the scene for over half a century, replacing them with an EU-wide association – the European Network of Transmission Systems Operators for Electricity.

Thus ended an era in European electric collaboration. The power sector's discursive hegemony on economy and high reliability was definitely challenged, as was its associated decentralized model of transnational governance. Enter the EU perception of "transnational vulnerability," its claim that only EU-level organization can make Europe's power system sound and secure, and its persistent equation of "Europe" with the EU polity and territory in matters of electric power as well. By then, vulnerability challenges had already exploded in the Commonwealth of Independent States due to rapid liberalization of the former Soviet system. What is more, this system was increasingly externalized as "non-Europe" as EU discourses on "Europe" became hegemonic. While the implications for Europe's actual material infrastructure remain to be seen, the stage seems set for reinventing electrical Europe on the EU level.

The dynamics of electric EU-ropeanization

We read the entrance of the EU and its direct forerunner organizations as a political drama in three acts.

In the first decades of experimenting with new forms of supranational governance in continental Western Europe, energy had been claimed as a major arena for political integration. Indeed, two of the three European communities – the European Coal and Steel Community (1951) and the European Atomic Energy Community (1957) – were related directly to primary fuels. Electricity infrastructure had been considered for the third community, the European Economic Community (1957), but was ultimately bypassed. This is remarkable because, at the time, integration theorists and politicians from the six participating states – Belgium, the Federal Republic of Germany, France, Italy, Luxembourg, and the Netherlands – saw transnational infrastructure as a producer of integration spillovers and thus a major candidate for common policy.[127]

The reason for this bypass was suggested in the influential Spaak Report preparing the 1957 Treaties of Rome. According to the report, electricity and gas infrastructure differed from other potential policy domains in their "technical and economic specificities," making them less well-suited candidates for a common policy; they were well dealt with by specialized sector organizations.[128] Thus when coal issues led the three communities to jointly set up an Interexecutive Working Group on Energy in 1961, they foregrounded energy source problems – such as security of supply in the case of oil and diminishing coal production – rather than infrastructure issues. By the way, despite a number of attempts, a common fuel policy did not take off either; it was repeatedly frustrated by member states' concerns for domestic coal market protection. Of these failing proposals, a 1964 Protocol of Agreement on Energy Policy intended to introduce fairer competition between energy sources, a wider diversification of oil supplies, and prices as low and stable as possible. In 1967 the Working Group was replaced by a Directorate-General for Energy, which developed Guidelines for a Common Energy Policy, seeking secure primary fuel supply and low and stable prices. Here electricity was mentioned briefly as a candidate for common regulations on open access and tariffs. Neither was implemented as energy remained "an extremely sensitive area of national sovereignty"; not even the oil crises of the 1970s inspired a Community community energy policy.[129] The result relevant to us here is that international electricity infrastructure governance was organized outside the European Communities framework in the more voluntary and broader membership organizations that we discussed in the previous sections. Interestingly, the same happened with transport and communications infrastructure.[130] In addition, in terms of perceived vulnerabilities, electricity issues seemed negligible compared with concerns about fossil-fuel energy security and miner employment.

The Second Act, in which electricity became a policy target, opens with the emerging concept of an Internal Market in the early 1980s, formalized in a European Commission White Paper by 1985 and the Single European Act by 1986. By now the Communities also included Denmark, Ireland, Britain, Spain, and

Portugal. Dissatisfied with de facto trade flows, the aim was to reinvigorate the economic integration process by combating internal frontiers. The White Paper listed some 300 legislative measures that could reduce physical, technical, and tax barriers to trade. The Single European Act set a target date for a liberalized common market by 1992 and defined steps accordingly. It included a target date (1992, later postponed) for realizing a common energy market, meaning an internal and liberalized common electricity market. The Treaty on the European Union (1992), finally, added EU involvement in the planning and financing of a "Trans-European Network"; by 1994 the first priority interconnection lists were compiled, including a number of transnational power lines.[131]

Importantly, EU electricity policy-making aimed at economic integration and (neo)liberalization, not reliability management. EU spokespersons and documents rarely questioned the reliability of electricity infrastructure and the power sector's decentralized governance model; the perception that Europe's electric power infrastructure was vulnerable still had not taken root. The 1988 European Commission policy document "The Internal Energy Market" praised Europe's highly interconnected electric power system and recognized that international exchanges were managed well by sector organizations such as the UCPTE and NORDEL without government interference.

Instead, EU electricity policy targeted perceived economic and political vulnerabilities. The European Commission itself was concerned chiefly with social and economic cohesion and with making Europe more competitive. Note that in this context "Europe" was identified with EU internal market integration: newspeak of "the costs of non-Europe" referred to internal fragmentation and barriers to trade hampering European economic performance, which was deemed problematic in an ever more competitive world and emerging economic recession.[132] Thus the cost of non-Europe in the energy sector is affecting our economic performance... The potential benefit of "more Europe" would be twofold: a reduction in costs as a result of greater competition and a reduction in certain unit costs as a result of the effect of scale and the optimization of investment or management.[133] To counter "non-Europe" in electricity, the European Commission prioritized "economic and competitive aspects of electricity," leading to governance issues such as monopoly control, the common carrier principle in which users would be able to purchase electricity from any power producer instead of being tied to the producer in their specific supply area, and open competition between power producers. The envisaged beneficiaries were large electricity-intensive industries, which had been lobbying for these principles, but also small users without substantial political representation. Electricity system reliability was mentioned only as a sector-specific concern, not as a primary target; security of supply still exclusively denoted the availability of primary fuel. Even in the next step, the formulation of the Trans-European Network program for electricity infrastructure, reliability and its governance were not problematized.

In addition, the new push for liberalization and Europeanization did not much affect the power sector's perception of high reliability. Initially, the UCPTE (soon renamed UCTE, dropping the "P" for production following the separation of production and transmission activities) was alarmed by the new developments: competitive pressures might jeopardize system security and increase the possibilities of blackout, for the common carrier principle might complicate international coordination.[134] Unable to withstand or block EU policy, the organization engaged in a debate with the European Commission to accommodate its concerns about EU policy. The result was positive: "the UCTE believes that the new deregulated market environment is compatible with an adequate level of system reliability."[135] New technologies geared to the new situation were explored, improved, and introduced.[136] For instance, by 2000, Wide Area Monitoring Systems (WAMS), as a supplement to earlier monitoring technology, offered real-time information about grid conditions in over 30 key nodes in the UCTE network. Such augmented monitoring was accompanied by innovative WAMS. In addition, UCTE security rules were tightened, in particular through a security package in 2002. Existing rules were sharpened and systematized in the eight policies of the *UCTE Operational Handbook*.[137] As a result, on the eve of the major blackouts of 2003 and 2006, many stakeholders, analysts, and politicians still considered continental Europe's electric power system to be extremely secure. The UCTE system adequacy forecast for 2003–5 and other documents noted that although cross-border power flows were increasing and the system was operated near its limits in some locations, so "the security of the UCTE system as a whole seems to be not at risk."[138]

In the Third Act, the "Italian Blackout" of 2003 inspired EU policy-makers to challenge the high-reliability consensus head on. The ground was prepared by several other large blackouts that same year – the northwestern blackout in Canada and the United States, in London, and in Sweden and eastern Denmark. The 2006 European blackout underscored the transnational nature of present-day power grid vulnerability.

Interestingly, these events did not change the high-reliability discourse in power sector organizations such as the UCTE. The Italian blackout might confirm that there was little slack in the system at some points, not least where Italian reserve generation capacity and load-shedding programs were concerned. Yet the disturbance was contained everywhere except in Italy. Besides, in Italy itself, supply was restored within five hours in northern Italy and ten hours in the entire mainland. The UCTE found "no fundamental deficiencies in the existing rule-setting of the UCTE system."[139] The existing decentralized governance mode also remained unquestioned: "The blackout and subsequent investigation has cast no doubt on this [decentralized] model in principle. On the contrary, the lack of a grid operator's empowerment and independence could be identified as a potential security risk."[140] In the next year, UCTE members again succeeded in running their systems in "a highly secure and reliable manner"; a year later the adequacy forecast for 2005–15 did not anticipate any major risks either, predicting a "reasonable security margin" by 2010.[141] The UCTE interpretation of the "European blackout" of November 4, 2006 follows the same line of interpretation: most consumers remained unaffected, while supply to most of those affected was back online within 30 minutes and to all within two hours.

Yet the UCTE president, Martin Fuchs, observed how, following the Italian Blackout, the "security of supply issue has come to largely dominate the discussion in terms of energy policy. Transmission system operators' functions and activities have never before been a matter of such considerable interest to politics and public."[142] Electricity infrastructure vulnerability quite suddenly became a key concern of EU policy-makers and entwined with other policy initiatives; it became an integral part of the movement to extend EU influence into the domain of transnational electricity infrastructure governance that we summarized above. Why did this happen? In our interpretation, this concern resonates well with the rapid emergence of what EU analysts term an EU "security identity" associated with an emerging "protection policy space."[143] In the last decade or so, EU policy-makers increasingly focused on transboundary threats, from disaster response and counterterrorism to food safety and avian influenza. Moreover, member-state governments were increasingly inclined to grant the EU powers in such matters of transnational citizen protection, thus contributing to a qualitative as well as a quantitative change in the formal European integration process. We expect, pending further research, that this context made EU policy-makers sensitive and receptive to transnational electricity disruptions such as the 2003 and 2006 blackouts. Either way, unprecedented policy measures followed, not least the EU's Third Legislative Package (then still in draft), including plans for an EU-wide electricity infrastructure regulatory agency. Notably, in March 2006 – half a year before the 4/11 blackout – member states had still rejected the notion of such agencies.[144] The interconnection of energy networks itself was inscribed into the Treaty of Lisbon, the amended "European Constitution" that came into force in 2009.

This EU pressure was stepped up even further after the rejection of the proposed European Constitution by French and Dutch voters in 2005. In response, the new European Commission charm offensive foregrounded the leading role of the EU in combating climate change, thus adding yet another layer of vulnerability and urgency to legitimate EU interference.[145] Facing these combined pressures of economic, security, and ecological vulnerabilities, the UCTE and other sector organizations' interpretation of economical, clean, and high-reliability performance and adequate transnational sector governance were no longer politically convincing. Moreover, important electricity producers recognized new business opportunities, such as foreign expansion and green subsidy schemes, and this supported ongoing political developments. In the realm of electricity infrastructure, international sector organizations followed the European Commission's suggestion to merge into the EU-wide European Network of Transmission System Operators for Electricity (ENTSO-E). Accordingly, the old mesoregional organizations were terminated in 2009. While continuing to contest the EU notion of transnational electric vulnerabilities, this new infrastructure organization implicitly copied and implemented the EU version of electrical Europe institutionally and discursively: "We are the European TSOs. We are ENTSO-E... [with an EU mandate] to ensure optimal management of the electricity transmission network and to allow trading and supplying electricity across borders in the Community."[146] As for the infrastructure hardware, the new organization inscribed the aim of an "interconnected European grid" in its mission statement. One of its first activities was to publish a call for projects developing a roadmap towards a pan-European supergrid to counter Europe's various electricity threats.[147] Based on the findings of this chapter, we perceive this initiative as confirmation that yet another round of interpreting and negotiating Electrical Europe is currently taking place.

Epilogue

November 25, 2005. Heavy snowfall causes electric power interruptions throughout the Netherlands. Supply to the town of Haaksbergen (25,000 inhabitants) near the German border is interrupted for between 30 and 61 hours.[148] The Dutch Royal Air Force flies in emergency generators to serve elderly homes and husbandry farms. Local entrepreneurs blame the responsible network company Essent Netwerk BV and quarrel about damages. The Dutch parliament is shocked and demands a thorough inquiry into the adequacy of the Dutch power grid.[149] This inquiry concluded that the Haaksbergen failure could not jeopardize the rest of the Dutch system, but the town itself is vulnerable: it is located at the end of a transmission line. Connecting Haaksbergen and other towns in a similar position into a ring structure to secure supply from two sides would cost €90 million annually, while annual profits would amount to only €4 million.

A year later the 2006 European blackout passes nearly unnoticed in the same parliament.[150] In contrast with EU politicians, Dutch MPs are not impressed. After all, many more faults happen locally, particularly in low- and medium-voltage distribution networks. Indeed, for consumers and small businesses, the blackout of November 4, 2006 accounted for less than 2 per cent of the annual average power outage per consumer per year.[151] Events such as the Haaksbergen local blackout seem much more disruptive and important. In 2010, Haaksbergen gets its second cable connection.[152]

We started this chapter with the so-called European blackout of November 4, 2006. At first glance this event seemed to represent a remarkable historical irony. Historical actors set up large transnational synchronous power pools that facilitated, next to power exchanges, immediate mutual support and system stabilization: a disturbance anywhere in the system would instantaneously be counteracted by all other machinery in the pool. Transnational electric interdependency thus reduced much electric vulnerability. Yet it also produced a new vulnerability in the form of cascading blackout, as the November 2006 events demonstrated: today a disturbance in northern Germany can turn off lights in Portugal or Tunisia within seconds. On closer inspection, however, this displacement of electric vulnerabilities proved subject to diverging interpretations: while the geography of the blackout signaled a new form of transnational vulnerability to EU policy-makers, to power-sector experts its successful containment and quick repair confirmed the secure state of the European power supply. Electric vulnerabilities, in short, are subject to interpretation, contestation, and negotiation in concrete historical and institutional contexts.

Our subsequent investigation confirmed that vulnerability perceptions were key, yet moving, targets in the shaping of Electrical Europe: electrical interdependencies and vulnerabilities were framed differently in the eras of isolated power systems, interwar electric nationalism and internationalization, postwar reconstruction, and ongoing electric EU-ropeanization. One implication is that present-day EU, state government, or power-sector claims about electric vulnerabilities should not be taken at face value. Rather, these should be related to their respective institutional logics. Another implication is that any measure to reduce present-day vulnerabilities will undoubtedly be criticized in the future for producing new vulnerabilities of its own. The currently celebrated promise of European "smart grids," for instance, may facilitate better real-time control of transnational power flows and fluctuations, and also balance out disturbances caused by new unstable renewable energy generators like wind or solar power. Yet simultaneously, smart grids heavily increase the dependency of electric power supply on information and communication technology infrastructure that can fail or be hacked.[153]

It is thus in the context of such ongoing EU-ropeanization of electricity supply that we end this chapter with the 2005 Dutch Haaksbergen event. After discussing magnificent electrotechnical collaborations spanning from Ireland to Siberia and from Norway to North Africa, the Haaksbergen incident is a welcome reminder that in the age of pan-European and global systems, the local remains a crucial unit of design, use, concern, identification, and vulnerability. Moreover, borders still matter: high-capacity power lines have pierced the Urals and the Mediterranean as electrical borders of Europe, yet the proximity of Haaksbergen to the Dutch-German border meant that the town was situated at the end of a transmission line. The primary grid crosses borders, but lower-level transmission lines usually do not, even in countries that have been at the heart of Europe's electrical integration project. Far from being a homogeneous space, Electrical Europe is a complex, multilayered entity of interwoven local, microregional, national, mesoregional, and transcontinental systems, transcending borders but not erasing them. Europe's electrical vulnerability geography follows suit: local failures are frequent, while rare transnational failures, such as the 2006 European blackout, provide a glimpse of the selective geographical extension of these complex systems. The importance of local vs. long-distance failures is interpreted and weighted differently in EU policy-making, national governments, power companies, and local communities.

It is a key task of transnational history to highlight and interrogate such entanglements between international, national, and local processes, not to obscure or erase them. To further this sort of inquiry, the following chapters zoom in on the interpretation and building of electrical Europe and its vulnerabilities "from below."

Acknowledgements

We are much indebted to Wil Kling, Johan Schot, and Geert Verbong; the participants of the EUROCRIT program; and two anonymous referees for critical comments on earlier drafts of this chapter. This project was funded by the European Science Foundation and the Netherlands Organization for Scientific Research (NWO) under Dossier no. 231-53-001.

Notes

1. This event is well documented in UCTE 2007. For further analysis, see Van der Vleuten et al. 2010a.
2. Van der Vleuten et al. 2010b.
3. *International Herald Tribune*, November 5, 2006.
4. Because of a spelling mistake in the *International Herald Tribune*, we here cite the same quote in *BBC News*, November 6, 2006.
5. European Commission, "Energy Commissioner Andris Piebalgs Reacts to Saturday's Blackouts," press release, November 6, 2006.
6. European Commission, "Blackout Last November Calls for Increased Cooperation Between TSOs Says Commissioner Piebalgs," press release, December 19, 2006.
7. "German-Triggered Blackout Exposes Fragile European Power Network", *International Herald Tribune*, November 5, 2006.
8. UCTE 2007, p. 6.
9. http://www.ucte.org, consulted on August 17, 2004. UCTE 2007, p. 12. The organization claimed to make "historical contributions to the ideal of a united Europe" for a long time. See UCPTE 1971, p. 1.
10. Using 4/11 for shorthand, we acknowledge that the infrastructure-related terrorist attacks of September 11, 2001 (9/11 in US notation) and March 11, 2004 (also known as 3/11) were, of course, much graver.
11. Nye 1998.
12. Misa and Schot 2005.
13. For a further discussion, see Van der Vleuten et al. 2010b.
14. For the development of these research questions and their embedding in different literature, see Van der Vleuten and Kaijser 2005; Van der Vleuten and Kaijser 2006; Van der Vleuten et al. 2007; and Kaijser et al. 2008.
15. Highlights in comparative, nation-centered electricity historiography include Millward 2005 and the all-time classic of Hughes 1983. For recent international electricity historiography, see Lagendijk 2008 and Hausman et al. 2008. On transnational technological history, see Van der Vleuten 2008.
16. League of Nations, Advisory and Technical Committee for Communications and Transit, Procès-verbal of the second session, held at Geneva, March 29–31, 1922. UNOG Archives, C.212.M.116.1922.VIII, Annex 7: "Report to the president of the advisory and technical committee on communications and transit on the requested action by the League of Nations for facilitating the cession by one country to another of electric power for operation of railways of international concern", p. 33. Also see Lagendijk 2008, pp. 39 and 61ff., and Schipper et al. 2010.
17. Cioc 2002, pp. 131–132; Lee 1991, p. 203; Kleisl 2001, pp. 25–26; Rathenau 1985, p. 7.
18. Rüegg 1954; Kittler 1933, p. 141; Kaijser 1997, p. 6; Van der Vleuten 1998, pp. 139–140 and 150–155.
19. Legge 1931, pp. 68 and 77; Halacsy 1970, p. 145; Maier 2006, pp. 134–135.
20. Lagendijk 2008, p. 45.
21. Klingenberg 1912, pp. 731–735; "Elektrische Grosswirtschaft unter Staatlicher Mitwirkung", *Elektrotechnische Zeitschrift* 35 (1916), pp. 81ff, 119ff, and 149ff. These arguments were picked up and propagated in neighboring countries, such as Van der Vleuten 1998 and 1999.
22. Van der Vleuten 1998.
23. Haas 1926, p. 989; Legge 1931, p. 5; Laporte 1932, p. 113.
24. Lagendijk 2008, pp. 56–57. Compare Millward 2005.
25. Van der Vleuten 1998, p. 129.
26. For example, in the Netherlands, Doyer 1916. In Denmark, Angelo 1917. For Britain, see Hughes 1983, pp. 289–291.

27. Hughes 1983, pp. 289–291, 319–323, and 350–362.
28. Lenin 1920/1965.
29. Coopersmith 1992, pp. 150–152 and 160; Coopersmith 1993.
30. Lagendijk 2008, p. 56; Hausman et al. 2008, p. 148.
31. Van der Vleuten 1999.
32. Varaschin 1997, pp. 100–101; Gugerli 1996, p. 288ff.
33. Hunt Tooley 1988, p. 59.
34. "Verhandlungen über die Wasserkraft und Elektrizitätswirtschaft mit den Landesvertretungen Salzburg" (undated), Deutschösterreichische Staatsamt für Handel und Gewerbe, Industrie und Bauten, Z.26064 III 1919, OeStA, Box 2184H, Folder 425–1.
35. For example, in the Netherlands and Denmark, see Doyer (1916) and Angelo (1917).
36. Compare Schot et al. 2008.
37. Hertner 1986; Hausman et al. 2008, pp. 52ff and 97ff.
38. Persoz et al. 1992; Fells 1998; Lagendijk et al. 2009.
39. Here we follow Gall 2006. See also Gall 1998 and Voigt 1998.
40. Sörgel 1932, pp. 118–119.
41. Trieb and Müller-Steinhagen 2007, map on p. 213. Compare the DESERTEC concept and program, see www.desertec.org/EN/concept (consulted on February 14, 2010).
42. See, for example, Higgins 2008, pp. 42–46.
43. Viel 1930; Oliven 1930; Schönholzer 1930. Compare Gall 2006 and Maier 2006.
44. On micro-, meso- and macroregions, see Troebst 2003.
45. Schönholzer 1930, p. 385. For a discussion about utopianism vs. the feasibility of Sörgel's Atlantropa-project, see Gall 2006.
46. Anastasiadou 2011.
47. For example, Pegg 1983.
48. Oliven 1930, p. 875.
49. Coudenhove saw European unity as a countervailing force to the rise of the Soviet Union and the United States. Coudenhove-Kalergi 1931, p. 4. On the European security of maintaining empires, see Adas 1989, pp. 385ff.
50. Schönholzer 1930, p. 385. The emphasis and quotation marks are as in the original.
51. Sörgel in 1938, cited in Gall 2006, p. 114.
52. This was the office of Albert Thomas, the director of the International Labour Organization.
53. League of Nations 1930.
54. Lagendijk 2008, pp. 69ff.
55. Hughes 1983; Bouneau 1994, pp. 794–795; Vanneste 1998.
56. Segreto 1994, pp. 68–71; Morsel 1994.
57. Maier 2006, pp. 131 and 138–149.
58. Maier 2006, p. 144. Allied bombings targeted electrical installations in only 0.04 per cent (RAF) and 0.05 per cent (US Air Force) of cases. See United States Strategic Bombing Survey 1945, pp. 83–85.
59. UCPTE 1987, p. 5; RTÉ 2004, p. 256.
60. UCPTE 1978, pp. 6–9.
61. For this distinction, see, for example, Troebst 2003.
62. Roberts 1989. This argument still echoes today – for example, De Bruijne et al. 2007.
63. UCPTE 1977, p. 103.
64. UNIPEDE 1955, pp. 126–127.
65. P. Facconi cited in the preface of UCPTE 1971.
66. UCPTE 1976.
67. UCPTE 1976, p. 167.
68. CEEC 1947, p. 10. See also Lagendijk 2008, p. 125ff.
69. CEEC 1947, p. 11.
70. OEEC 1950, p. 24.

71. For a detailed discussion, see Lagendijk 2008, pp. 144ff.
72. UCPTE 1952, p. 5.
73. UCPTE 1952, p. 4. For a detailed overview, see, UCPTE 1976.
74. UCPTE 1976, pp. 153 and 188.
75. Verbong 2006.
76. Steklov 1960, p. 138 and Högselius 2006, p. 249.
77. Steklov 1960, pp. 136–137 and Lebed 2005.
78. Djangirov et al. 2002, p. 1, and Bondarenko et al. 1992.
79. "1921–2002 gody," http://www.so-ups.ru/index.php?id=925 (consulted on March 2, 2010). Michel et al. 1964, p. 208, and Sagers et al. 1982.
80. Sagers et al. 1982, p. 291.
81. Venikov et al. 1989, p. 19.
82. Michel et al. 1964, p. 217; Sagers et al. 1982, p. 301, and Bondarenko et al. 1992, p. 386.
83. Myrdal 1968, p. 626.
84. Neumann 1995, pp. 118 and 127.
85. Khrushchev 1959, p. 8.
86. Lebedev (General Secretary USSR Committee for the USSR participation in international power conferences) to Sevette, December 1, 1964, UNOG Archives, registry fonds GX, file 19/6/1/15-32212. Also see Bondarenko et al. 1992, p. 388.
87. Johnson 2006.
88. "Note sur la constitution de l'Union Franco-Ibérique pour la coordination de la production et du transport de l'électricité," April 4, 1963, Historical Archives of the European Union, fonds OECD, file 1157.8, EL/M(63) 1, Annex II, HAEU.
89. SUDEL 1984 and Lagendijk 2008, pp. 181–183. On the special case of Yugoslavia, see Lagendijk and Schipper, forthcoming.
90. Kaijser 1995 and 1997. The United States also tried to forge more cooperation between these countries through the Marshall Plan. See Lagendijk 2008, pp. 145ff.
91. Wistoft 1992, p. 87.
92. Wistoft 1992, p. 88 and Van der Vleuten 1998.
93. Fridlund 1998, pp. 185–190.
94. Persoz et al. 1992, pp. 62–65.
95. "Vorschläge der DDR für die Arbeit der Ständigen Kommission für Elektroenergieaustausch zwischen den Teilnehmerländern des Rates für gegenseitige Wirtschaftshilfe und für Ausnutzung der Wasserkräfte der Donau," Berlin, April 10, 1956, BArch, fond DE 1, file 21753.
96. These steps included a quadrilateral (GDR, Poland, Hungary, and Czechoslovakia) energy conference in Budapest in 1956, a second one with load dispatchers and related experts, and lastly one in December that laid the groundwork for what would become the CDO. Ministerium für Kohle und Energie, Berlin, "Durchzuführende Massnahmen die sich aus dem Schlussprotokoll der vierseitigen Energiekonferenz in Budapest vom 26.1–10.2.56 ergeben," March 20, 1956, BArch, fond DG 2, file 14218; "Protokoll der Beratung der Arbeitsgruppe über Fragen der Vorbereitung eines Vorschlages über Grundprinzipien des Dispatcherbetriebes, der mit dem Austausch von Elektroenergie zwischen den Teilnehmerländern des RfgW im Zusammenhang steht," Budapest, December 1957, BArch, fond DG 2, file 14218.
97. Savenko 1983, pp. 9–14, 33ff, and 57–58; UNECE 1963, p. 69; UNECE 1964, p. 25; Kaser 1965, pp. 58 and 81; Maximov 1963; and Persoz et al. 1992.
98. Kaser 1965, p. 81.
99. Thiry 1994, p. 4.
100. Riccio 1964, p. 4, and Gicquiau 1981, p. 153.
101. Lagendijk 2008, pp. 184–190.
102. Lagendijk and Schipper, forthcoming; Allmer 1985 and Lagendijk 2008, p. 190ff. See also Chapters 4 and 5, this volume. On Austria–Czechoslovakia links, see Savenko et al. 1983, pp. 120–121. An earlier PhD thesis focused on this, see Schneider 1994.

103. Persoz et al. 1992 and Hammons et al. 1998.
104. Hammons et al. 1998 and Feist 2004, pp. 1226–1228.
105. Högselius 2006.
106. Granadino et al. 1999 and Zobaa 2004, pp. 1401–1403.
107. Van der Vleuten and Lagendijk 2010a.
108. De Heem 1952; UCPTE 1959, pp. 130–138; and Cahen et al. 1964. See also UCPTE 1976.
109. UCPTE 1965 and 1966.
110. UCPTE 1966, pp. 6–7.
111. UCPTE 1990, p. 22.
112. For example, UCPTE 1986, pp. 39–43.
113. UCPTE 1990, p. 20.
114. Semenov 1997.
115. Venikov et al. 1989, p. 19. Compare Makarov et al. 2005.
116. "Protokoll der Beratung der Arbeitsgruppe über Fragen der Vorbereitung eines Vorschlages über Grundprinzipien des Dispatcherbetriebes, der mit dem Austausch von Elektroenergie zwischen den Teilnehmerländern des RfgW im Zusammenhang steht," Budapest, December 1957; Anlage, Entwurf. "Grundprinzipien zur Erarbeitung einer Vereinbarung über Fragen der Betriebsführung im Falle eines gemeinsamen Betriebes der Energiesysteme," BArch, fond DG 2, file 14218.
117. "Einschätzung der 21. Tagung der Ständigen Kommission Elektroenergie des RGW," Sofia, December 14, 1965, BArch, fond DC-20, file 19589.
118. Ibid. Compare Voropai et al. 2005.
119. Working Group 37.12 1994, p. 2.
120. Semenov 1997, p. 3.
121. "Gas Crisis Revives Memories of Communist Era in Bulgaria," *EUbusiness*, January 7, 2009, http://www.eubusiness.com/news-eu/1231354021.63. (consulted on April 30, 2010).
122. Thanks to Nadzeya Kiyavitskaya for observations on Byelorussian blackout patterns in the 1980s. Their interpretation in a system-management context is ours.
123. Bondarenko et al. 1992, p. 384.
124. Riccio 1964, p. 4; Savenko et al. 1983, p. 118; UCPTE, "Protokoll über die Sitzung der UCPTE-ad-hoc-Gruppe Ost-West-Verbund am 30. Mai 1994 in Wien," May 30, 1994, p. 4, UCTE Archives; Houry et al. 1999, p. 638; and Centrel, "Charter of Centrel," October 11, 1992, p. 1, UCTE Archives.
125. "Italy Slowly Comes back to Light," *BBC News*, September 28, 2003.
126. Commission of the European Communities 2007.
127. Schot et al. 2011 and Schot 2010.
128. Comité Intergouvernemental créé par la Conférence de Messine 1956, p. 126.
129. The quote is from Kohl 1978, p. 111. See also Lucas 1977; Hassan et al. 1994; and Commission of the European Communities 1968.
130. Laborie 2006; Henrich-Franke 2008; and Schipper 2008.
131. Padgett 1992; Schmidt 1998, p. 191; Commission of the European Communities 1987 and 1988.
132. Commission of the European Communities 1987, p. 6.
133. Commission of the European Communities 1988, p.6.
134. UCPTE 1998, p. 15.
135. UCTE 2001, p. 25.
136. Kling 2002; Kling 1994; and Breulmann et al. 2000.
137. UCTE 2004.
138. UCTE 2002, p. 5.
139. UCTE 2004, p. 11.
140. UCTE 2007, p. 10.
141. UCTE 2004, p. 5, and UCTE 2002, p. 5.
142. UCTE 2002, p. 4.

143. Boin et al. 2006, p. 405.
144. Stephen Castle, "EU Summit Fails to Address Protectionism Fears," *The Independent*, March 26, 2006.
145. For further interpretation of the current transition in European energy regimes, see Van der Vleuten and Högselius 2012.
146. See www.entsoe.eu (consulted on April 20, 2010).
147. ENTSO-E 2010.
148. Directie Toezicht Energie 2006, p. 20.
149. "Vragen van het lid Hessels aan de minister van Economische Zaken a.i. over de langdurige stroomstoring in Haaksbergen afgelopen weekeinde (Mondelinge vragenuur)," in *Handelingen van de Tweede Kamer der Staten Generaal*, vol. 27 (2005–2006), pp. 1852–1854.
150. "Vragen gesteld door de leden der Kamer, met de daarop door de regering gegeven antwoorden (657)," in *Aanhangsel van de Handelingen van de Tweede Kamer der Staten-Generaal*, 2006, pp. 1411–1413.
151. Ministry of Economic Affairs 2007.
152. Directie Toezicht Energie 2010.
153. Misa 2011.

Part II
Negotiating Neighbors

Introduction

In Part I we investigated the emergence and governance of Europe's transnational infrastructure vulnerabilities from a pan-continental perspective. We discussed the processes of creating – or failures to create – large territories of collaboration and trust in European energy relations for the cases of two major energy carriers, electricity and natural gas. Importantly, we found that international organizations existed but were comparatively weak contributors to this process; by contrast, state governments and their alliances with major energy companies (whether state or privately owned) proved to be central builders of Europe's energy geographies. For this reason, Part II continues to inquire about Europe's critical energy infrastructure from the perspectives of selected countries. Moreover, we have selected three countries situated on Europe's major Cold War border, the Iron Curtain: in the cases of Finland, Bulgaria, and Greece, attempts to establish reliable energy linkages and cope with different kinds of energy criticalities took place in a particularly charged political environment.

These three countries are located in regions of Europe that are not easily defined and tend to overlap in complex ways. This makes them intriguing to study from a critical transnational infrastructure point of view. All three countries are young nation-states formed in the nineteenth and early twentieth centuries. Prior to that they had belonged to the Swedish and Russian (in the case of Finland) and Ottoman (in the cases of Bulgaria and Greece) empires. During the Cold War – the period at center stage here – Finland was a militarily neutral capitalist country with strong but tense relations with the Soviet Union. Bulgaria was one of the closest Soviet allies within the Communist bloc, but with historical ties to the South and West, as well as with neighbors that were not as integrated with the Soviet Union as Bulgaria itself – from communist Romania and Yugoslavia to capitalist Greece and Turkey. Greece, for its part, regarded itself as the very origin of Western culture, it was a capitalist country and a NATO member, but it remained geographically situated in the southeastern periphery of capitalist Europe, facing communist and Muslim neighbors that some perceived to be the very antitheses of the West. In addition, all three countries directly faced the Iron Curtain, though from different sides. Therefore our case selection allows us to study the role of one of Europe's most prominent internal barriers in the geography of Europe's infrastructure vulnerability.

Against this background, the ways in which system-builders in these countries were able to create reliable energy relations and minimize domestic vulnerabilities were far from clear at the outset. As we shall see, their technical and economic priorities did not necessarily match the political and military logic of the time. Indeed, the seeming contradictions between technology, economics, and politics were more often than not resolved, paving the way for far-reaching transnational networking. Explaining how this was possible is a major theme in this part of the book. To do so, the chapters go beyond the standard diplomatic and foreign-policy histories that have dominated both national and European historiographies of Finland, Bulgaria, and Greece. They delve into the additional dimension of international networks of infrastructure, people, and communities of practice in the field of energy.

In Chapter 4, Karl-Erik Michelsen analyzes the uneasy energy relations between Finland and the Soviet Union. He focuses first on dams and hydropower stations in a river that the Soviet Union seized from Finland during the Second World War, and next on the Finnish purchase of two Soviet nuclear power plants in the 1970s. The Kremlin exerted considerable pressure to make Finland choose Soviet reactors, but Finnish engineers changed the overall reactor design by adding Western safety components. These reactor purchases implied increased technological dependency on the Soviet Union, including the supply of nuclear fuel, while at the same time reducing Finland's dependency on electricity imports from that country. Hence the content of Finland's energy vulnerability shifted but not its geographical direction.

Ivan Tchalakov, Tihomir Mitev, and Ivaylo Hristov demonstrate in Chapter 5 that Bulgaria experienced an even stronger dependency on the Soviet Union. In the late 1940s and early 1950s, Soviet leaders coerced Bulgaria into importing Soviet coal for generating electricity in large-scale power plants, and abandoning the course of medium-sized power plants based on indigenous energy sources. The vulnerability of power shortages, endangering the industrialization program of the Bulgarian Communist Party, was substituted by import dependency on Soviet coal. In the following decades, ultra-high-voltage power lines were also to facilitate direct electricity imports from the Soviet Union via Romania, which served as a transit country. In addition, Bulgaria received Soviet technical assistance to develop its electricity system and a nuclear power infrastructure; like Finland, Bulgaria purchased Soviet export reactors. The effect on the country's energy vulnerability was similar to the Finnish case: dependencies on imports of electricity and coal were replaced in part by a new dependency on nuclear fuel and know-how. Indeed, nuclear power in combination with new power plants using domestic lignite (built when the Soviet influence on Bulgarian energy policy was relaxed) made Bulgaria self-sufficient in electricity and even a leading electricity exporter, selling power to Romania, Turkey, Yugoslavia, and, more recently, Greece. Bulgaria became a Balkan electricity hub on which its neighbors became dependent.

In Chapter 6, Aristotle Tympas, Stathis Arapostathis, Katerina Vlantoni, and Yiannis Garyfallos show that Greece provides a counterexample to the Finnish and Bulgarian stories. Greece deliberately chose not to build any nuclear power plants – although it long debated a nuclear program – but instead to import large amounts of electricity from abroad. The Greek state power company had ambitious plans in the late 1970s to build a nuclear reactor not far from Athens, but a major earthquake not far from the intended plant alerted the general public to the risks involved. This forced the government to abandon all nuclear plans and to focus on expanding power plants based on domestic lignite and on electricity imports. This import of electricity was possible thanks to power lines that Greece had built in previous decades to its Cold War enemies in the north: Albania, Yugoslavia, and Bulgaria.

The chapters in Part II demonstrate how vulnerabilities resulting from (planned or existing) transnational infrastructure were perceived at the national level, and how these were dealt with in the given domestic political, economic, and cultural settings. Furthermore, they demonstrate how the three countries decided to – or were forced to – concentrate on coping with some vulnerabilities rather than with others. In addition the three chapters discuss how "Europe" was perceived and built by national players. In Finland, the radically different safety cultures in nuclear engineering strengthened a Finnish perception of the Soviet Union as something completely outside "Europe" – it was explicitly perceived as "non-Europe." The insistence on adding Western technology to Soviet reactors came to symbolize Finland's technological, political, and cultural distance from the Soviet bloc, where similar reactors emerged in very different ways. In the case of Bulgaria and Greece, transnational electricity relations and interconnections instead produced a postwar energy region that largely coincided with the European part of the former Ottoman Empire.

4
An Uneasy Alliance: Negotiating Infrastructures at the Finnish-Soviet Border[1]

Karl-Erik Michelsen

Prologue: The quest for electricity cables across the Gulf of Finland

In spring 2006, the Russian power company United Power Ltd announced plans to install a 1000 MW submarine cable under the Gulf of Finland. The high-voltage cable – commonly referred to as the "sea cable" – would connect the town of Kernovo near St. Petersburg with the Finnish coastal town of Kotka. Although the announcement was a surprise, the plan itself had been conceived much earlier in the mid-1990s. More precisely, the sea cable had been inspired by a Russian presidential decree issued on May 7, 1995, in which President Boris Yeltsin had declared energy to be a strategic component of Russia's industrial and foreign policy. Accordingly, state-owned energy companies were encouraged to explore business opportunities beyond Russian borders, and this was precisely what United Power sought to do.

United Power was owned by Russian energy group BaltEnergo, which in turn was a subsidiary of the state-owned holding company, Rosenergoatom, Russia's nuclear energy operator, the owner of ten nuclear power plants in Russia with a total of 31 reactors. The company had made plans to build several new reactors in Russia, China, Iran, and other countries in the near future, and the sea-cable initiative was a further component of Rosenergoatom's international strategy.

United Power's announcement opened the door to speculation that quickly gained traction in the Finnish media. Journalists identified controversial individuals who were either financing or consulting for the company, which for its part seemed to be neither Russian nor Finnish. This was typical of Russian business at the time in its transition from the socialist to the free-market system. Many links seemed to connect United Power to the political elite of the Kremlin, although the company vigorously denied these accusations. Rather than pursuing any political interests, it claimed, the project would result in significant economic benefits for both Russians and Finns. According to András Szép, chairman of the board of United Power,

It is mutually interesting, in both Russia and Finland, to have this additional possibility to trade in electricity. We know that. We have conducted our own study on the market effects and feasibility of the project. And the figures were fully confirmed by an independent study made by the VTT Technical Research Center of Finland on behalf of the Finnish government, which determine that the market impact will be around 6–8 per cent, which means savings for the Finnish nation and industry amounting to more than EUR 200 million a year.[2]

However, it was not economic arguments but the perceived technological dependency that decided the fate of the sea-cable project. The cable would be able to transmit more than 8 TWh of electricity annually from Russia to Finland, but this was only one side of the coin. According to Sergei Averianov, managing director of the Leningrad Nuclear Power Plant, which cooperated closely with United Power on the sea cable project,

> We decided very early on that we would install a two-way cable that can transmit electricity in both directions. It will therefore be a bridge. It can fulfill the need for electricity and it can utilize power stations on both sides of the border. That is the real benefit of the cable.[3]

This was exactly what the Finnish government was afraid of. Although Finland already imported more than 10 TWh of electricity annually from Russia, the national energy policy emphasized self-sufficiency and self-control of electricity production. The sea cable seemed to endanger these goals. United Power was going to purchase electricity from the Russian northwestern grid, which was the main supply artery for St. Petersburg, yet supply from this grid was by no means guaranteed. The reason was that demand for electricity in Russia's second largest city was growing fast and the northwestern grid had thus come under heavy stress. In particular, the electricity supply in northwestern Russia was threatened by problems at the Leningrad nuclear power plant, the main power plant in the region. It had four 1000 MW RBMK (Chernobyl-type) reactors, but these were already 30 years old and the plant's operating license was under threat. Plenty of cash from both Europe and Russia had been poured into modernizing the plant but, despite these improvements, the production capacity had to be reduced by nearly 30 per cent as a result of the perceived safety risks.[4]

Against this background, the Finns saw a risk that electricity would be exported from Finland rather than imported from Russia. In spring 2006, the Finnish national transmission company, Fingrid, decided to launch a all-out attack on United Power. The 1000 MW high-voltage cable, it was argued, would create serious risks at the location where it was to be connected to the national grid. Because Finnish high-voltage lines were not designed to tolerate extreme load peaks, expensive improvements would have to be made to increase the carrying capacity of the grid should the sea cable be built. Fingrid's estimate of the required investments rose to €1.5 billion.[5]

United Power responded by pointing out that electricity for the submarine cable was not to come directly from the Leningrad nuclear power plant but rather from the northwestern grid as a whole, which had many other power sources. The company also promised to build two 900 MW thermal power stations near St. Petersburg to secure the production of electricity in all circumstances. Last but not least, the 1000 MW cable could be split into two 500 MW cables and connected to the Finnish national grid at different locations as a means of further reducing the stress on the Finnish system. Szép showed a considerable amount of goodwill when addressing the different alternatives:

> We will accept the possibility of diverging – to land two cables at different landing points geographically – and will even consider their separation in time – meaning one cable in, say three years, and the second in a later phase, according to when the main grid is ready, and its capacities.[6]

Finland's minister of trade and industry, Mauri Pekkarinen, had the final word in the controversy. Fingrid and the major power companies lobbied the minister not to grant the Russian company the license necessary to build the connection, and they were successful. According to Pekkarinen,

> It's a question of self-sufficiency. What happens in the middle of winter when the temperature drops way below zero and it's freezing cold? What happens if Russia cuts her exports of electricity when the temperature goes below −22 °C? They've already had problems and it's not even winter yet.[7]

The license was thus not granted to the Russians by the government.

The sea-cable episode would perhaps have been quickly forgotten had not something odd happened at the same time on the other side of the Gulf of Finland. Estonian power company Eesti Energia A/S and two other Baltic power companies, Latvenergo and Lietuvos Energija, were preparing to lay a 350 MW submarine cable that would connect the Baltic electricity grids with that of Finland. This project, although very much like the one proposed by United Power, was never associated with any risks or vulnerabilities. On the contrary, both Fingrid and the Ministry of Trade and Industry strongly supported the connection. As the president of ABB, Fred Kindle, put it at the opening ceremony of the "EstLink 1" cable,

> This is more than an energy project. It brings the EU closer to the goal of creating a European electricity network, and extends the benefits of a reliable power supply with low environmental impact.[8]

Why, then, was it possible to connect a cable from Estonia to the Finnish national grid, while an almost identical one from Russia would create risks and vulnerabilities? Helena Raunio, writing for the engineering journal *Tekniikka ja talous* (Technology and Economy), answered the question as follows:

> The sea cable is an entirely different story depending on where it comes from. EstLink from Estonia will be ready by November 2006, but the other cable from

Russia doesn't even have a schedule. Transnational infrastructures are governed by rules of their own. One needs to know who is financing the projects and who the proposed partners are. Economic rationalities are important, but there are other factors, too. Transnational infrastructures are also about fear, trust and suspicion. The sea cable episode shows that the old phrase is still valid: The fear of Russia is the mother of all wisdom.[9]

Trust, distrust, and the cultural construction of criticality

This chapter investigates transnational infrastructures at the border between Finland and the former Soviet Union. As the sea-cable story demonstrates, decisions to build large-scale infrastructures are not necessarily based on economic calculations alone. Rather, political and cultural aspects play a major role in the decision-making process. The Finnish authorities did not trust the Russian energy system – despite the fact that United Power promised to carry out seemingly necessary improvements and investments to ensure reliability – and they tried to find politically correct "excuses" to reject United Power's proposal. They chose to support the Estlink project, which was economically and environmentally as problematic as the offer made by United Power. The difference lay in the cultural and political relations between the countries involved. Finland and Estonia shared a similar ethnic and cultural heritage, and there had always been a strong political bond between the neighbors.[10]

Transnational infrastructures transfer vital raw materials and energy over long distances and across a multiplicity of borders. It is the wish of every country to build cross-border connections with neighbors who are friendly and cooperative. Yet some neighbors are neither. This is where the concept of trust comes in. Sociologists, psychologists, and political scientists have demonstrated how trust acts like glue in international relations. It holds people, institutions, and organizations together and it makes collaboration feasible despite cultural differences.[11]

Technical aspects constitute risks for transnational infrastructures. Power stations and transmission lines can break down and computers can go dead without warning. But organizational and social factors must also be taken into account, because infrastructures are operated, managed, and governed by professionals who represent a variety of institutional and corporate cultures. If partners share negative historical experiences and if there is a lack of trust, collaborative governance of transnational infrastructures can be difficult or even impossible. This makes these infrastructures vulnerable to accidents due to mismanagement and poor governance.[12]

Trust is associated with social capital, which is an instantiated informal norm that promotes cooperation between two or more individuals. Shared historical experiences can shape informal norms and produce social capital, but the result can also be the opposite. If historical experiences are negative, social capital and trust can be lost. This takes place at different levels of society. Everyday routines create trust in families and well-managed projects make workplace personnel more trusting. At the level of the state, democracy and open access to decision-making increase trust and loyalty among people and social classes. In centrally governed states where power is concentrated in the hands of one party or one leader, this process is weak and societies suffer from a lack of trust. According to Francis Fukuyama,

> This thing occurred in the former Soviet Union after the Bolshevik Revolution, where the Communist Party consciously sought to undermine all forms of horizontal association in favor of vertical ties between Party-State and individual. This has left post-Soviet society bereft of both trust and a durable civil society.[13]

Russia's relations with other European nations are still lacking in trust, even though the communist government collapsed 20 years ago. The reason is obvious. Two world wars, the Cold War, genocides, and ethnic cleansing have produced enough bad historical experiences to erode trust far into the future. In general, there is still fear of expansive powers that use ideology and military strength to enhance their economic and political agendas. Although European integration has increased cooperation between nation-states, there is still a lack of trust and mutual respect even between member states of the European Union (EU). At the EU's external border the lack of trust becomes even more evident. The EU accuses Russia of energy imperialism and of being an unreliable supplier of gas, oil, and electricity. Meanwhile, the EU is viewed in Russia as an "energy NATO" that restricts competition and creates consumer cartels. There are similar problems and accusations at the EU's southern and southeastern borders as well.[14]

In his report to the European Council for Foreign Affairs, Pierre Nöelle, director of the Energy Policy Forum, summarizes the problem in EU–Russia relations as follows:

> Current attempts to use direct diplomacy to solve Europe's problem with Russian gas are unlikely to succeed because the EU and Russia have divergent interests. Europe wants to de-politicize the EU–Russia gas relationship in order to integrate Russian gas imports into a competitive pan-European gas market and to maximize the volumes it can import from Russia. But Russia – or its current leadership, at least – wants precisely the opposite: to keep the politics in the gas relationship. A depoliticized EU–Russia gas relationship would be a disaster from the Russian leadership's point of view.[15]

Finland and Russia share a long and troubled history. There have been peaceful periods when collaboration and cooperation have enhanced economic growth on both sides of the border, but the positive experiences have been overshadowed by wars and conflicts that have destroyed trust and confidence. Finnish-American historian William Copeland introduced the concept of an "uneasy alliance" in his 1973 book of the same title. The concept describes the unique nature of Finnish-Russian relations over the past two centuries. Russia and the Soviet Union have influenced Finnish political culture but, compared with the Baltic States, Finland has enjoyed more freedom and independence. In return, Finland has accepted the geopolitical facts and refrained from criticizing Russian and Soviet domestic and foreign policies. The uneasy alliance or "bounded sovereignty" maintained the status quo in Finnish-Russian/Soviet relations during the early twentieth century. However, the Winter War of 1939–40 and the Continuation War of 1941–4 broke the structure that had been created in previous decades. Finland survived the war as an independent nation, but afterwards the Soviet Union enclosed Finland in an iron grip. Trust and loyalty never returned, even though Finland officially pledged peaceful coexistence with the Soviet Union. When the Cold War ended, many believed that the "new" Russia would denounce the communist regime and become a liberal Western nation. These hopes were dashed by Vladimir Putin, who navigated Russia into the new millennium while retaining old doctrines. Power was centralized inside the Kremlin and opposition voices were muted by censorship and harsh discipline.[16]

Finland is therefore a perfect place to investigate how political, ideological, and cultural factors affect transnational infrastructures. Today there are three high-voltage transmission lines and one natural gas pipeline entering Finland from Russia. Together these annually bring more than 50 TWh of energy across the border, constituting a substantial proportion of Finland's overall energy needs.

For centuries the cultural boundary between East and West has passed through Finland; and, as a result, Finland is neither a Western nor an Eastern country but a mixture of both. Technologically, however, it has always gravitated towards the West and the technological culture in the country is predominantly of Western origin. The Soviet Union, on the other hand, preferred technologies that reflected communist ideologies and proletarian values. When these two very different technological styles collided at the Finnish-Soviet border, a difficult political and cultural process took place. Finnish engineers disliked Soviet technologies and, when given a choice, never accepted Soviet constructions. Nonetheless, Finland was politically forced to collaborate with the Soviet Union, and economically it was necessary to agree upon transborder energy infrastructures. As the sea-cable episode demonstrated, it was up to politicians to make the final decision. Whenever pressure from the East was weak, Finnish politicians rejected the offers made by the Soviets and later the Russians.

However, during the Cold War, Finland became an immensely important ally to the Kremlin because it was the only "Western country" that collaborated with the Soviet Union. Therefore the Kremlin often used ruthless political and ideological pressure to force Finland to accept Soviet technologies and build cross-border infrastructures. In this way the Soviet Union was able to demonstrate to the West how its good-neighbor policies would bring economic benefits for Finland and possibly also for any other countries willing to cooperate with Moscow.

This uneasy alliance had dominated the building of cross-border infrastructure on the Finnish-Soviet border from the early 1940s, and, as the sea-cable episode illustrates, the same pattern persists even though the Soviet Union has collapsed and Finland has become a member of the EU. It is the task of this chapter, therefore, to investigate in greater detail the origin and evolution of this pattern. This is done through two case studies. The first looks at the struggle for hegemony between Finland and the Soviet Union during and after the Second World War. The war paved the way for Soviet occupation of large tracts of Finnish territory on the Karelian Isthmus, which had been of major industrial importance to Finland before the war. In the chaotic situation that resulted from this occupation, electric power stations, dams, transmission lines, factories, and harbors changed hands. When the border was moved, the region's infrastructure and natural resources were divided between the former enemies. After the war, the Soviet Union and Finland had to negotiate regarding the use of common resources and ways to manage and govern previously national infrastructures that now had become transnational. A critical point in negotiations concerned several hydropower stations along the Vuoksi River, which connects Lake Saimaa and Lake Ladoga. Before the war the river was inside Finnish territory, but the new border made it a shared resource.

The second case study focuses on the Finnish nuclear power program, which took shape in the 1960s and 1970s. Finnish nuclear scientists and engineers, and also the power industry, were seeking the transfer of Western technology to Finland. Regarding this as a challenge to Soviet nuclear hegemony, the Kremlin pressured Finland to purchase a nuclear power reactor from the Soviet Union. The Finns considered Soviet technology unsafe and dangerous, believing that the Finnish energy system would be put at risk if an unsafe nuclear power reactor were connected to the grid.

These two case studies and the sea-cable episode illustrate how the concept of uneasy alliance or bounded sovereignty has shaped the building of transnational infrastructure on the Soviet-Finnish border. The relationship between the two neighbors has been troublesome, but despite the political and ideological differences, transnational infrastructures have been built across the contested border.

Political scientists, sociologists, and historians of technology have investigated the relationship between politics and technology for a long time. They have discovered not only that politics shapes technology but also that technology shapes politics. At the Finnish-Soviet border the interaction between politics, ideology, and technology has been complex, and it is very difficult to pinpoint the direction of influence. This chapter shows that politics and technology are interwoven into a seamless web and that the outcome of transborder infrastructure projects cannot necessarily be predetermined.[17]

Negotiating with the enemy

BBC News has put up a website showing how European borders have changed during the past two centuries. Clicking on a small box makes the great empires of the nineteenth century disappear and small independent nation-states appear in bright rainbow colors over the eastern and central parts of Europe. At the next click, the rainbow colors are replaced by light pink, representing the satellite states of the Soviet Union during the Cold War. Finland, the Scandinavian countries and Western Europe are all painted in grey, although light pink would have been equally appropriate for Finland.[18]

The Finnish-Soviet border was contested after Finland left the Russian Empire in 1917. After a bloody civil war, Finland wanted to distance itself from the ideologically dangerous socialist Russia. Very little traffic was allowed to pass across the border and instead Finland turned towards Western Europe, which represented Western democratic values. Finnish industry had a difficult time finding new customers in the West, but by the end of the 1920s, Finnish timber, pulp, and paper companies were well established in Western European markets.

The geopolitical balance in Northern Europe changed during the 1930s, when Germany pushed aggressively towards Poland and the Baltic States, and the Soviet Union strengthened its military presence along the western border. For Finland the most important issue was the security of the city of Leningrad, which was located only 30 km from the Finnish border. Finland and the Soviet Union had signed a non-aggression pact, but it was effectively discarded in August 1939, when Vyacheslav Molotov and Joachim von Ribbentrop signed the Soviet-German non-aggression pact.

One week after the pact was signed, Germany and the Soviet Union invaded Poland. Soon after that the Soviet Union demanded military and naval bases along the Baltic Sea coast and in the Baltic States. Similar demands were sent to the Finnish government but, in contrast with its Estonian, Latvian, and Lithuanian neighbors, Finland refused to allow the Red Army to set foot on Finnish territory. As a result, the Winter War broke out between Finland and the Soviet Union in November 1939.[19]

It was expected that the Red Army would wipe out the poorly equipped Finnish army in a few weeks, but the impossible happened. The Finns put up tough resistance and, with the help of an exceptionally cold winter and much snow, the invasion was halted. This lasted until February 1940, when the Soviets finally broke through the defense lines and forced the Finnish government to settle for peace. The peace terms were devastating. Finland kept its independence but the Karelian Isthmus and large territories along the eastern border were ceded to the Soviet Union. The loss of the city of Vyborg and two other towns, plus all factories, power stations, warehouses, harbors, and farms, slashed a deep wound in the Finnish economy, society, and culture. Finland emptied the ceded land and more than 400,000 Karelians were evacuated to the other side of the new border.

Once the initial shock faded away, the Finnish government analysed the new situation. More than 25,000 soldiers were dead and another 45,000 were injured. The Finnish-Soviet border had moved almost 200 km westwards, putting what had been one of the country's most important industrial regions in the hands of the Soviets. The war had destroyed the Finnish economy and almost half a million refugees were struggling to find a place to live. Finland suffered from shortages of food, electricity, and fuel. Sweden was the only foreign country that could help. Germany, the long-time friend, was now allied with the Soviet Union and the war in Western Europe cut off connections to the outside world.

The new Finnish-Soviet border gave rise to several controversies. The Soviet foreign minister, Molotov, had used a thick pen when he drew the new boundary on the map during the peace negotiations in Moscow. The line left room for interpretation; naturally, it was the Kremlin that had the last say in the dispute. The Finnish government hoped that the Enso-Gutzeit pulp and paper mill and the Enso community – both of which were located just on the new border – would be saved, but these hopes were crushed by Molotov. The Soviet Union did not forego the opportunity to take over the largest pulp and paper mill in Finland.

This was another devastating blow to the Finns. Enso-Gutzeit was not only the biggest paper company in Finland but also the symbol of a modern industrial state. The company's main mill at Enso was located on the banks of the Vuoksi River, very close to the Imatra rapids, which had been harnessed in the 1920s to provide for Finnish electricity production "forever." Together they formed the nucleus of the Vuoksi River Valley industrial region, which was to be the Finnish version of the German *Ruhrgebiet* or America's Tennessee Valley Authority.[20]

The Vuoksi River Valley did indeed have the potential to become an important industrial region. The biggest waterway in Finland, it discharged the waters of the Saimaa Lake District into Lake Ladoga. The Karelian forests provided raw materials for the pulp and paper industry, and the power of the river could be utilized in electrochemical and metallurgical processes. Moreover, the Saimaa Canal, which was located close to the river valley, provided access to the Gulf of Finland and to European and global sea routes.

The nature of the Vuoksi River changed dramatically in 1929 when the tumbling and foaming waters of the famous Imatra rapids were harnessed and almost 125 MW of electricity was directed to the grid. Another power station, Tainionkoski, was built a few kilometers upstream, adding 70 MW of electricity. A third project took shape in the middle of the 1930s at Rouhiala, where private power companies prepared to produce as much as 100 MW of hydroelectricity. The fourth and final leg of the series of rapids in the Vuoksi River was the twin rapids of Enso and Vallinkoski, located a few kilometers downstream from Imatra and just upriver from the Enso pulp and paper mill. Enso-Gutzeit launched the construction project in 1937 and the hydropower station at Enso was almost complete when the Winter War broke out.[21]

The electrification of Finland thus advanced rapidly through the exploitation of the Imatra and other rapids in the Vuoksi River. The Finnish government established a state-owned utility, Imatran Voima, to operate the Imatra power station and to manage the 560 km high-tension transmission line (the Iron Lady line) that fed electricity from Imatra and the other Vuoksi River power stations to population centers in the southern and southwestern parts of Finland.[22]

This dynamic development came to an end, however, through Soviet occupation of the Karelian Isthmus and the lower end of the Vuoksi River in March 1940. Finland lost about a quarter of its installed hydropower capacity. The Rouhiala and Enso power stations (the latter of which was still under construction) were left on the Soviet side of the border, whereas the Imatra and Tainionkoski power stations remained in Finland. The national grid had to be reorganized because the connections from Rouhiala were interrupted. However, it was not immediately apparent what the Soviet Union was going to do with the power stations that were now on Soviet territory.

It did not take long to find out what the Soviet plans were. Only a few days after the peace treaty was signed, the Red Army took over the Rouhiala power station. In addition, special troops were assigned to build a 110 kV high-voltage line that was intended to connect the former Finnish power stations to the municipal grid of nearby Leningrad. The Soviet authorities ordered the Rouhiala Power Company to immediately hand over all technical and hydrological documents. Meanwhile, the Finnish crew that was still in place could keep the power station running and the electricity could be directed across the new border to the Finnish transmission grid.[23]

The Finnish authorities adopted a tough but collaborative attitude towards the former enemy. The peace treaty was considered unjust and even disgraceful, but the neighbors were forced to cooperate. Finland sent to Moscow experienced negotiators who knew the rules of the game. Instead of refusing to cooperate, it was better to stay calm and nudge negotiations in a direction that would provide Finland with the best outcome.

A.E. Kotilainen, the president of Enso-Gutzeit, set a good example, although he had every reason to be outraged with the Soviets, who had taken his mill and occupied his hometown. At the first meeting with the Soviet side in spring 1940, he pledged collaboration and goodwill:

> The best way to build trust between our delegations is to tell things as they are. We have brought our best experts here and we are ready to give all information to you. I hope that you appreciate the fact that we have been very open and our proposals are based on the genuine will of Finland to handle these issues in a positive way.[24]

The first negotiations focused on the Rouhiala power station. The lack of a high-voltage transmission line between the station and Leningrad gave the Finns some hope. The country was in desperate need of electricity and nothing could compensate for the loss of Rouhiala. For this reason, the Finnish delegation tried to persuade the Soviets to allow this power station to feed electricity across the border to the Finnish national grid until the high-voltage line to Leningrad was completed. To demonstrate goodwill and collaboration, the Finnish delegation handed over all technical and hydrological documents and agreed to pay for the electricity that had been transmitted from Rouhiala to Finland since the end of the war.[25]

At first the strategy seemed to work. The people's commissar, Stepanov, promised to respect international law and consider the Finnish proposal. One option was to use the Finnish national grid as a gateway for transmitting Rouhiala electricity to the Hanko naval base (approximately 100 km to the west of Helsinki), which had been leased to the Red Army. The naval base consumed approximately 1.5 MW of electric power and it was technologically feasible to use Rouhiala hydroelectricity to satisfy the needs of the Hanko base.[26]

The collaborative spirit disappeared, however, when the new Enso power station that was under construction was put on the agenda. The rapids that would supply this power station were located, after the peace treaty, partly in Finland (Vallinkoski rapids) and partly in the Soviet Union (Enso rapids). However, the construction of the new power plant was designed to increase the height difference of the Enso rapids by building an unusually high dam (16 m). When completed, this would raise the water level so that the Vallinkoski rapids 4 km upstream would be combined into a new and more powerful "waterfall," to be located on what was now Soviet territory.[27] The construction of the new dam had not been completed by the Finns when the Winter War broke out. The Soviet foreign minister, Molotov, made clear, however, that he intended to have the dam completed and that the Soviet Union would take all of the electricity. The Finns would thus not be able to exploit the Vallinkoski rapids, even though these were still on Finnish territory. The Finnish authorities appealed to international law for compensation but in vain. The Soviet Union interpreted the law differently and refused to pay.[28]

This left Finland with three options. First, it could build a new power station that would utilize the Vallinkoski rapids' elevation drop and feed electricity to the national grid. Second, it could sell the rapids to the Soviets and thus eliminate the problem. Third, it could continue negotiations until a long-term solution could be agreed upon for cross-border collaboration. The first option was ruled out because it would challenge Soviet hegemony at the border. The second option was equally impossible because it was against the Finnish constitution to sell property to a foreign nation. The only option left was to continue negotiations until a consensus was reached. Finland and the Soviet Union were now connected to each other by the Vuoksi River and the electric power network that stretched across the border.[29]

The Finnish delegation approached Molotov again in November 1940. Its new proposal was based on a transborder collaborative scheme that, according to the Finns, was now necessary to ensure the rational use of the Vuoksi waterway's hydropower resources. According to Finnish calculations the capacity of the Vallinkoski rapids was approximately 500 million kWh annually, representing a little more than 70 per cent of the total production of the new, large Enso power station. It was fair, therefore, for the Soviet Union to compensate Finland for the loss of energy. Instead of asking for financial compensation, however, the Finns proposed a permanent transborder connection that would deliver to Finland 45 per cent of the electricity that was produced in the new Enso power station.[30]

It took six months before Molotov responded to the Finnish proposal. The Soviet Union regarded the Vallinkoski rapids as part of the Enso power station and therefore, Molotov explained, Finland had no legal right to claim compensation. This was the final contact between the neighbors, because one month later Adolf Hitler ordered the Wehrmacht to invade the Soviet Union. The Finnish army followed the German invasion, and with the help of the German war machine the Finns pushed the Red Army out of the occupied territories. By the end of 1941 the Vuoksi River valley was again part of Finland and the power stations that had been lost were reconnected to the national grid. The Red Army had blown up the nearly completed Enso power station while retreating, but it was subsequently rebuilt by the Finns. However, after the defeat of Germany in the battle of Stalingrad, the tide turned and in the summer of 1944 the Red Army launched a massive offensive against the Finnish army in the Karelian Isthmus. However, as they were again not able to penetrate deep into Finland or crush the Finnish army, peace negotiations were held in Moscow and Finland exited from the Second World War in September 1944. The border returned to where it had been settled in March 1940.

But the question of cooperation in the Vuoksi River valley refused to go away, even though the Iron Curtain had sealed the border between Finland and the Soviet Union. The Rouhiala and Enso power stations (the latter having eventually been completed in 1942) were connected to the Leningrad grid, while Tainionkoski and Imatra fed electricity to the Finnish national grid. There was very little contact and practically no cooperation across the border. Nevertheless, the loss of two big hydropower stations was a serious blow to the Finnish energy system. As Finland was committed to paying massive war reparations to the Soviet Union, every drop of falling water was needed to provide energy for the country's power-hungry industry.

In this situation, the Finns launched a renewed attempt to obtain compensation for the loss of the Vallinkoski rapids. The question was put on the table in 1948 when the Soviet Union sent a delegation to Finland to negotiate the regulation of the flow and the discharge of the Vuoksi River. The Soviet authorities wanted to increase the minimum depth at the Enso power station and to obtain guarantees that the average discharge of the river would not fall below 540 m^3/s. The Finns were ready to accept this rate but they opposed a raising of the water level at Enso because it would expand the mass of water in Lake Saimaa and cause serious flooding. In addition they again made clear that they were dissatisfied with the loss of Vallinkoski hydropower. According to their latest calculations, Finland lost about $8 million annually because Vallinkoski energy was supplied to Leningrad rather than to the Finnish national grid. Also annoying was that the Soviet Union had raised the water level at Enso without consulting the Finns. A higher water level downstream decreased the production of electricity upstream. For example, during 1946 and 1947, Imatran Voima lost almost 10 million kWh due to the selfish behavior of the Soviets.[31]

It took more than a decade before transborder collaboration between Finland and the Soviet Union took a new turn. In 1960, Enso-Gutzeit, Imatran Voima, and the Soviet foreign trade organization, Mashinoexport, signed an agreement restoring the 110 kV connection from the Enso power station to Imatra. One of the four turbines of the station was designated to deliver 22 MW of electricity to the Finnish national grid. The connection was managed by the Leningrad power network operator, Lenenergo. Although it did not compensate for the Finns' loss of Vallinkoski power, it was nevertheless a strong signal of the peaceful coexistence that had been advocated by the Finnish and Soviet governments since the death of Stalin in 1953. Finland had already opened up access to the national grid on its western border – through a connection to Sweden that was operational from December 1958 – but now the eastern border had also been opened up for electricity imports.[32]

The new connection across the eastern border revived old arguments about the use of the Vuoksi waterway's hydropower resources. If electricity was transmitted from the Enso power station back to Finland, it made sense to coordinate the discharge of the Vuoksi River jointly. In the Soviet Union, issues like these were decided at the highest political level, inside the high walls of the Kremlin. A joint Finnish-Soviet border water commission was established in 1964 to investigate, coordinate, and regulate the use of common resources along the 1000 km Finnish-Soviet border. In addition, the commission tried to increase accessibility to waterways located on both sides of the border. From the Finnish point of view, the permanent joint commission was an important instrument that could be used to enhance trust and confidence, but also to control the use of common resources and transborder infrastructures. Without such an instrument the Soviets could continue dictating the terms of cooperation and the use of hydropower resources in the Vuoksi River, and also in other rivers and lakes along the long border.[33]

The border water commission did not banish distrust and tension from Finnish-Soviet relations, but it helped to establish a permanent contact between the former enemies. Finnish foreign policy during the Cold War followed a doctrine of "pragmatic neutrality" that advocated friendly and cooperative relations with both East and West. The main objective of energy policy was to enhance self-sufficiency and to utilize domestic energy resources as much as possible. Nevertheless, transborder connections were allowed in designated locations where they did not threaten the stability of the national grid. The Enso-Imatra connection was considered "safe" because it utilized hydropower resources of the Vuoksi River and the connection delivered power one way to the Kaukopää pulp and paper mill in Finland. The connection was also technologically safe because the Finns had designed and constructed the Enso power station so that it was technologically compatible with the Finnish standards. The connection to the Swedish grid followed a different logic, being established to balance the production of hydropower during periods of low precipitation.

This long, painful process was the first real encounter that prompted the Finns and the Soviets to sit down to negotiate the management and governance of a transborder connection. It goes without saying that the parties did not trust each other and that very little could be done to enhance trust during the war. However, the border water commission and the transborder connection from Enso to Kaukopää demonstrated that former enemies were able to establish permanent transborder connections and manage them jointly.

The negotiation process at the Finnish-Soviet border followed unique rules. Soviet delegations were bound to the Kremlin's orders, which were more politically than economically oriented. The Soviets were more interested in scoring political victories than gaining economic benefits. For the Finns the situation was the opposite. After the war they had to become accustomed to living and collaborating with a politically aggressive neighbor; in order to secure the supply of electricity, certain compromises had to be made. Although the rules and patterns of the uneasy alliance were understood on both sides of the border, they were never openly admitted. The Soviets tried to increase political and ideological influence in Finland, while the Finns tried to contain the pressure by allowing transborder connections to be built which could be controlled and managed jointly. The critical point was the trunk line, which became a symbol of independency and sovereignty.

Containing criticality

Finnish energy policy, based on self-sufficiency and only limited transborder connections, was put to the test during the Cold War, when political and economic integration advanced on both sides of the Iron Curtain. At the same time, Finnish society was undergoing rapid industrialization and urbanization, leading to greatly increased consumption of electricity. It was estimated that domestic energy resources would not be able to meet demand from the early 1970s; beyond this, Finland would become increasingly dependent on imported electricity, oil, gas, and coal.[34]

This was not a pleasant scenario. International trade in fossil fuels was plagued with ideological and political tensions, which Finland wanted to avoid to the greatest possible extent. The Soviet Union was a major supplier of coal, oil, and gas, so these raw materials were already included in the bilateral trade agreement between Finland and the Soviet Union. Coal and oil could also be purchased from Poland and the Middle East, but both regions were politically unstable and constant threats of war and terror made this option the least favorable.[35]

How then could Finland meet its demand for electricity without getting involved in political and ideological struggles? This was the central issue of Finnish energy policy from the mid-1960s, when all domestic hydropower resources had been harnessed. New sources of electricity had to be found, domestically or abroad. This is when atomic energy stepped in as the ultimate solution to the country's energy problems.[36]

The peaceful use of atomic energy became available in 1953 when the US president, Dwight D. Eisenhower, announced to the United Nations (UN) General Assembly that his country was going to declassify thousands of scientific and technological documents concerning nuclear fission. In addition, he pledged that the United States and other members of the nuclear family would create an atomic bank from which politically reliable countries could obtain enriched uranium to be used in scientific and technological experiments. The Soviet Union was not mentioned but the Kremlin quickly responded to the challenge by offering similar cooperation to politically "credible" nations.

In August 1955 the "Atoms for Peace" program was officially inaugurated. Scientists, engineers, and policy-makers from all over the world came together in Geneva to share information and discuss the future of atomic energy.[37] The Geneva conference gave wings to an atomic enthusiasm that spread rapidly throughout the world. It was estimated that the price of atomic electricity would be next to nothing and that uranium would replace oil, gas, and coal as the main source of thermal power. These optimistic expectations were somewhat overshadowed by fears of nuclear war and technological failures in the power plants that could cause health hazards and environmental catastrophes. But most countries still considered it of great importance to advance nuclear power.

Finland sent a small delegation to Geneva to find out what the Atoms for Peace program might offer a small country that had been isolated from the international community since the end of the war. Although atomic energy promised inexpensive and almost limitless access to electricity, there were a number of political and ideological issues that needed to be sorted out before Finland could participate in a transnational program. The Atoms for Peace program was coordinated by the UN, and the Finnish application for membership of this organization was still pending.[38]

Finnish atomic energy policy emphasized education, technological research, and international cooperation. A small country could educate a critical mass of scientific and technical experts, but this had to be done by networking and collaborating with more advanced countries and with the help of international organizations, such as the International Atomic Energy Agency (IAEA) and the Organisation for Economic Co-operation and Development (OECD). It was suggested that a first full-sized nuclear power plant should be completed in Finland by the early 1970s. At that time the national demand for electricity was expected to exceed 20 million KWh.[39]

But the road to atomic paradise proved to be rough and windy. The atomic enthusiasm that inspired the first Geneva conference faded away as the Cold War escalated in the late 1950s and early 1960s. The civilian use of nuclear power became a battlefield where the Soviet Union and the United States fought for technological hegemony. Both superpowers signed bilateral deals with allies for nuclear reactors and supplies of uranium. The Finnish authorities tried to balance between East and West and, as a result, neither of the superpowers established a dominant foothold in Finland. Finnish nuclear scientists and engineers studied in the United States, Sweden, and Denmark, while policy-makers, diplomats, and politicians paid visits to Soviet nuclear facilities in Moscow, Novo-Voronezh, and Obninsk. The head of the Finnish nuclear power program, Professor Erkki Laurila, defined the Finnish position in the hegemony game in 1962 as follows:

> We should buy a test reactor that suits nuclear technologies already existing in Scandinavia. This way we would get a "ticket" to the ongoing research and development process in the West. From a technical point of view the Soviet standard reactor is not bad, but if we purchase the reactor from the Soviet Union it sets in motion a process that we cannot control. It is better therefore, to turn to the West and acquire the test reactor from there.[40]

Most European countries had no choice but to accept an offer of nuclear cooperation from the United States or the Soviet Union. Those who received the offer from the United States were considered lucky, because US nuclear technology was regarded as safe, reliable, and efficient. The situation on the other side of the Iron Curtain was very different. The Soviet Union had developed nuclear technology from the early 1950s, but the Soviet technological style was clumsy and crude, and the nuclear facilities and power plants were fully integrated with the top-secret military-industrial complex. Moreover, as Professor Laurila pointed out,

Soviet technology was beyond Finnish control. If a test reactor was ordered from the Soviet Union, Finland would become dependent on Soviet nuclear technology for the foreseeable future. This would have major implications when the nuclear power program moved to the next phase. Finland feared that it might be forced to adopt Soviet nuclear technology and that this would jeopardize the security and reliability not only of the nuclear power reactor as such but also of the national electricity system as a whole.

This same concern was hidden between the lines of the document that formally opened the international bidding for a 300 MW Finnish nuclear power station. The document had been written in 1965 by state-owned power company Imatran Voima, which had been charged with managing the nuclear power project in Finland. Imatran Voima needed an international company that could not only design and construct a safe and reliable nuclear power station but also manage the project. Invitations were mailed to 11 nuclear manufacturers around the world in July 1965.[41]

Heikki Lehtonen, president of Imatran Voima, was convinced that this was the right approach. The company welcomed the participation of all competent manufacturers in the international bidding. The final decision was to be based on objective technological and economic criteria. No political or ideological arguments were to be taken into account. This was the first time this kind of open bidding had been organized in the field of nuclear power. As Lehtonen wrote to his chief engineer, Tauno Rask, "I'm sure everyone will participate – except the Soviets, perhaps. At least, this is what we're counting on here."[42]

By the time the deadline arrived, eight manufacturers had confirmed their participation. The missing three were the manufacturer from the Soviet Union, General Atomics from the United States and Britain's General Electric. General Atomics did not have the right kind of reactor and General Electric was occupied elsewhere. Imatran Voima had sent two invitations to the Soviet Union, because the company did not know which organization or institution would be the one to participate in the bidding. No reply had been received from Moscow, and Imatran Voima thus went ahead with the selection process. The company had promised to analyze the tenders and choose two or even four candidates for the final round.

This is when everything started to go wrong. The Soviet Union had, indeed, sent a reply but it was addressed not to the power company but directly to the Finnish government. This mysterious letter raised concerns at Imatran Voima. The company had hoped that the Soviets would stay out of the race and that the first nuclear power station could be ordered from the West. Now these hopes faded, and political and ideological dimensions entered the bidding.[43]

Imatran Voima accepted the tender from Moscow, even though the official deadline had expired. It turned out to be both technologically and economically inferior and could not compete with the much stronger tenders of the Western companies. The initial analysis revealed that the five best offers came from Westinghouse, Siemens, AEG, ASEA, and the Canadian General Electric.[44]

Imatran Voima had hoped that it would be able to escape the ideological and political tensions plaguing the civilian use of nuclear power by allowing all nuclear manufacturers to participate in the bidding. These hopes were quickly crushed when it became clear that none of the Western companies would be able to deliver enriched uranium without permission from the IAEA and possibly also from the US government. The Soviet Union also had a surplus of enriched uranium, but the Soviets were refusing to sell uranium fuel for reactors other than the ones they had designed and constructed themselves.[45]

Despite this troublesome information, Imatran Voima decided to go ahead as planned and choose the best three contenders for the final round. They were AEG from West Germany, Westinghouse from the United States, and General Electric from Canada. These three had the most advanced reactor technology, well-designed safety systems, and a cost-effective production record.[46]

This result did not please the Soviet Union, however, which demanded that its tender should be reconsidered and readmitted to the competition. This in turn angered Canada's General Electric, and the company decided to pull out of the competition. This opened the door to further speculation and, by early 1967, the Swedish company ASEA, and the British manufacturer UKAEA, re-entered the bidding with new tenders. Imatran Voima tried to reorganize the competition but in vain. A hegemonic struggle between East and West had taken over the bidding and it was no longer possible for anybody to predict the final outcome.

This question could not be decided by Imatran Voima alone. The Finnish government could not passively watch the country being drawn into the political storm. Imatran Voima had had good intentions, but now it was time to blow the whistle and take a time-out. The government ordered Imatran Voima to inform all participants that all tenders had been rejected and the bidding cancelled. Furthermore, there was no reason to launch a new round of bidding and no order should be placed on the basis of the current tenders. The nuclear power project could be reconsidered when and if the political climate calmed down. Finnish energy policy was no longer looking at nuclear power as a future solution. Instead the focus was on coal, oil, and gas.[47]

This was a tough decision for Imatran Voima. The company's atomic energy team had spent almost four years investigating tenders, without result. Imatran Voima lost much of its reputation as a credible international partner in high-technology projects, and many engineers left the company for better opportunities elsewhere. Worst of all, the Western media ridiculed both the company and the Finnish government for their allegedly gutless and soft attitude towards the Soviet Union.

The nuclear moratorium came to an end in the summer of 1969 when the Finnish government succumbed to political pressure and agreed to send a delegation of experts to Moscow to negotiate nuclear collaboration. Although no agreements were supposed to be signed, everybody on board understood that this undertaking would decide the future of the Finnish nuclear power program. The Soviet Union had already informed Finland that it would not be allowed to purchase enriched uranium for Western reactors. But Soviet uranium was available, and the Kremlin promised long-term contracts if the Finns accepted the terms.[48]

Summer days in Moscow can get hot and humid when continental weather from Siberia sets in. Local people know how to avoid the heat by escaping to their "dachas" outside the city, but this privilege was not given to the small group of men standing in Red Square. They were all dressed in black suits, white shirts, and black bow ties. The Finnish delegation had chosen Red Square as the place to decide their final negotiating strategy, because it was close to the Kremlin and there were no listening or spying devices.[49]

The Finnish delegation tabled its agenda, which included three very difficult demands. First, the reactor and the reactor building had to be covered by a gastight steel containment. Second, the reactor and all of its components had to be designed according to US ASME standards. Third, the Soviets had to provide unrestricted access to all manufacturing units, including the uranium-enrichment plants.[50]

The Soviet delegation could not possibly accept this agenda. The head of the delegation, the academician Andranik Petrosyants, gave the Finns a long summary of Soviet achievements in the field of nuclear technology. The Soviet Union had been the first country to develop power reactors that could be used for civilian purposes. In addition, the Soviets had developed safety standards that exceeded those in the West. The safety record of the VVER-type reactor – the Soviet light-water reactor – was spotless and no accidents or critical events had been reported. There were already several reactors of this type installed in the Soviet Union and the Council for Mutual Economic Assistance (COMECON) countries. Soviet scientists and engineers had assessed the possible risks, and the necessary precautions had been taken to secure the safety of the reactors and power plants. The Soviet Union used its domestic standards when designing nuclear reactors and components, and there had been no complaints. The last point on the agenda was simply impossible, because the Soviet constitution did not allow foreign citizens any access to high-technology factories and research institutions.[51]

The Finnish delegation was determined not to give in, however. The containment that was to cover the reactor vessel and protect the nuclear power plant in the case of catastrophic accident was the key issue. If the Soviets refused to redesign the reactor, the deal was off. The same applied to the ASME standards. It was impossible to accept design standards that were incompatible with the rest of the standards used in Finland. Quality control was a normal procedure in the West and open access to manufacturing units, the Finns argued, increased transparency and trust.

The Moscow negotiations lingered for weeks without much progress. Imatran Voima tried to make the Soviets give up the deal by slowing down the negotiations and refusing to accept any compromises. This frustrated the Soviet delegation, which had received strict orders from the Kremlin to persuade and pressure Imatran Voiman to sign the contract. The Soviet Union needed a foreign reference, and if Finland refused to accept the VVER reactor, who else would then think of buying a Soviet-designed reactor? On the other side of the table, this was exactly what the Finnish negotiators were trying to avoid. Imatran Voima opposed the Soviet nuclear power station and the company had no desire to become a part of the Soviet technological sphere.

The deadlock was broken in political negotiations that took place behind the scenes. The agreement was made but the parties were not Imatran Voima and

Technopromexport – instead they were the Finnish and Soviet governments. What was decided inside the Kremlin walls was a great and unpleasant surprise to Imatran Voima. The company was named as the main contractor for the first nuclear power station in Finland and the role of Technopromexport was reduced to being a subcontractor, which would deliver two VVER reactors, reactor vessels, two steam turbines, and certain parts of the primary circuit to Finland. It was left to Imatran Voima to redesign the power plant, adapt it so that it matched Finnish requirements, manage the project as a whole and find subcontractors – in Finland and abroad – for those parts that Technopromexport did not provide.

Even after the agreement, there remained more questions than answers. How should Imatran Voima redesign a Soviet-designed nuclear power station when it had no previous experience in the field of nuclear energy? How could it find a manufacturer that could design and construct the massive steel containment that would cover the VVER reactor? And finally, how much would all of this cost, and would this kind of transnational project ever result in a safe, durable, and efficient nuclear power station?

These and many other questions were at least partly answered on January 21, 1977, the day when the first reactor of the Loviisa nuclear power plant, built on the shores of the Gulf of Finland, went critical for the first time. Imatran Voima had worked for almost eight years to accomplish the project, which was popularly referred to as "Project Eastinghouse." Time and money had been spent and the project was almost four years behind the original schedule. Finally, everything was in place. The Finnish president, Urho Kekkonen, and the Soviet premier, Alexei Kosygin, attended the opening ceremonies. Those Soviet engineers who saw the power plant for the first time could not believe that it was, in fact, a Novo-Voronesh-type power plant. The reactor building was made of reinforced concrete and the reactor hall was covered by a massive steel containment that sealed the reactor from the outside world. There were no broken windows or dirty floors, but instead the reactor hall and other facilities were spotless and painted in neutral colors. The only reference to the origin of the VVER reactor was found on the steel lid that covered the reactor vessel. The bright red lid was adorned with the famous letters CCCP.

The Loviisa nuclear power plant became a truly transnational facility containing high-technology components from both sides of the Iron Curtain. The Soviet Union delivered the reactors and turbines, while the safety technology was purchased from the United States. The instrumentation came from West Germany and the process computer was manufactured in Finland. Thousands of construction workers came from the Soviet Union, Finland, and other European countries. The project was managed by Imatran Voima with the help of several consulting companies from Switzerland, the United States, Sweden, and West Germany. This was an exceptionally difficult task during the Cold War, when the transfer of high technology across the Iron Curtain was subject to strict control.[52]

The obvious question was: Why all this trouble? Finland needed additional power-production capacity to ensure self-sufficiency, but nuclear power was deeply political and connected Finland to the ideological struggles of the Cold War. As a result, when the first nuclear power station was plugged into the national grid, the Finnish energy system became dependent on the Soviet Union. The uranium fuel was shipped across the border and the nuclear waste returned. In addition, the VVER reactor and Soviet technology connected Finland to its eastern neighbor. These connections had to be managed and governed in order to minimize risks and maximize sovereignty.

The question was all about trust – as it had been in the Vuoksi River case and would be later in the sea-cable episode. Nuclear power was a new and dangerous source of energy, and specialized skills were needed to control the reactors and their production of electricity. Finnish engineers did not trust Soviet nuclear technology, but the risks were contained by the new construction and additional safety technologies from the West. The Soviets never understood the lack of trust; several times they tried to convince the Finns that Soviet technology was safe and reliable. Antti Vuorinen, head of the Finnish Radiation and Nuclear Safety Authority, ended this dialog when he repeated what Lenin had said several decades before: "Doveryay, no proveryay" – "Trust, but verify."[53]

Fear, distrust, and the quest for control at the Finnish-Soviet border

In 2006 the Finnish authorities refused to grant United Power permission to build a 1000 MW submarine cable from Russia to the coast of Finland because the national grid was too weak to tolerate the massive load of electricity in one geographical location. As mentioned above, United Power criticized this decision because it was not based on objective calculations. The Finnish authorities and the national grid company, Fingrid, never seriously considered the second proposal, which would have divided the 1000 MW cable into two 500 MW cables, to be landed on the northern shore of the Gulf of Finland at two different locations. Instead of playing a fair game, the Finnish government decided to finance the EstLink project, which was technologically analogous to the one offered by United Power.

It is no secret that the Finnish government and Fingrid disliked United Power's offer. They also feared that a sea cable from the Leningrad nuclear power station would make the national grid vulnerable and Finland too dependent on the Soviet energy system. These fears were grounded in the long history of energy collaboration with the Soviet Union. Finland had developed special relations with its mighty neighbor during the nineteenth and twentieth centuries, and this uneasy alliance resulted in fruitful economic collaboration, but also political and ideological pressures. The Soviet Union was an aggressive and expansive superpower that jealously guarded its geopolitical interests in Europe and other parts of the world. In 2006, even though the Soviet Union had been dissolved more than a decade earlier, the memory of this volatile collaboration refused to go away.

The case of the sea cable and the two other episodes discussed in this chapter demonstrate how energy collaboration between Finland and the Soviet Union was plagued by distrust and fear. Although common natural resources and more than 1000 km of common border provided a perfect framework for close and profitable collaboration, there were only a few cross-border connections built during the nineteenth and twentieth centuries. Instead of advocating collaboration, Finland tried to preserve its energy independence and only those connections that were considered necessary were allowed to cross the border. The national grid was the most precious component of the national energy system, and no foreign connection was allowed to plug into it without permission from the Finnish government.

This chapter has demonstrated how the political, ideological, and cultural context shaped criticality at the Finnish-Soviet border. The Finns did not trust the Soviets and vice versa. The lack of trust resulted from wars and geopolitical pressures that continued until the collapse of the Soviet Union in 1991. The bad blood between the neighbors stemmed from political institutions, but also from the Soviet technological style, which was perceived as alien by Finnish engineers and managers.

Current scholarship of critical infrastructures focuses on critical events and breakdowns of complex technologies. This chapter suggests that the definition of criticality can be enriched if the focus is placed on the cultural and political context that shapes the design, construction, and governance of transnational infrastructures. None of the transborder infrastructures which crossed the Finnish-Soviet border became critical, although all elements of risk were present. Soviet engineers and managers had a "theoretical" approach to risk and criticality. If calculations showed that no risk existed, there was no need to build expensive backup systems and protective structures. Their Finnish colleagues approached risk and criticality from a different angle. Theoretical calculations were made, but practical considerations played the major role in the final decision-making. The two approaches clashed when transborder connections were planned. Finnish politicians could not say "no" to the transborder connections that were initiated by the Kremlin, but Finnish engineers and managers were able to contain the risk by redesigning Soviet technologies and by creating joint management institutions to govern common resources.

Notes

1. I would like to thank my colleagues Mikko Kohvakka, Pekka Rautio, Virpi Kaisto, and Kristiina Korjonen-Kuusipuro at Lappeenranta University of Technology for their creative comments and collaboration.
2. Manuscript of the TV program *MOT*, "Venäläistä sähköpeliä," May 22, 2006. http://lotta.yle.fi/motweb.nsf/sivut/ohejelma?opendocument&pagei
3. Ibid.
4. Raili Leino, "Sosnovy Bor jatkaa remonttia," *Tekniikka ja Talous*, May 19, 2005. See also "Proposed undersea cable would double electricity imports from Russia," *Helsingin Sanomat*, December 31, 2005.
5. "Future of the high voltage sea cable is still open," *Nordicum. Scandinavian Business Magazine*, 3/2006, p. 47.
6. "Russian company sweetens offer of large undersea electric cable," *Helsingin Sanomat*, June 9, 2006. See also the manuscript of the TV program *MOT*, "Venäläistä sähköpeliä," May 22, 2006.

7. Arja Haukkasalo, "Pekkarinen: Päätös merikaapelista syys-lokakuussa," *Energia-lehti*, August 16, 2006.
8. "The first power link between Estonia and Finland was inaugurated on 4 December 2006". www.nordicenergylink.com
9. Helena Raunio, "Merikaapeli on eri kaapeli," *Tekniikka ja talous*, September 12, 2006.
10. See, for instance, the speech by the president of Finland, Ms. Tarja Halonen, in the Estonian parliament (Riigikogu), May 5, 2010. http://www.tpk.fi/Public/default.aspx?contentid=191689&culture=fi-FI
11. See, for example, Luhmann 1979 and Giddens 1984.
12. Lundestad 2003, pp. 11–12.
13. Fukuyama 1999.
14. Sergei Karaganov, "Time to Back off: Russian-European Relations,", *New York Times*, May 15, 2007.
15. Nöelle 2009.
16. "Tasavallan Presidentin (Urho Kekkonen) turkkilaisille sanomalehdille myöntämä haastattelu," March 30, 1977, Finnish National Library, digital archive, UKK927638. https://oa.doria.fi/bitstream/handle/10024/9294/TMP.objres.4098.html?sequence=1. See also Baker and Glasser 2005.
17. See, for example, Winner 1986.
18. See http://news.bbc.co.uk/2/sh'. red/spl/hi/europe/02/euro_borders/html/8.st
19. Laaksonen 2005.
20. Michelsen 1999.
21. Kristiina Korjonen-Kuusipuro, "Voimaa Vuoksesta," *Tekniikan Waiheita* 3/2007, pp. 5–7.
22. Myllyntaus 1991, pp. 95–96.
23. Johan Nykopp, "Rouhialasta rauhanteon jälkeen Suomeen saadun sähkövirran maksu," September 27, 1940. See also Johan Nykopp, "The Price of Electricity Produced and Delivered from Rouhiala since the End of the War," April 9, 1940, Archives of the Finnish Foreign Ministry (UM), 45/7 D 1940.
24. A.E. Kotilainen, "Moskovan neuvotteluista," June 28, 1940, UM, 45/7 1949.
25. "Selostus eräistä Rouhialan voimaitosta ja sen venäläisille luovuttamista koskevista seikoista," April 29, 1940, UM, 45/7 D 1940.
26. Johan Nykopp, "Rouhialasta rauhanteon jälkeen Suomeen saadun sähkövoiman maksu," October 8, 1940, UM, 45/7 1940.
27. Stora Enso Ltd, *Company History*, http://www.funduniverse.com/company-histories/Stora-Enso-Oyj-Company-History.html
28. Aaro PakaslahtI, "Suomen ja Neuvostoliiton suhteiden uudelleenjärjestelyä II," November 15, 1940, UM, Poliittisia tiedotuksia 93. See also "Enso-Leningrad, Uuden korkeajännityslinjan rakentaminen," October 9, 1940, UM, 45/7 1940.
29. "Enson-Vallinkosken voimalaitosta käsittelevän sekakomission suomalaisten jäsenten muistio," (undated), UM, 45/7 1940.
30. Aaro PakaslahtI, "Suomen ja Neuvostoliiton suhteiden uudelleenjärjestelyä II," November 15, 1940, UM/Poliittisia tiedotuksia 93, UM, 45/7 1940.
31. "Eräiden rajavesistöjen käyttöä koskevien oikeudellisia kysymyksiä selvittelevän toimikunnan mietintö", Committee report, Helsinki, 1962, pp. 9–12. See also Uolevi Raade, "Koskee Vuoksen voimalaitoksia," October 8, 1947, UM, Rajavesikomitea 1947.
32. The agreement between Finland and the Soviet Union was signed in 1960, but electricity started to flow across the border in 1963.
33. "Suomen tasavallan ja Sosialististen Neuvostotasavaltain liiton välinen rajavesistöjä koskeva sopimus," April 24, 1964, UM, Rajavesikomitea.
34. Bror Nordqvist, "Finland's Future Needs and Possibilities for Power and Heat," presentation at the First Geneva Conference on Peaceful Use of Atomic Energy, August 8–20, 1955, Imatran Voima Archives (IVOA), AEN 1965–7, FHA.

35. Kuisma 1997.
36. Komiteamietintö B76:1966, *Atomienergiakomitean mietintö*, Helsinki, 1966.
37. Laurila 1967, p. 243.
38. Michelsen 2010.
39. Laurila 1982, p. 112.
40. Erkki Laurila, "TKK:n reaktorilaboratorio", April 9, 1962, Archives of the Finnish Ministry of Trade and Industry/Atomic Office (KTM/Atomitoimisto), HS1, VNA.
41. Imatran Voima, "Invitation to Bid," July 15, 1965, IVOA, Kirjeistö 1965, A-J, FHA.
42. Pentti Alajoki to Tauno Raskila, IVOA, Kirjeistö 1965, O-T, FHA. See also Heikki Lehtonen, "Feelers from Finland," *EuroNuclear*, Vol. 2. No. 10/1965, p. 526.
43. Michelsen and Särkikoski 2007, pp. 88–90.
44. Ibid., pp. 90–92.
45. Ibid., pp. 115–116.
46. Heikki Lehtonen, "Teknopromeksportin atomivoimalaitostarjouksista," July 25, 1967, IVOA, Juhani Santaholman kokoelma, Atomikierros 1968, FHA.
47. Ministry of Trade and Industry to Imatran Voima, July 25, 1968; "Hallintoneuvoston pöytäkirjat," July 26, 1968, IVOA, Juhani Santaholman kokoelma, Atomikierros 1968, FHA.
48. Erkki Laurila, "Atomivoimalaitoksen hankintaa koskeneista neuvotteluista Moskovassa," July 17–23, 1969, KTM/Atomitoimisto, Hj10, VNA.
49. Numminen 2000.
50. Erkki Laurila, "Atomivoimalaitoksen hankintaa koskeneista neuvotteluista Moskovassa," July 17–23, 1969, KTM/Atomitoimisto, Hj10, VNA.
51. Michelsen and Särkikoski 2007, pp. 146–147.
52. Numminen and Laine 1983, p. 243.
53. Michelsen and Särkikoski 2007, pp. 238–245.

5

Bulgarian Power Relations: The Making of a Balkan Power Hub

Ivan Tchalakov, Tihomir Mitev, and Ivaylo Hristov

Prologue[1]

In early 1983, Todor Bozhinov is appointed as the new Bulgarian minister of the electric power industry. Eager to demonstrate his professional managerial capacities, he introduces a number of economic and organizational changes. These have recently been discussed in the Council for Mutual Economic Assistance (COMECON), and are aimed at increasing the productivity and efficiency of member states' power systems. Some of the measures directly affect the reliability of electricity supply. For instance, the customary summer maintenance and repair works, during which many generating units lie idle, are to be drastically reduced. Also, two full months of coal emergency reserves are to be cut to the equivalent of two weeks, thus freeing significant financial and material resources. About a third of the Bulgarian coal supply comes from Soviet Ukraine via Black Sea ports, but this supply line seems quite reliable.

However, the winter that follows is unexpectedly cold. In early January 1984, most Soviet Black Sea ports are frozen. The Ukrainian coal supply to Bulgaria is disrupted. The limited emergency reserves of lignite at Bulgaria's biggest electric power complex, Maritsa East, are quickly spent, and a number of thermal generators are shut down. The remaining generators work at extremely high loads and many break down, partly due to the reduced repair works in the previous summer. For almost two months the Bulgarian economy and population face severe electricity cuts – two out of every four hours. At the time, one of us is living on the 11th floor of a (typical) apartment building, where power cuts also deprive the family – including a six-month-old baby girl – of heat and water because of the failure of the electric steam and water pumps.

January 1984 becomes known as the country's most serious electricity crisis since the early 1950s. Bozhinov is dismissed and his predecessor is reinstated. Still it takes more than six months for Bulgaria's power supply to be fully restored, and by then the current minister has been hospitalized with symptoms of severe exhaustion. The events of that winter help to legitimize the expansion of nuclear power in Bulgaria as a way to strengthen the domestic energy supply and reduce dependence on coal and electricity imports. They also serve to boost the establishment of a large hydropower pump storage complex as a way to improve power-system reliability, and the construction of new high-voltage power lines linking Bulgaria to the Soviet power system and several other countries in the region, thus further increasing Bulgaria's electrical integration with its neighbors and strengthening the country's role as a Balkan power hub.

Introduction

The events of January 1984 and their aftermath illustrate how the Bulgarian electric power supply developed in the context of broader energy exchanges between a number of countries, not least in cooperation with COMECON and in bilateral exchanges with non-COMECON Balkan partners. Such exchanges provided advantages and helped Bulgaria to develop from structural energy poverty into a prominent Balkan power hub. Yet they also implied new vulnerabilities, as in the case of the sudden disruption of coal imports. Moreover, these energy and power issues had a sharp political edge and constituted key elements in domestic quarrels as well as in Soviet-Bulgarian relations.

This chapter investigates Bulgaria's electrification in the framework of the Eastern European socialist project. We outline the technological and institutional principles in building the Bulgarian electric power system, as well as the patterns of electric and energy integration with other COMECON countries – and especially the Soviet Union. From this perspective we address two major research themes. First, the Bulgarian case allows us to study several aspects of Europe's infrastructure integration.[2] As a COMECON partner, Bulgaria was involved in efforts to organize electricity systems in socialist countries under a single "umbrella" – the Integrated Power Systems *(Obedinenie energeticheskie Sistemi)*, thus reproducing Europe's Cold War division in the electrical sphere.[3] On the other hand, as we shall see, other power lines – as well as flows of electricity, machinery, energy sources, and skilled workers – crossed the capitalist-socialist boundary and to some extent negated Europe's Cold War fragmentation. Indeed, we argue that Bulgarian efforts to link up with its socialist partners as well as its capitalist neighbors eventually gave rise to a new role for Bulgaria as a key player in southeastern European electric power exchanges across supposed socialist-capitalist borders.

Second, we ask how Bulgarian perceptions of vulnerability informed and shaped the processes of infrastructure integration. As socialist system-builders, Bulgarian power authorities needed to address and reconcile several forms of vulnerability in technical (domestic power shortages, import dependencies) and political (harsh domestic conflicts, managing Soviet-Bulgarian relations) spheres. As we shall see, vulnerability considerations greatly informed the historical development of Bulgaria's domestic power supply and its embedding in transnational energy and electric power relations, inspiring such measures as introducing ever more powerful generating units; diversifying electricity production into hydropower, thermal, and nuclear units; improving the reliability and stability of domestic transmission lines; and continuous investment in interconnections with its COMECON and Balkan countries.

The political, economic, and technological relations between Bulgaria and its (socialist and capitalist) neighbors during the Cold War period were not static. For the purpose of this chapter, we distinguish three main periods or stages in

Bulgarian electrification, each one characterized by different perceptions of criticality and vulnerability; different developments in electricity infrastructure; and different patterns in international relations. The first period covers the dramatic and turbulent period from the communist takeover in 1944, when access to electric power was limited, to the early 1960s, when Bulgaria was by and large electrified. Structural power shortages during the Soviet-style modernization process were perceived as the main criticality and driver. In the second period, Bulgarian power authorities sought to meet the ever-increasing power demand by deepening cooperation with the Soviet Union and other COMECON countries, thereby producing a new form of energy vulnerability – import dependency. The third period, which started around 1975 and ended with the demise of communist Bulgaria in 1989, is characterized by concerns about system stability and new vulnerabilities following the Bulgarian nuclear age, as well as the country's emergence as a Balkan power hub.

Communism and its criticalities (1944–65)

December 1948 – the victorious Fifth Congress of the Bulgarian Communist Party. The party chairman, Georgi Dimitrov, proclaims from the tribune: "Through industrialization, electrification, and the mechanization of agriculture, our country should achieve in 15–20 years what other countries have taken a century to achieve. To this end a strong energy basis has to be created by using the country's thermal and hydropower potential." The technological expertise is to come from the Soviet Union, and Dimitrov hence adds: "for the Bulgarian people the friendship with the Soviet Union is of critical necessity, in the same way as sun and water are necessary for every living being."[4]

One year later, in 1949, the so-called People's Court (Naroden sad) tries several top-level government officials, who had been appointed by the Communist Party after 1945 to electrify the country. Originally created as an instrument of terror against the former bourgeoisie, the People's Court now turns against enemies inside the Communist Party and its allies. Traycho Kostov – member of the Communist Party Politburo, minister of electrification and vice prime minister in the 1946 government – is sentenced to death and hanged. His successor in the 1946–7 government, engineer Manol Sakelarov, is also sentenced to many years in prison, where he dies in 1954. Deputy ministers of electrification in the same government, the engineers Lyubomir Kairyakov and Marin Kalburov, are also sentenced – the first dies a few years later in prison, while the second is cleared and set free in 1956. Their official crime is "deliberately impeding electrification of the country by building wasteful small water power stations and holding up the techno-economic collaboration with the Soviet Union."[5] *Concretely, Kostov's sin is his insisting that in the field of electric power, Bulgaria should not change the patterns inherited from its historical tradition but follow them. By this he means that the large-scale introduction of relatively small power stations based on local sources of hydropower and thermal power have to be favored over the centralization of power generation in a few large utilities proposed by Soviet experts.*[6]

This harsh start to the communist era in Bulgarian electricity history illustrates that electrification was a major concern in government programs to build postwar, socialist Bulgaria. It also shows that the personal stakes for individual system-builders could be exceptionally high; pushing electrification strategies that fell out of favor might not only lead to financial and employment sanctions but could become a matter of life and death for their supporters. Personal security became part of the equation. Moreover, it is noteworthy that the object of disagreement was a choice between two electrification strategies that are well known from Western histories: electrification could take the form of "planned systems" to radically break with past practice, or of "evolving systems" following the lines of "inherited," existing patterns.[7] To understand this conflict we start by briefly addressing the legacy of pre-communist electrification efforts.

The electrical communist revolution

Up to 1944, Bulgaria was one of the least developed countries in Europe in terms of electrification, with merely 13 per cent of all settlements being electrified. According to the index of consumed energy per capita, Bulgaria was last among all European countries.[8] Its modest domestic generating capacity of 130 MW consisted of 60 per cent thermoelectric units and 40 per cent hydroelectric generators.[9] The system was fragmented: the first national electrification plan was launched only in 1941. Only after the "the golden age of Bulgarian electrification," as the period from 1950 to 1970 is sometimes called, would Bulgaria be electrically integrated and fully electrified.

In terms of transnational relations, pre-communist Bulgaria was oriented towards Western models of electrotechnical development, not least the German model of public electrification. Western European technology and know-how (predominantly from Germany, Austria, Switzerland, Czechoslovakia, and Hungary, and to a lesser extent from Belgium, Britain, and some other European countries) played an important role. Western European engineers came to Bulgaria, and Bulgarian engineers were trained in Central and Western European schools (Vienna, Prague, Karlsruhe, Munich, Darmstadt, Paris, Grenoble, and Toulouse were some of their favorite destinations).[10] Accordingly, it was from this direction that the first ideas of large-scale electrification reached the country. Besides importing the notion of a national power grid, German and Austrian engineers helped to map the hydropower potential of the country and design several major dams.[11]

The first major changes after 1944 were of a discursive and institutional nature. After the occupation of Bulgaria by the Soviet army in September 1944 and the establishment of a "people's democratic government" dominated by the Communist Party, economic development was redirected under the famous slogan of Lenin, "Communism is Soviet power plus the electrification of the whole country," which was repeatedly cited in communist propaganda. The Bulgarian people were bombarded with this slogan from their earliest school years on. The catchword of "socialist industrialization" implied a pro-Soviet, anti-Western economic policy that forced Bulgarian engineers to cut off existing ties with Western Europe and the southern Balkan countries, and to reorient themselves toward their eastern allies.

The country followed the Soviet-inspired political priority of developing heavy industry and the related need for large amounts of electricity. Thus the main challenge for the Bulgarian power industry became to match the expected growth in industrial demand through a corresponding increase in power supply, and to enable access to electricity in all locations. The main concern was to ensure "availability" and "connectivity," and this also defined the criticality of the emerging system.[12]

As in other Eastern European countries, the Bulgarian energy sector was reorganized following the Soviet model. Electrification became a public task with the Act on Industry, Bank and Mine Nationalization and Electrification (1947) and the Electrical Industry Act (1948), and particularly the Act of the Elprom Syndicate (1948), the Water Syndicates Act (1948), the Act on the Construction of the Rositsa and Topolnitsa Dams (1948), and the Act on Heat and Electricity Supply Joint Ventures (1948). These acts nationalized private power stations and related transmission networks, and they suspended the previous patterns of public–private cooperation in the field of electric power. This was considered to be the only way toward a rapid electrification of the country and to building a unified power system. A new state enterprise, Energoobedinenie, would be responsible for the generation, transmission, and distribution of electric power. Another agency, Elprom, was to absorb electrotechnical manufacturers. An electrical power construction trust, Energostroy, was responsible for constructing, upgrading, and maintaining power plants, transmission, and distribution networks. Finally, the state bureau Energohydroproject would work on electric power design and R&D.[13] All energy companies and institutions were placed under the administration of the new Ministry of Electrification, Water, and Mineral Wealth (Ministerstvo na Elektrifikaciata, Vodite i Melioraciite). The idea to nationalize and centrally manage the energy sector was developed in the framework of the first Two-year Plan for Economic Development (1947) and was aimed at relieving the continued strain on developing heavy industry caused by energy shortages.[14]

The power industry was nationalized more smoothly than other sectors. There were three main reasons for this. First, German philosophies of electrification had already introduced and paved the way for the notion of public electrification and publicly owned facilities. By 1944, privately owned electric power stations provided only 17.5 per cent of all electricity. Second, the majority of Bulgarian engineers, graduates of Central and Western European engineering schools, favored public intervention in the energy sector and did not oppose nationalization. Third, during the early years of the communist period, the Bulgarian authorities preserved the pre-communist technical intelligentsia (with a few exceptions, which will be discussed below), suddenly in such great demand due to the new plans for the large-scale industrialization of the country. An additional motive for protecting this group of engineers was that, at least in the field of electric power, the majority of engineers favored the public ownership of power facilities and transmission networks. As a consequence, many leading Bulgarian engineers, who had previously worked in the country's private and public electric power enterprises, now became involved in the development of the nationalized electricity infrastructure.

Technological choices

There was, however, no consensus about the choice of a specific technological trajectory. As noted above, under the first ministers of electrification, Traycho Kostov and Manol Sakelarov, the "inherited" pattern of electrification was seen as most appropriate for the Bulgarian situation. Further electrification, according to Kostov and Sakelarov, should exploit the availability of relatively small, local hydropower resources. This would allow an electrification pattern in which Bulgaria could remain largely independent from other countries, since power sources as well as technological expertise were available domestically.

When Georgi Dimitrov, a former chairman of the Third Communist International who had just returned from the Soviet Union, was appointed prime minister in 1948, he found these early efforts of communist electrification ill-organized and fragmented. The recent construction of some 30 thermal and hydropower plants, transformer substations, and new transmission lines after 1944 was dismissed as highly unsatisfactory. Moreover, Dimitrov considered the small scale of these power plants to be "a mistake." In fact, this view was suggested by Stalin's fear of the strong nationalist stand of Traycho Kostov, who championed Bulgaria's relative independence from the Soviet Union. As Richard Staar points out, Stalin was shocked by the developments in Tito's Yugoslavia and became suspicious of all Bulgarian communists who had spent the war in their own country rather than in the Soviet Union. Thus Kostov was accused of being a Titoist confronting Stalin, and was executed after an arranged show trial.[15] The accusations of reliance on relatively small power stations based on national resources were justified only in the light of the Soviet-imposed rapid industrialization, which favored heavy industry. Because the country lacked the necessary resources for this, however, industrialization took longer, incurring a high price and remaining largely inefficient.

The trial against Kostov marks a turning point in the history of Bulgarian electrification. Henceforward, a "policy of accelerated electrification" was pursued, the Soviet model of building large and powerful electric power stations was adopted, and intensive collaboration with the Soviet Union took center stage. As Dimitrov made clear, the country needed large-scale technology. Since Bulgaria lacked the specialists and scientific expertise to build very large power plants, close cooperation with the Soviet Union was necessary. Bulgarian power engineers were to step back from their leading positions.

Electrification did indeed accelerate. The early 1950s saw the creation of a Bulgarian national grid. This happened in connection with several large (thermal and hydropower) projects initiated with the help of the Soviet Union. The effort was strengthened by the Second General Plan for Electrification, which was developed in collaboration with the Soviet Ministry of Electrification and launched in 1954. Furthermore, in 1955, Bulgaria signed an agreement with the Soviet Union

Table 5.1 Installed capacity in the Bulgarian electricity system, 1955–65 (MW).

Year	Thermal	Hydro	Nuclear	Total
1955	298	134	–	432
1960	380	475	–	925
1965	1357	770	–	2,127

Source: Spirov 1999.

for "mutual help" on nuclear technology (see further below). Other exchanges included the provision of specialists, consultation, and machinery from the "Big Soviet Brother." Until the department of electric power at the Technical University of Sofia (Sofiiska Politehnika) became a full electrotechnical faculty in the mid-1950s, Bulgarian students were sent to Soviet technical colleges to study power and heat engineering.

As a result of all of this, by the mid-1950s about 70 per cent of all Bulgarian settlements had been electrified. Electrification of the country was virtually complete by 1960. The total installed capacity had increased almost sevenfold to 900 MW (up from 130 MW in 1944; Table 5.1).

Continued electricity shortages and the quest for transnational interconnection

The accelerated development of heavy industry, coupled with the process of rapid urbanization, continued to demand ever-increasing amounts of power. From 1948 to 1960, Bulgarian industrial output increased by a factor of six, and the annual industrial growth rate of 16 per cent was one of the highest in the world. The share of industrial workers in the total workforce increased from 9.5 to 22.7 per cent. By 1960, large industrial enterprises provided some 34 per cent of GDP. This massive industrialization process came with a mass exodus of people from rural to urban areas, and the urban population grew from 25.7 per cent in 1946 to 64.8 per cent in 1985.[16] Although the electric power system grew at an impressive pace, the demand for electricity to power industrialization and urbanization grew even faster. Bulgarian officials gradually realized that the Second General Plan for Electrification of 1954 would fail to meet the new demands. Top-level politicians interpreted the country's vulnerability in the electricity sector as a permanent electric power shortage holding back the development of heavy industry. Pending further action, power authorities coped with the situation by rationing electricity – by sector, region, and time of day.

In terms of long-term planning, the problem, as described by the officials, was seen to lie in Bulgaria's failure to develop a strong electricity industry based on local resources.[17] As centralization and reliance on larger units, pursued from the late 1940s, had proved insufficient, new strategies were needed. Again, solutions to the problems were sought in the direction of stronger cooperation with, and thus dependency on, the Soviet Union and COMECON. These transnational relations now entered a new phase. Besides continued exchanges of knowledge, machinery, and resources to strengthen electricity production in Bulgaria, efforts were stepped up to cooperate on a transnational power grid.

The construction of transnational power lines had thus far been negligible. In 1950, Bulgaria acquired its first cross-border power line, a 60 kV cable across the Danube River connecting the Bulgarian town of Russe with the Romanian town of Giurgiu (Gurgevo) on the opposite bank. This link, however, was of only local importance and preceded the construction of an integrated national grid by several years.

More far-reaching transnationalization efforts were planned in the framework of COMECON. This was created after a meeting between Bulgaria, Hungary, Poland, Romania, the Soviet Union, and Czechoslovakia in January 1949. The main tasks of the meeting were

> to achieve greater economic cooperation between countries with a people's democracy and the Soviet Union...the meeting considered the necessity to establish a Council for Economic Assistance by representatives from the countries participating in the meeting, on the basis of equal representation, with the task of exchanging economic experience to provide technical assistance to each other, for mutual assistance with materials, supplies, machinery, installations, and such.[18]

During its first years, this cooperation took the formal shape of bilateral agreements with the Soviet Union. The outcome in terms of transnational power lines, however, was judged to be disappointing, so that for the time being it was uncertain whether the socialist countries would be able to link up their electricity systems.[19]

In 1956, however, the COMECON established a Commission for Electric Power Exchange and Utilization of the Hydropower Potential of the Danube, a body that was transformed into the Standing Commission on Electric Power (Postoiannaia komisia po elektroenergia) in 1959. Guided by this commission, cooperation on delivery of power among the member states began in earnest. An already existing power line between Hungary and Czechoslovakia became the basis for future extensions. In 1960, Poland and the German Democratic Republic (GDR) joined their grids to these two countries. The next step was the creation of a first link between the electricity systems of the Soviet Union and its Eastern European satellite states. This succeeded in 1962, when the connection Mukachevo (Ukraine)–Schaioseged (Hungary) was taken into operation.[20]

The same year an agreement was signed in Moscow between the COMECON member states (the Soviet Union, Poland, East Germany, Bulgaria, Romania, Czechoslovakia, and Hungary), aimed at the creation of an integrated power system (IPS) involving most of the socialist countries. This was part of a broader COMECON effort to unify the economic plans of all of its member states.[21] The IPS aimed for the creation of a transnational electric transmission network between COMECON countries, and to strengthen mutual assistance among the partners. Bulgaria was eventually connected to the COMECON grid in 1967 via a 220 kV line to neighboring Romania (Boichinovtzi–Craiova). In joining the IPS, Bulgarian engineers and communist leaders adopted a strategy to import large amounts of electric power. In this way they aimed to combat the domestic power crisis. To a lesser extent they also hoped to use the interconnected system as a way to prevent blackouts in case of sudden, unexpected capacity shortages.

Reversing the salient (1965–75)

> It seems today that Bulgarian power system is very well balanced – approximately 30 per cent nuclear, 55 per cent thermal and 15 per cent hydropower. But to tell you the truth, I have no impression that one was consciously looking for such proportions, since they naturally came from solving the acute problem where we could supply equipment for the power stations... The sources were second in order, although the Varna thermal power station was built after the Soviet minister of coal industry signed a contract with Bulgaria that guaranteed twenty years of regular coal supply from one of the Ukrainian mines. I remember how proud our Minister of the Power Industry Konstantin Popov was – he was waving the document everywhere, this was a big achievement...[22]

Thus Nikita Nabatov, a longtime investment manager in the Bulgarian power industry, remembers the constant struggle to solve domestic power shortages. This quote illustrates that in our second period (1965–75) the earlier crisis in terms of electricity shortages had largely been solved. The outcome seemed rational, balanced, and favorable: the new system was based on an impressive domestic power production capacity, with diverse energy sources and solid integration into transnational power grids. Yet Nabatov remembers the dynamics of this process as a continuous "acute problem solving," struggling to take into operation new capacities in the system. Moreover, the quote touches upon changes in the criticality of the Bulgarian power system; the enduring energy crisis was finally solved but, in its stead, Bulgarian power authorities now needed to manage new vulnerabilities stemming from increased dependence on the Soviet Union.

Coal and transnational relations

As for the expansion of domestic generating capacity, with the organizational and technological help of the Soviet Union, East Germany, Czechoslovakia, and some other COMECON members, several powerful thermal power plants were constructed during the late 1960s and early 1970s at Varna, Russe, Svishtov, Vidin, and Gabrovo, and some smaller ones at other locations. These plants were designed to be fuelled by rich anthracite coal from the Donbass industrial region in Soviet Ukraine, which was imported to Bulgaria by way of the Black Sea and the Danube.

The new plants were built in close cooperation with the Soviet Union. The new power station at Bulgaria's ancient Black Sea port city of Varna, for instance, was jointly designed by the Bulgarian R&D institute Energoproekt and the

Moscow-based institute Teploenergoproekt. Most of the equipment was manufactured at the Taganrog machine-building plant in Ukraine. By the end of the 1960s the station possessed three 200 MW units, which made it one of the leading thermal power stations in the Balkan region. Later it was expanded to 1200 MW. Equipped with its own specialized coal port, the station was considered to be an important means of compensating for the lack of high-quality coal sources in Bulgaria. Apart from its convenient position on the Black Sea, its location also facilitated the improvement of electricity supply in northeastern Bulgaria, which up to then had remained underdeveloped in terms of electric power production. At the same time, however, the operation of the plant was linked to a vast dependency on the Soviet coal supply.

Equally important developments took place in southeastern Bulgaria. Back in 1953–4, Soviet specialists N. Mikhailov and S. Tzigankov from the Moscow All-Soviet Institute of Heat Engineering had already devoted intensive efforts to studying the Maritsa East lignite coal basin in this region. However, the properties of the coal were not the most favorable: it contained a lot of water and sulfur, and its heat value measured only 4000–7100 kJ/kg. As a result, the proposed power station was expected to work at the margin of profitability. Moreover, it would have to be supplemented by drying plants for the coal, which would consume part of the heat produced.[23]

Nevertheless, the study suggested that a large thermal electric power station should be constructed near the Bulgarian lignite fields. In September 1960 the First Komsomolska thermal power plant (now Maritsa East 1) was brought online. With several 50 MW units, it was a large plant even at this time. Later the plant was enlarged by adding two 150 MW power units, which transformed it into the biggest power plant in the country. In 1965 it produced 30 per cent of all electric power in Bulgaria. In 1969 a second lignite power plant, Maritsa East 2, was inaugurated, consisting of four units with a total capacity of 660 MW.

The remarkable increase in thermal-power facilities based on low-quality lignite with high levels of sulfur compounds soon worsened the environmental situation in the southeast of the country. For more than two decades, acid rain and heavy dust during windy periods became part of the lives of the local population. The situation did not improve until the 1990s, when most of the power-station chimneys were equipped with filters to capture the dust and the most dangerous compounds from the exhaust gases.

The First Komsomolska power station played a prominent role in the development of thermal power in Bulgaria. As Professor Nikola Todoriev put it later, "it was not only a pioneering plant in lignite coal utilization in Bulgaria, but it became a real research center for the Bulgarian power industry and its main center for training specialists in thermal power production. Most of our best experts in the field have been trained there." In the early 1960s, experimenting with technologies adopted in the First Komsomolska, Professor Yakimov, in collaboration with his disciples Ivan Tchorbadjiiski and the abovementioned Nikola Todoriev, set out to develop an improved version of the original burning technology that eliminated the necessity of drying the coal and significantly increased the heat yield.

According to Hristo Todoriev, son of Nikola Todoriev and also a professor of electric power engineering at the Technical University of Sofia, the initial idea that sparked these innovation activities came from the lignite coal power plant at Kozani in northern Greece, whose construction had begun in the 1950s. This power plant, located not far from the Bulgarian border, was built by the German electricity giant RWE (see Chapter 6). RWE used more modern and efficient fan mills for the coal than the Bulgarians did. To test the applicability of the technology to the less favorable properties of the Maritsa East lignite coal, in the mid-1960s, several trains loaded with Bulgarian lignite coal were sent through the Iron Curtain to Kozani for an experiment and burned by a team of Bulgarian, (West) German and Greek engineers.[24] Returning to Sofia with the data, Todoriev and I. Tchemedjiev developed a technology that replaced the existing hammer mills for coal with highly efficient fan mills, opening the door to the construction of larger power-generation units. As it turned out, these innovations radically decreased costs and rendered Bulgarian lignite power plants among the most efficient at the time. Apart from making the coal-drying plant unnecessary, the energy needs of the thermal plant proper decreased from 14 per cent to 10 per cent.

For our story, it is important to stress that this innovation project counterbalanced some of Bulgaria's dependence on the Soviet Union. First, it utilized a domestic energy source. Second, the new innovations made it possible to further expand the use of this fuel in the electricity system. Subsequent power stations such as the Bobov Doll thermal power plant (1971–1975) in southwestern Bulgaria were designed to operate with domestic lignite as well as imported hard coal.[25] In that way the domestic lignite and coal could substitute for imported coal in case Soviet coal imports were disrupted.

According to interviewed witnesses, the regained autonomy in thermal power production raised concerns among the Soviet leadership. The Bulgarian Black Sea and Danube power plants, which still used imported Soviet coal, were seen to "calm" these concerns. Balancing these two types of vulnerability – import dependency and the risk of provoking Soviet concern – became an explicit political strategy: when Todoriev was appointed the new minister of the power industry (Ministerstvo na elektrifikaciata) in 1976 he expressed his fears that further development of Maritsa East power complex might give rise to problems in Bulgaria's relations with the Soviet Union. However, Bulgarian party leader Todor Zhivkov replied: "Your duty is to defend Bulgarian interests; there are other people in the Politburo who take care of the Bulgarian-Soviet friendship."[26]

An important aspect of the cooperation with Greece in connection with the development of new lignite technologies was that it inspired further knowledge and specialist exchanges across the Iron Curtain and the accumulation of domestic expertise. At least in the field of electric power, the Iron Curtain turned out to be permeable. In the late 1960s the Greek-Bulgarian exchanges were followed up by a visit to Bulgaria by a delegation of US power engineers, who wished to learn more about the experiences gained by the Bulgarians in burning such low-quality coal. A few years later a Bulgarian delegation returned the visit.[27] Out of these visits in the early 1970s a Research Laboratory for Power Technology was established at the Technical University of Sofia, financed under a special contract with the UN Development Program. For almost two decades this laboratory served as the main channel for scientific exchange with the West. Bulgarian engineers thereby also developed extensive relations with relevant research centers in West Germany, France, Spain, and other capitalist countries.[28]

In terms of technological style, participants in the lignite technology project remember that deviations from the original Soviet design were not easy to push through, given the undisputed prestige of Soviet experts, practical problems, and political sensitivities. Yakimov, who had lived in the Soviet Union in the 1930s and 1940s, including a period imprisoned in a gulag, had to use his political capital from the pre-Second World War period and his prestige as the first Bulgarian professor of heat technology to convince the Bulgarian political leadership of the advantages of improving on the Soviet designs used in the first version of lignite-fueled power production at Maritsa East. There were practical complications, too – the proposed innovations necessitated further changes in the construction of the combustion chambers, which were produced in the Soviet Union. Still, the political sensitivity of this seemingly technical problem was revealed when in the mid-1960s Nikola Todoriev, Yakimov's successor, was told to report on the proposed design changes at a meeting with the Politburo of the Soviet Communist Party. Todoriev realized that he would risk serious prosecution in the case of failure. Eventually the technology worked well, which paved the way for him to be appointed minister of the power industry in 1975.[29]

Bulgarian nuclear relations

In parallel with the development of lignite power plants, Bulgaria took its first steps into the nuclear era. Nuclear power was seen as another response to the country's vulnerability in terms of energy shortages and its dependence on imported coal, and it was also a matter of national pride.

Nuclear developments in Bulgaria started through bilateral agreement with the Soviet Union, signed in 1956. It foresaw the construction of an experimental nuclear reactor, IRT–1000, near Sofia. The reactor was built by Soviet specialists and went critical for the first time in 1961.[30] This was followed in 1966 by another bilateral Soviet-Bulgarian agreement on the construction of a full-scale nuclear power plant in the country.[31] For the next three years, however, no concrete steps were taken to construct such a plant. This was partly because the project was considered to be controversial. The minister of the power industry, Konstantin Popov, was hesitant because of the lack of qualified personnel, experience, and sufficient funding. Moreover, he believed that the technology was still not mature enough.[32]

A turning point came in late 1969. Without informing Popov, a number of high-level representatives at the Ministry of the Power Industry, supported by a Politburo member who was also deputy chairman of the Council of Ministers (Ministerski savet) and minister of building works (Ministerstvo na stroitelnite raboti), Pencho Kubadinski, ordered the beginning of construction for the first reactor. Ground was broken on October 14, 1969. At the plenum of the

Communist Party held a month later, the defenders of the nuclear idea thus presented the other officials with a *fait accompli*.[33] The debate about whether or not Bulgaria should have a nuclear plant came to an abrupt end.

The plant was similar to the ones built in the GDR (Greifswald) and in the Soviet Union (Novovoronezh), which had been constructed a few years before the Bulgarian one.[34] In Bulgaria the power plant was built at Kozloduy on the Danube, just at the Bulgarian-Romanian border. According to the agreement between the Soviet Union and Bulgaria, the Soviet Union provided a loan to pay for the plant at a highly favorable interest rate of only 2 per cent.[35] In response to the urgent need for nuclear specialists, between 1969 and 1975, more than 150 students were sent to the Soviet Union to study nuclear engineering, sponsored by the Bulgarian government. Some of them attended courses at the Moscow Energy Institute (Moskovskii Energeticheski Institut) for four years as regular students, while others were sent to Novovoronezh nuclear power plant for one year or six months of training. Another group of 40 specialists went to the GDR to attend seminars at the nuclear plant there. After 1971 a directive from the Council of Ministers obliged all nuclear specialists to attend educational courses at the Bulgarian experimental reactor.[36] In 1973 the first group of students that graduated in Moscow returned to Bulgaria, where the preparations for running the first reactor were under way.

The Bulgarian nuclear power plant used mainly Soviet technology, and for this reason the project was carried out by the Russian state firm Teploelektroproekt.[37] The Bulgarian R&D and design institute Energoproekt played an auxiliary role, designing some subsections of the blocks. However, Bulgarian specialists understood that Soviet nuclear technology lagged behind that of the West and that there were security risks involved. In that sense it was also a risky decision for Bulgaria to establish the nuclear power plant. With its signature of the Non-Proliferation Treaty in 1972, the country's opportunities for international collaboration increased. However, the International Atomic Energy Agency (IAEA) did not significantly influence the construction of the nuclear power plant.[38] For example, most of the IAEA experts who inspected the Kozloduy plant were in reality Soviet specialists. Western authorities rarely visited Eastern European nuclear sites.

The plant's first block was taken into operation in 1974 and the second in 1975. Before long the first accidents occurred. The danger of a sudden shutdown of some of the big 440 MW nuclear power units was new to the system and added a new notion of vulnerability to the Bulgarian energy grid.[39] This was coupled with the new threat of radioactive pollution in case of accidents in the reactors' active zone. Hence with the launching of the Kozloduy nuclear power plant the inherited notion of criticality was radically redefined; from then on, the problem of nuclear safety gradually emerged as a key issue.

Expansion of transnational links

The third element in overcoming the domestic energy crisis, in addition to the expansion of lignite and nuclear power, was the strengthening of transnational electricity links so that power could be imported. In 1965 a 220 kV transmission line from the Soviet Union through Romania to Bulgaria was inaugurated, the purpose of which was to import Soviet electricity.[40] In 1972 a more powerful, 400 kV transmission line to the Soviet Union was built, crossing the border in the northeastern part of the country.[41] This connection made it possible for the country to import significant amounts of energy from the Soviet Union. In 1974 this amounted to 15 per cent of total domestic consumption. In the years that followed, the import of electricity directly from the the Soviet Union decreased and stabilized at about 4.5 TW/h annually.

In conclusion to this section, the criticality and vulnerability aspects of the second period could be framed along two lines. According to the first line, during the period 1965–75 the notion of connectivity/availability was still dominant. This was true despite the fact that during this period the Bulgarian electricity system grew to become one of the largest in the Balkan region. The growth in consumption was absorbed through the establishment of some specific industries (such as petrochemical, steel production, and electrified railways) and the growing use of electric power for heating the large, newly built apartment houses in the cities (Tables 5.2 and 5.3).

According to the second line, the period marked the strengthening of interconnections and interdependencies with the Soviet Union and COMECON. In this context an important new trend was that Bulgaria's dependence on Soviet specialists, technologies, and raw materials was balanced by the development of indigenous technologies and the increased utilization of local resources. Yet the successful electrification and accelerated development of the Bulgarian electricity system undoubtedly owed much to the close cooperation with the Soviet Union, and most of the power plants were built with the support of Soviet specialists and R&D institutes, while relying on Soviet technology and equipment.[42]

Table 5.2 Specific consumption of electric power in Bulgaria, Greece, and Yugoslavia, 1939 and 1970 (kWh/person).

Country	1939	1970
Bulgaria	42	2298
Greece	55	1117
Yugoslavia	75	1278

Sources: Bulgarian Power Industry. Annals and Energetika.

Table 5.3 Installed capacity in the Bulgarian electricity system, 1970–5 (MW).

Year	Thermal	Hydro	Nuclear	Total
1970	3264	814	–	4075
1975	4312	1720	880	6912

Source: Spirov 1999.

Although the system managed to provide electricity for the national economy, it thus remained critically dependent on Soviet technology and resources.

Grid optimization and the emergence of a Balkan power hub (1975–89)

> On the evening of March 4, 1977, a powerful earthquake struck the region of Vrancha, southern Romania. The epicenter of the earthquake was only 300 kilometers from the nuclear power plant Kozloduy. The chief engineer on duty was Borislav Dimitrov, who immediately reacted and shut down the plant. Most of the citizens from the nearby town arrived to provide any necessary first aid.[43]

In the third period analyzed in this chapter, starting in the mid-1970s, the problem of the stability of electric infrastructures became crucial. We call this last stage the "optimization" of infrastructure. This period saw a continuation of the trend, discussed above, of Bulgaria's increasing emancipation and autonomization of the system vis-à-vis the Soviet Union, and the country's transformation into a key regional actor in the Balkans. Through the Kozloduy nuclear power plant, Bulgaria became the 11th country in the world to use nuclear power commercially, and the seventh to launch operation of a 750 kV transmission line. If we take into consideration the residential area and the population of the country, it is not surprising that Bulgaria encountered rich opportunities to become a power hub in the Balkans. This was made possible through three pillars of the system – the Kozloduy nuclear power plant, the Maritsa East thermal power complex, and the Chaira pumped storage hydropower plant, together with the strong electric power transmission lines connecting it with the COMECON countries, and at the same time, the newly built links with neighboring countries outside COMECON (Table 5.4).

The process of "optimization" can be said to have started in 1977, triggered by the earthquake near Kozloduy. The technology at the nuclear power plant survived the earthquake without significant damage, but the event caused the government

Table 5.4 Installed electric power and electricity consumption in Bulgaria and other countries, 1989.

Country	Population (millions)	Installed power (GW) Total	TPP	HPP	NPP	Consumption (TWh)	kW/ person	kWh/ person
Belgium	9.9	4.1	7.2	0.1	5.5	66.5	1.42	6680
Greece	10	9.0	6.7	2.3	0	35.5	0.9	3550
Denmark	5.2	9.1	6.7	2.3	0	32.8	1.9	6370
Netherlands	15	17.6	16.9	0.25	0.5	81.1	1.17	5410
Bulgaria	8.6	11.7	6.9	2.0	2.8	45.9	1.36	5340
EC	327	435	252	81	102	1806	1.33	5521

Source: Spirov et al. 1998, p. 344.

to set up a financial plan to improve the plant's safety system. This improvement process, as it turned out, opened up a unique opportunity for Bulgarian energy organizations to gradually revive their institutional links to Western countries – for example, by purchasing know-how and components from Germany, Japan, and the United States. Also, Western standards for better seismic indexes were imported, although this delayed the completion of the third and fourth reactor blocks at Kozloduy to 1980 and 1982, respectively. A criterion when working out the original plans had been that the reactor blocks had been able to withstand an earthquake with a maximum strength of 5.4 on the Richter scale. The new hydro shock absorbers that were imported from Japan, however, complied with the worldwide standard of 7.5 Richter. The 1977 earthquake had measured 7.2. It took a long time to deliver and install these hydro shock absorbers, and the engineers needed time to deploy the imported devices.

The strengthened contacts with Western industry were linked to improved technological capabilities in Bulgaria itself. Imports of advanced Western equipment were combined with a decreasing number of Soviet nuclear specialists working in Bulgaria, and with a growing role for Bulgarian industry. The Bulgarian electrotechnical industry now produced most of the equipment, such as high-voltage switchgear, circuit breakers, on-load tap changers for large transformers, high-voltage asynchronous electric motors, and synchronous electric motors for nuclear power production.[44] Indeed, the industry started to export electrical equipment to foreign markets and the production of appliances under license of Western companies was launched.[45] Moreover, Bulgarian engineers now contributed to electrification projects abroad – in Lebanon, Pakistan, Syria, Vietnam, Kuwait, and later Nigeria, the Soviet Union, Nicaragua, Malta, Jordan, Iraq, Iran, Egypt, and Bangladesh.[46]

Also, as we have seen, the large Maritsa East lignite-thermal complex was an important step in strengthening the country's competencies and autonomy in the field of electric power. It was explicitly considered as providing the country with greater autonomy, while it also enabled export of technological know-how. The Maritsa East thermal power plant contributed to the development of thermal technology in Romania, Czechoslovakia, Spain, and other countries.[47] In the 1960s, as we have seen, the three thermal power plants in the Maritsa East complex produced a third of all electricity in Bulgaria. By the 1980s, however, the balance between local and imported raw coal had been inverted, so that just 30 per cent of the coal used for electricity production was imported.

On the other hand, this build-up of domestic technological capabilities did not bring collaboration with Soviet partners to a full stop. In the early 1980s another two state-of-the-art nuclear reactors were commissioned from the Soviet Union. With an output of 1000 MW each, they were much more powerful than the earlier four 440 MW reactors, and they were the first 1000 MW blocks to be built outside the Soviet Union. These immense reactors went critical for the first time in 1987 and 1991, respectively. However, the planning and operation of these new blocks took place largely without Soviet help, a fact that gave rise to great pride in both Communist Party and engineering circles.[48]

The new 1000 MW nuclear blocks significantly increased the Bulgarian power supply. In stark contrast with the earlier periods of constant structural shortages of electricity, Bulgaria now had more generating capacity at its disposal than it could use domestically. After the introduction of the two 1000 MW reactors at the Kozloduy nuclear power plant, the total capacity of this plant grew to 3760 MW. In addition, a second, brand-new nuclear power plant with two 1000 MW reactors (at Belene) was being planned.[49] However, the installation of the new very large power blocks gave rise to new demands for stabilization of the power system. Solutions to this problem were devised that rested on the development of new national and transnational capacities. Nationally, the efforts were concentrated on a new pumped storage hydropower plant, Chaira. The transnational capacities, for their part, took the form of new cross-border transmission lines. These are discussed below.

Hydropower cascades, balancing the system

As part of the strategy for emancipating the Bulgarian electricity system, Bulgarian system-builders embarked on the construction of powerful hydropower cascades with pumped storage technology. The first of these were the facilities of Dospat-Vacha and Belmeken-Sestrimo,[50] built in the Rila-Rhodopes mountains and collecting waters by way of a long headrace and tunnels. The plants were developed in consideration of the growing nuclear capacities and the specificities of lignite coal burning technology at Maritsa East, where the low heat yield meant that no significant variations in power production could be tolerated. Similar to nuclear power plants, the thermal power plants were unable to respond rapidly to daily variations in power consumption. The new hydropower cascades, where electricity production was more flexible (see Chapter 8, this volume), were designed to optimize the operation of the thermal and nuclear base load plants, and to improve key technical parameters of the Bulgarian grid system. Significantly increasing the efficiency of the entire system, they also marked the next attempt to emancipate from the Soviet Union in the field of electric power generation.

From the early 1980s the first two hydropower facilities in the Rila-Rhodopes mountains were no longer seen as sufficient to stabilize the system. Against this background, Bulgarian engineers proposed and designed a new, much larger pumped storage hydropower plant in order to increase the capacity of the existing Belmeken-Sestrimo cascade. It was to be used as a source of quick replacement capacity for any lost generating capacity, thus providing for flexibility of operation and reliable handling of peak loads in the overall electricity system of Bulgaria and the region. The Chaira plant was designed by specialists from the Energoproekt Institute. In contrast with the case of the Maritsa East thermal power complex, however, Bulgarian system-builders took the decision not to develop the technology on its own but to rely on already existing foreign technology, albeit with the participation of Bulgarian enterprises. In the 1980s a contract with Toshiba was signed for the first (of four) 200 MW pump turbines and motor generators. The main equipment was to be delivered and installed by Japanese engineers, while the remaining auxiliary equipment and devices were to be produced by the domestic electrotechnical industry and, accordingly, handled by Bulgarian engineers. The last three pump turbines were completed only after the fall of the communist regime.[51]

Transnational connections: From COMECON to non-COMECON partners

The Kozloduy nuclear power plant, the Maritsa East thermal power stations, and the Chaira pumped storage hydropower plant were all important steps through which Bulgaria increased its relative independence from Soviet energy resources and technology. Equally important for stabilizing the system and countering vulnerability, however, was the strengthening of the national high-voltage transmission grid as well as the development of new transnational connections.

As mentioned in the previous section, Bulgaria had been connected with the Soviet Union through a 400 kV line since 1972, allowing Bulgaria to import large amounts of Soviet electricity. In the 1980s this link was expanded to 750 kV as part of a wider effort to strengthen the COMECON's IPS.[52] The connections with Romania were also strengthened through the construction of a new 400 kV line across the Danube in 1986, linking the Bulgarian nuclear power plant at Kozloduy with the Tzantzareni substation in Romania.

The 750 kV transmission line went from a nuclear power plant in southern Ukraine through Romania to a substation at Varna in Bulgaria. Construction began in 1982 and the line was inaugurated on May 4, 1988. Romania expected 3–4 billion kWh of imported power from the Soviet Union and for that reason agreed to build the facilities on its territory. Bulgaria had to pay Romania for the use of its territory to transport electricity. The agreement between the two countries also included flows of electricity through other connections, which reduced the losses.[53] This scattered the Soviet flow somewhat, but still there were no regular flows from the satellites to the Soviet Union. These existed only in the case of blackouts. Much more important than preventing blackouts in the Soviet Union, however, was the fact that the new, immensely powerful link was designed to stabilize the Bulgarian power system, which was soon expected to have several new 1000 MW nuclear reactors in operation. The Soviet-Romanian-Bulgarian connection was to serve as a regulator for Bulgaria's power system in case of the emergency shutdown of a 1000 MW nuclear reactor. The new line also provided greater opportunities to import electricity from the Soviet Union. The project gave rise to pride among Bulgarian engineers, since there were only six other countries in the world that operated such powerful transmission lines. The section on Bulgaria's territory was built by local engineers.[54]

Through cooperation within the Standing Commission on Energy within the COMECON and the operation of the IPS, electricity production in the European COMECON member states increased by 6.7 times in the period 1955–80, and consumption per capita exceeded the average world level by 2.0 times.[55] Towards the end of the 1980s, Bulgarian electricity consumption amounted to 5300 kWh per capita, which was roughly at the same level as England, Denmark, or France, and by far outstripped that of its Balkan neighbors.[56]

During its 25 years of existence, the IPS of the COMECON managed to put in operation 4 transnational 750 kV lines, 20 lines at 400 kV, and 10 lines of 220 kV, connecting about 170,000 MW of installed capacity. Eventually it created a reliable interconnected system comprising a territory of 1,628,000 km² with a population of more than 150 million (Figure 5.1).

The system played an important role in strengthening local economies and improving stability. But it was also a source of instability since it did not work with precision. There were frequent complaints about the lack of more sophisticated technical support and better coordination between the members. The automation devices produced by COMECON members were not able to satisfy the demands

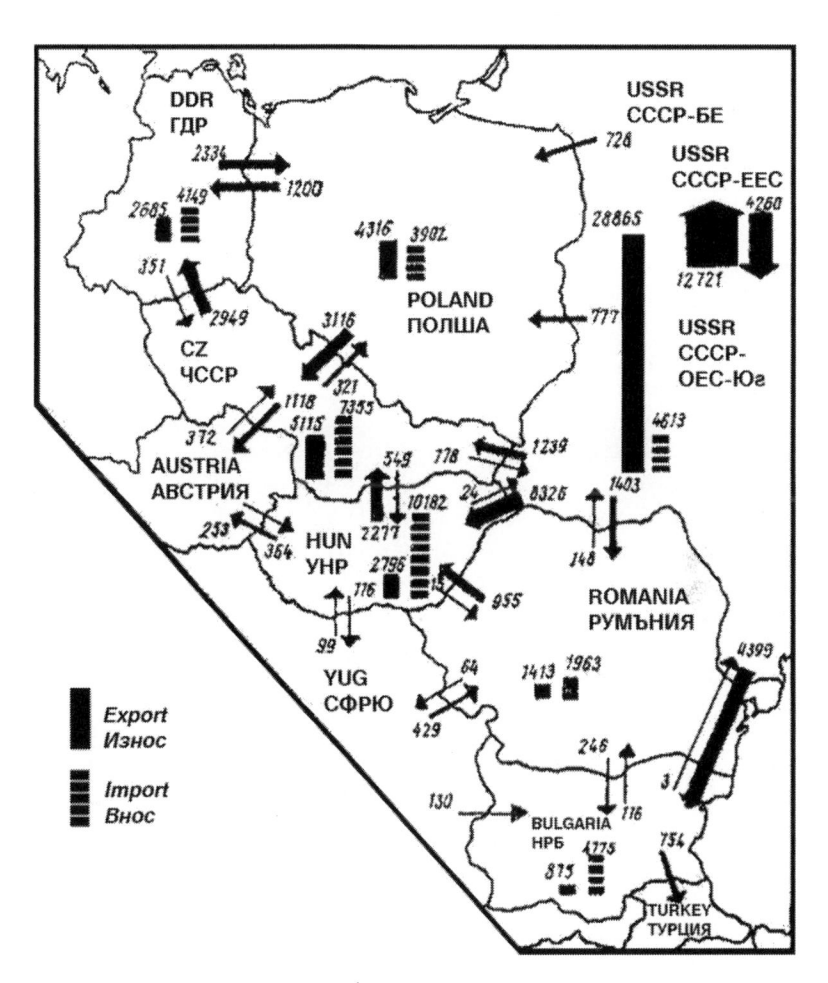

Figure 5.1 United electricity systems of the European COMECON countries.
Source: Griniuk 1981, p. 53.

of the grid. They often collapsed, resulting in irregularities in power transmission. The problems increased with the establishment of 750 kV lines. Studies reveal that the situation improved only from the mid-1980s, when the first computer technology was introduced into the monitoring and management of the system. From this perspective, Bulgaria, through its links with other COMECON countries, was involved in a vulnerable cooperation.

In parallel with its involvement in COMECON cooperation, Bulgaria proceeded to develop links to non-COMECON countries. Since the Bulgarian electricity system, as noted above, was more powerful than that of any of its Balkan neighbors, the latter showed an interest in linking up with Bulgaria as a means to improve the stability of their domestic systems, and possibly to import electricity from Bulgaria as well. Bulgarian actors, for their part, saw a possibility not only to further increase the stability and efficiency of the country's system but also to become an electricity hub in the Balkans. This idea was supported by a study financed by the UN Economic Commission for Europe, carried out in the late 1970s. The study argued that the country could become an electric power center in the Balkans, and it was met favorably by the top political leadership.[57] The idea was also indirectly supported through discussions within the COMECON, whose Standing Commission presented a report in 1980 dealing with the prospects for connecting the COMECON grid with the UCPTE grid in Western Europe. Bulgaria, sharing borders with three non-socialist countries, supported the idea.[58]

The actual development of connections to non-COMECON countries began back in 1975 through the construction of a 400 kV link to Turkey. The purpose of this line was to export Bulgarian lignite electricity. From 1975 to 1985, one of the 210 MW units at Maritsa East 3 was permanently connected to the Turkish power system. Since Turkey had different electricity standards, the Turkish region bordering on Bulgaria was, as a matter of fact, isolated from the rest of the Turkish territory, with its electricity system synchronized with the Bulgarian system. In 1985, however, an international crisis erupted when Bulgaria unexpectedly cut exports in order to cover domestic needs. According to H. Hristov, deputy director of the plant at that time, this was the first blow in the relations between the two countries in the field of electric power. H. Todoriev claims that instead of cutting the supply, Bulgaria could have shut down several 40 MW electric steel furnaces. However, the party leadership preferred to break the international agreement rather than to jeopardize planning targets with regard to steel production. During the following years, Turkey, lacking other options, continued to import Bulgarian electricity, but Bulgarian-Turkish electrical relations continued to suffer from irregularities. Eventually, Turkey built several natural gas power stations on its European territory, thus reducing its dependence on Bulgarian electricity deliveries.

The same year as the Bulgarian-Turkish connection was built, in 1975, a 220 kV connection was established with Yugoslavia (from Stolnik near Sofia to Nish in Serbia). This link was later upgraded to 400 kV. At the same time the first connection with Greece was designed, in the form of a 400 kV link. It went operational in the early 1980s, connecting the Bulgarian town of Blagoevgrad with

Thessaloniki. In the 1980s an additional, smaller connection with Yugoslavia was built. It took the form of a 110 kV link between the Bulgarian town of Petrich and the Yugoslavian town of Strumiza, now in independent Macedonia.

Hence, in the mid-1980s, Bulgaria was connected by powerful transmission lines to all of its neighbors, among them the Cold War enemies Turkey and Greece and the non-aligned country of Yugoslavia. Being seen as expressions of easing up European Cold War relations, these interconnections also brought significant economic gains for Bulgaria. The country was able to export part of its surplus electricity for highly valuable hard currency.

Conclusion: Becoming the power hub of the Balkans

The contemporary electric power sector of Bulgaria was established to a large extent as a result of cooperation with COMECON countries and especially with the Soviet Union. This meant giving preference to certain partners at the expense of others, while it also created a strong dependence on Soviet energy resources and technologies.

Yet this is a rather one-sided view, since although Bulgarian electrification took place in the framework of the larger process of communist industrialization, the country managed to develop a distinctive power infrastructure that eventually allowed it not only to emancipate itself from almost total dependence on the Soviet Union but also to become a key power actor in the Balkans (Table 5.5).

As we have seen in this chapter, from the late 1960s on, Bulgarian system-builders followed their own line in power development, which allowed Bulgaria to change the initial pattern of being a net importer of knowledge, technologies and specialists to becoming an exporter in these fields. There emerged a creative and original technical milieu in power-industry development, and far-reaching experience was gained through rapid development of thermal and hydropower, giving rise to a considerable R&D, design, and production potential.

In addition, something very important happened after the collapse of communism. Rapid integration with the winners of the Cold War began in the field of electric power, in particular in nuclear power. The improvement of safety conditions in post-communist nuclear plants was top priority: within only a few years, Western standards in operating such plants were introduced unconditionally, monitored by the IAEA and backed by international funding for improved equipment and (re)training. These measures improved the safety, stability, and efficiency of the Kozloduy nuclear power plant. Between 2004 and 2007, however,

Table 5.5 Installed capacity in the Bulgarian electricity system, 1985–95 (MW).

Year	Thermal	Hydro	Nuclear	Total
1985	6508	1975	1760	10,913
1995	6550	2440	3760	12,750

Source: Spirov 1999.

the first four reactors at this plant were shut down, as imposed during negotiations for European Union membership.[59]

At the same time, in the early 1990s, a large surplus of electricity emerged due to the economic crisis that followed the transition from an administrative to a market economy and the corresponding rapid decline in industrial production. This was especially pronounced in machine-building, electronics, and the electrotechnical and military industries, among others. As it turned out, neighboring countries were eager to swallow this surplus, and Bulgaria thus emerged as a major exporter of electricity (Table 5.6; Figure 5.2).

Looking just at the data, one could conclude that during the post-communist period the electricity sector became an oasis of development amid an extremely turbulent period in Bulgarian economy and society. This is partly true, especially up to the beginning of this century, when the organizational structure of the power system was preserved more or less intact and its operation remained in the hands of technocrats. However, it, too, was not spared by the shadow trends in post-socialist economies, including incompetent political intervention and corruption, asset-stripping, and especially the gradual erosion of human capital (dismantling of the R&D units, deterioration of technical education, and emigration of some of the best engineers and professors).

Table 5.6 Electricity consumption in Bulgaria, 1960–94, by sector (percentage of total).

Year	Industry	Households	Others	Total (GWh)
1960	53%	11%	36%	4685
1970	58.8%	13.2%	28%	19,616
1994	33.5%	25.6%	40.9%	28,860

Note: "Industry" includes metallurgy, chemical industry, light industry, and food-processing industry. "Other" includes agriculture, transportation, communication, auxiliary consumption, and transmission losses.
Source: Spirov 1999.

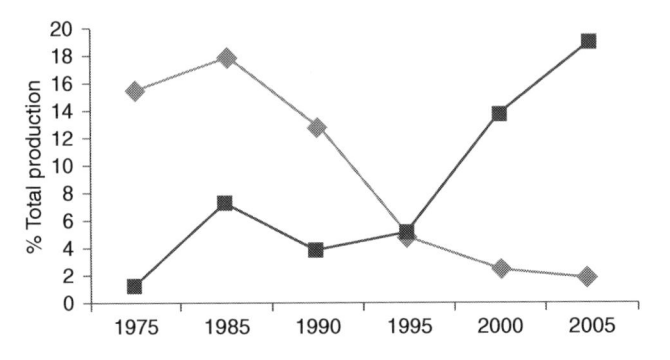

Figure 5.2 Bulgarian imports and exports of electricity, 1975–2005 (percentage of total production).
Source: UNIPEDE and EUROSTAT.

Let us conclude by reassessing the question of the extent to which one could speak of "Bulgaria as a power hub of the Balkans" and what kind of policies made this possible. There were three different yet interacting processes during the final decades of communist Bulgaria. The first, more specifically in thermal and hydropower, preserved and developed local traditions and continuously aimed to preserve some degree of autonomy, in terms of both resources and technology. The main evidence includes the establishment of strong design and R&D capabilities, which resulted in some original technology and decisions, such as the Maritsa East thermal complex and the Chaira pumped storage hydropower plant.

The second tendency – most pronounced in the field of nuclear power and to a certain extent in thermal power, especially at their earlier stages – was based on close cooperation and almost complete dependence on Soviet technology, resources and know-how. Although it also expanded the scope of its international relationships with Western partners from the late 1970s on, Bulgarian nuclear power remained closely bound to the Soviet Union during the entire period of the study (and even beyond).

The third tendency was to expand transnational links through transfers of technology and know-how, even across the Iron Curtain, and the diversification of transnational transmission lines in non-COMECON directions. During the final period discussed in this chapter, Bulgaria extended its transnational electricity links to the Soviet Union and Romania, linked up with Turkey and Greece, and strengthened existing links with Yugoslavia. This, eventually, allowed Bulgaria to become the most important electricity exporter in the Balkan region in the post-Cold War period.

The process of becoming a power hub was contested and negotiated in a number of instances and resulted from the Bulgarian national strategy for developing the power industry by tuning in to the long-term interests of the Soviet Union and COMECON, and because of the innovative solutions devised by Bulgarian engineers to compensate for insufficient local resources. By taking all of this into consideration we may conclude that Bulgaria's role as a power hub is a complex and multidimensional concept – it should indeed be considered as a hub distributing electricity flows, a hub of power capacities, a hub of technological competence, and a regional junction.

Notes

1. Compiled from interviews with experts in the power industry.
2. See Van der Vleuten and Kaijser 2005, pp. 21–48. See also Chapter 1, this volume.
3. Chapter 3, this volume.
4. Introduction by the editor. *Spianie Energia i Voda*, January–February 1951.
5. Spirov et al. 1998, p. 227.
6. Spirov et al. 1998; Georgiev et al. 2001.
7. Hughes 1983.
8. Georgiev et al. 1998.
9. Georgiev et al. 1998; Georgiev et al. 2001.
10. Spirov 1999.

11. Interestingly, a dam in the central northern region designed for over a billion cubic meters ranked among the largest in the Balkans; it was ultimately built by Bulgarian and Soviet specialists in the 1950s, thus constituting a technological continuity during a regime change.
12. This specific criticality pattern of availability and connectivity could also be framed as "user vulnerability". See Chapter 1, this volume.
13. Spirov 1999, p. 46.
14. Spirov et al. 1998, p. 45.
15. Staar, 1982.
16. Hristov 2000; Mladenov et al. 2009.
17. Georgiev et al. 1998; Georgiev et al. 2001.
18. In the same year, Albania also joined the council, leaving it in 1962. During the following decades, a few more states became part of the organization: East Germany in 1950, Mongolia in 1962, Cuba in 1972, and Vietnam in 1978. In 1964, Yugoslavia made an agreement with the organizational body to participate in its framework for shared value issues. See Bulgarian Academy of Sciences 1981, pp. 347–348.
19. See Marer 1976.
20. Savenko et al. 1983.
21. Neporojni 1978.
22. Author's interview with N. Nabatov, sustainable investment manager in the Bulgarian power industry.
23. Todoriev 1982, pp. 5–7.
24. According to Professor I. Tchorbadjiiski, Maritsa East lignite coal was also burned experimentally in Poland. Spending late 1960 in West Germany on an Alexander von Humboldt fellowship, Tchorbadjiiski placed several sacks of Bulgarian lignite coal in his car trunk, to be studied in the experimental burning chamber in the research lab there. "I prepared my PhD based on the results obtained from this fellowship," he recalled. Author's interview.
25. Todoriev 1982, pp. 17–18.
26. Todoriev 2001, p. 73.
27. Reminiscences of Nikola Todoriev, presented by his son, Hristo. As author of one of the key innovations in the new technology of lignite coal burning, the young associate professor, Nikola Todoriev, was personally appointed by then minister of the power industry, Konstantin Popov, to guide the US delegation. As a member of the group of Bulgarian experts returning the visit to the United States, he made several presentations on principles of the new technology.
28. Todoriev 1982.
29. An interview with N. Nabatov confimed these events. He mentioned, however, a dispute between N. Todoriev and Alexander Tzvetanski, head of another group of thermal power engineers, who opposed the direct burning of the coal. He insisted that such low-quality coal required preliminary enrichment to increase its yield. It is true, said Nabatov, that in direct fan mill burning the coal does not burn completely, which means that more coal is needed for any given amount of heat. Yet when the costs of enrichment are calculated, it is not clear which technology is better, and obviously Todoriev managed to prove that he was right. I still have some doubts, he said, that the final decision was rather political and administrative, and not purely technocratic. Author's interview with N. Nabatov, sustainable investment manager in the Bulgarian power industry.
30. See Bulgarian Academy of Sciences 1986.
31. Lambov et al. 1981.
32. Author's interview with former vice minister of electricity, Oved Tadjer. Nikita Nabatov confirmed that some of the advisors to the Bulgarian Communist Party leader, T. Zhivkov, were among the strongest supporters of nuclear power; he remembers a debate with two of them in the late 1960s, who tried to convince him about the efficiency and profitability of nuclear power stations.

33. Ground was broken officially on October 14, 1969, just a few weeks before the plenum. Author's interview with Oved Tadjer.
34. Materials about the role of the nuclear power plants in Eastern Europe come from COMECON and other archives based in Bulgaria. See bibliography.
35. See Ministry of Foreign Affairs of the People's Republic of Bulgaria and Ministry of Foreign Affairs of the USSR 1971, pp. 419–424.
36. Reports to the Council of Ministers, 1971–3, Bulgarian Central State Archive, DSO "Energy and Coal", Inventory 565, file 1, N 88.
37. Lambov et al. 1981.
38. "Initial negotiations between IAEA and Bulgaria for concluding the Non-Proliferation Treaty. Concluding the treaty", February 3, 1972, Ministry of Foreign Affairs Archive, Inventory 18, file 224.
39. Only two years after the establishment of the plant, a serious earthquake struck the region where it was built. The earthquake changed the safety policy of the Bulgarian government with regard to the new technology. The case is revealed in detail in the next paragraphs.
40. The line was also linked to the COMECON interconnected power grid, with its central dispatching organization in Prague. Information about the transnational links of Bulgaria's power system is from the journal *Energetika*, 1950–89.
41. In this particular period Moldavia was part of the Soviet Union.
42. TPP Sofia, TPP Maritsa 3, TPP Parva Komsomolska, TPP Maritsa East 2, TPP Dimo Dichev, TPP Varna, and almost all of the hydropower plants.
43. Announcement in the newsletter "First nuclear station – Kozloduy", no. 2, March 1977.
44. This production was of great significance for the development of the electric power sector as well as for other industrial sectors, such as the machine building, chemical, and metallurgical industries.
45. One of them was Swedish company ASEA (later incorporated with Brown-Bovery as ABB). See Spirov 1999, p. 92.
46. Spirov 1999.
47. Although it was never patented, in the 1980s the technology was sold to power plants in Spain and some other foreign countries. After the collapse of communism the Bulgarian firm TOTEMA still owned some of the licenses, and obtained new orders from countries like Australia and Indonesia. Author's interview with Nikolay Ivanov, manager of the power engineering company TOTEMA.
48. Author's interview with engineer Mitko Iankov.
49. The Belene nuclear power plant was the next project for establishing an atomic power station in Bulgaria. The project was started in the early 1980s. The station never went operational, however. During the political changes in 1989 the project was abandoned. This was also related to the advent of free public opinion in the state, which had been deeply influenced by the Chernobyl catastrophe. The project was frozen for more than a decade before it was again reconsidered as a potential perspective in the power field.
50. The full capacity of the Belemeken-Sestrimo cascade was reached after the completion of the Chaira pumped storage power plant in the late 1990s. With its 1599 MW, it is the most powerful hydroelectric cascade in Bulgaria today.
51. These were built in Bulgaria under Japanese supervision. Due to the financial difficulties even before the fall of communism, in the late 1980s, Toshiba vitiated the contract and construction was stopped. A few years later a compromise solution was found. Bulgaria and Toshiba concluded a new agreement, according to which the Japanese company handed over the documentation of the pump turbines and the motor generators that it had produced for Chaira. In this way, a decade later, the construction of the Chaira pumped storage hydropower plant was completed with a grant from the World Bank.

The Bulgarian manufacturers, Vapcarov (Pleven) and Elprom (Sofia), produced the other three hydro aggregates, according to Japanese know-how and documentation.
52. Austria also profited from this cooperation because it received electricity from Poland transported through Czechoslovakia, where it crossed the Iron Curtain.
53. "Protocol for the united power system of COMECON", Central State Archives, Sofia, fund 521, Inventory 4, file 228.
54. Sokolov 1988.
55. Neporojni 1981.
56. Todoriev 1988.
57. Todoriev 2001, pp. 73–74.
58. The report was discussed at the 57th meeting of the COMECON Standing Commission. Several future directions for possible development of the united systems were dealt with. One of the topics was the extension of the system and its connection to the Western European grid. The Bulgarian representatives proposed such an extension and also mentioned the already existing transmissions to Turkey and Yugoslavia. The last of these (Nish-Stolnik), according to their proposal, could be used to export power to Austria. "Protocol of the Energy Minister for the 57th meeting of the Standing Commission on Electricity Cooperation in COMECON," Moscow, 1980, Central State Archives, Sofia, Fund 521, Inventory 4, file 239.
59. The changes also helped with completion of the Chaira pumped storage hydropower plant, which further stabilized the system.

6
Border-Crossing Electrons: Critical Energy Flows to and from Greece

Aristotle Tympas, Stathis Arapostathis, Katerina Vlantoni, and Yiannis Garyfallos

"A major geological surprise": From the 1995 earthquake to a transnational history of electric power infrastructure

"Dance of the Richters," read the title of a half-page article in the May 12, 1995 issue of the popular Greek newspaper *Nea* (News). "There have been 30 earthquakes of over 4 on the Richter scale in the last 40 days in many areas of Greece... the intense seismic activity of recent days has caused uneasiness, but the seismologists reassure us that it is not an unusual phenomenon and that there is no risk."[1] The earthquake that hit Greece the following day was unusual on many levels. *Makedonia* (Macedonia), the newspaper with the largest circulation in Northern Greece, called it "a major geological surprise."[2] The 6.6 Richter quake had its epicenter at one of the few areas in Greece that was not considered seismogenic, near the city of Kozani. This was the largest city in the western part of the Greek region of Macedonia.[3]

The newspaper reminded its readers that the Kozani area was also "the energy heart of Greece, with 70 per cent of the country's electricity produced there."[4] According to *Nea*, "the preparedness of the Public Power Corporation (PPC) prevented a major blackout." More precisely, instead "of lasting several hours... [it] was limited to a few minutes of service interruption at many points in northern Greece and other areas of the country." These short blackouts were attributed to "anomalies" in power switches at the Kozani Energy Center and the High Voltage Center at Kardia. According to the Public Power Corporation there was no permanent damage to the four hydroelectric stations of the Kardia Energy Center. To cover "the loss of 950 MW from this interruption, three hydroelectric units were immediately put in operation." This journalistic account of the blackouts concluded by emphasizing that "Automatic machines and technicians had previously checked the safety of the installations and the dams, which is what they do on a permanent basis as well."[5] The Greek power system had functioned well and there was no further danger.

While confirming that the 1995 earthquake did not result in a devastating blackout, engineers from the Greek Public Power Company's (PPC) Departments of Production and Transmission Studies diverged from journalists in explaining why the Greek electric power network withstood the shock of the earthquake so well. In the narrative of the PPC engineers, too, the earthquake had caused an "anomaly" at the Kardia Voltage Center near Kozani. Immediately prior to the anomaly the total load of the Greek network had amounted to 4,215 MW, whereas the total production was 4,016MW. The Greek network imported 100 MW from Albania and 135 MW from the Former Yugoslav Republic of Macedonia (FYROM) to cover the gap. Due to the disruption of several thermoelectric and hydroelectric units in the Kozani area and the loss of the connection to Albania, the Greek network lost 1150 MW and was about to face a general blackout. The Greek and the Bulgarian networks were not yet synchronously connected when the failure took place. It was the connection to neighboring FYROM that saved the Greek network: with 835 MW (20 per cent of the total load) flowing in immediately after the anomaly, the blackout could be limited to a few areas in Thessaloniki and Attica that were intentionally cut off because 300 MW were still lacking. Once the general blackout was avoided, all consumers were reconnected within 10 minutes. Hence, "Despite the large size of the disturbance there were no further consequences for the Greek system."[6]

Thus, while Greek journalists emphasized national resilience to disturbances, the PPC engineers highlighted how Greece was rescued by its neighbors. And not just any neighbor: FYROM was considered to be a hostile neighbor by many Greek citizens and politicians. Since its declaration of independence from Yugoslavia in 1991, the country was called Macedonia by most of its citizens and several other countries, triggering Greek fears of territorial claims to the adjacent part of Greece that was also called Macedonia. Hence the Greek, and later the United Nations and European Union (EU), practice of calling the new country the Former Yugoslav Republic of Macedonia. Evidently, however, the well-known political tensions between Greece and FYROM did not prevent a rather unnoticed technical link between the two countries in the form of electric power lines, which proved its worth when the earthquake struck.

This largely unnoticed but critical link offers an example of what historians have called the "hidden integration" of Europe.[7] In the case of Greece, such hidden electrical integration began in the most adverse political environment. Two power lines between Greece and Yugoslavia, one line between Greece and Albania, and one line between Greece and Bulgaria were built despite the ideological and political divisions of the Cold War. For about three decades, Greece received critical amounts of electricity from its Cold War enemies, not from its Cold War allies. The Greek power authorities failed to build a direct link to Italy during the Cold War. This link would have helped Greece to avoid a dependency on countries belonging to the other side of the Cold War camp, but it was deemed technically and financially problematic. Neither did Greece connect to Turkey, a country belonging to the same ideological-political Cold War camp, but also a traditional Greek enemy. Electric power lines that connected the Greek to the Italian and the Turkish electricity networks were built only after the Cold War under the shared ideological-political orientation of economic neoliberalism.[8]

In this chapter we take a closer look at the historical shaping of such cross-border connections. On one hand, this allows us to inquire about suggestive contrasts

Table 6.1 Sources of electricity in Greece, 1961–2005 (percentages).

	Lignite	Oil	Hydro	Gas	Renewables	Import
1961–7	42%	32%	24%			2%
1968–73	35%	41%	22%			2%
1974–9	48%	37%	14%			1%
1980–5	54%	28%	13%			5%
1986–91	69%	20%	9%			2%
1992–2000	84%		8%			2%
2001–5	63%	9%	6%	16%	2%	4%

Source: Data reconstructed through archival research at the PPC Monthly Report (Minieo Deltio Ergasion, DEI).

between hidden infrastructure links and ideological-political divides between neighboring countries. In the following section in particular, we provide an overview of the various connections between the Greek and neighboring electric power networks, placing special emphasis on the first Greek-Yugoslavian line built in 1961. On the other hand, this transnational perspective also allows us to study the role of cross-border power lines in Greek electrification. For in existing accounts of Greek electrification, again, these cross-border links are largely absent. Before the Second World War, Greece lacked any long-distance electric power transmission lines, despite several proposals for a national power grid; local and regional power systems remained isolated and dependent on imports of British coal. In 1950 the Public Power Corporation was established as a state utility, purchased scattered local-private electricity companies, and set out to build a national transmission network. The most important transmission lines of this new national network originated at thermal power stations near rich lignite deposits, with hydroelectricity providing a substantially lower percentage of the country's electricity. Imported oil provided an additional important source of electricity generation in Greece throughout the postwar period (Table 6.1). Lignite obtained an even more dominant position in the 1980s, following the abandonment of plans to build one or more nuclear power plants. Kozani, situated centrally in the main Greek lignite region and near a river of considerable hydroelectric potential (Aliakmonas), became the electricity center of Greece.[9] In the last section we discuss how cross-border power connections played an important yet neglected role in this story. An important example is the abandoning of Greek nuclear power plans and the construction of Greek-Bulgarian power import links instead.

Let us start the story from the beginning, however, by taking a closer look at the border.

"The grand energy proletariat of the Balkans": Electric lines and national borders in engineering debates from the Second World War to the Civil War

On February 2, 2000, on the occasion of the World Wetlands Day, the prime ministers of three Balkan countries – Albania, FYROM/Macedonia and Greece –

signed a declaration establishing Prespa Transboundary Park. This transnational area included the Great and the Small Prespa Lakes, the highest tectonic lakes in the Balkans at an altitude of 853 m. There was a third lake 10 km to the west of the Great Prespa, Lake Ohrid. It lay about 150 m below Great Prespa and was fed by it. One of Europe's oldest and deepest lakes, Ohrid had been declared a World Heritage site by UNESCO in 1979. The Prespas and Ohrid, along with Lake Malik, form a group of lakes that were known as the Desaretian lakes. In contrast with the rapid degradation of wetlands worldwide, these lakes still represented a healthy ecosystem. Prespa Park hosted the largest breeding concentration of the critically endangered Dalmatian pelican, significant populations of the endangered pygmy cormorant, and seven rare heron species, including the great white egret. It was also a rare breeding ground for the ferruginous duck. In addition, Prespa Park hosted the only breeding colony of the great white pelican in the EU. It goes without arguing that these Balkan lakes represented a critical transnational European natural resource.[10]

Several decades earlier, Greek engineer Th. I. Raftopoulos had explicitly argued that the Desaretian lakes were unique in Europe and of critical importance, but not because he thought that they might one day be the last European breeding resort of the great white pelican. When he looked at the Desaretian lakes, Raftopoulos saw an abundant supply of electrons, not a habitat of rare birds. In his vision, these lakes ought to become a critical hydroelectricity infrastructure (Figure 6.1). "One would have to try hard to find in Europe a hydraulic force comparable to that of the Desaretian lakes," wrote Raftopoulos in an article published in a 1943 issue of *Industrial Review* (*Viomihaniki Epitheorisis*). "If Sweden had such a case," he continued, "it would stop boasting about its plan to use Lake Vänern as an energy reservoir: one centimeter of the water of this Swedish lake corresponds to a depository of 5 million kWh, whereas one centimeter of water in our lakes yields 8.5 million kWh."[11]

Both the use of the present tense ("yields") and the possessive antonym ("ours") were rather problematic in this case. First, none of the seven hydroelectric plants that Raftopoulos envisaged was ever erected. The Desaretian lakes never gave a single kilowatt hour of electricity to Greece. The Raftopoulos 1943 proposal to turn the Desaretian lakes into a critical energy-generation infrastructure stimulated engineering debates over the proper configuration of the future national power network of Greece in the mid-1940s, but it was silently abandoned by Raftopoulos and the rest of its promoters only a few years later. Second and related, the Desaretian lakes never became exclusively Greek. Great Prespa was shared by Greece, Albania, and Yugoslavia, and Small Prespa by Greece and Albania. Ohrid was completely outside the Greek borders and was shared by Albania and Yugoslavia. As for Malik, it was located fully within the Albanian borders.

Raftopoulos was a key member of a team set up at the National Bank of Greece early during German occupation (in July 1942) in order to develop plans for the future Greek electric power network. There had been earlier (interwar) engineering proposals for hydroelectricity generation plants and for long transmission lines to bring electricity from these plants to the cities. But these plans were blocked by

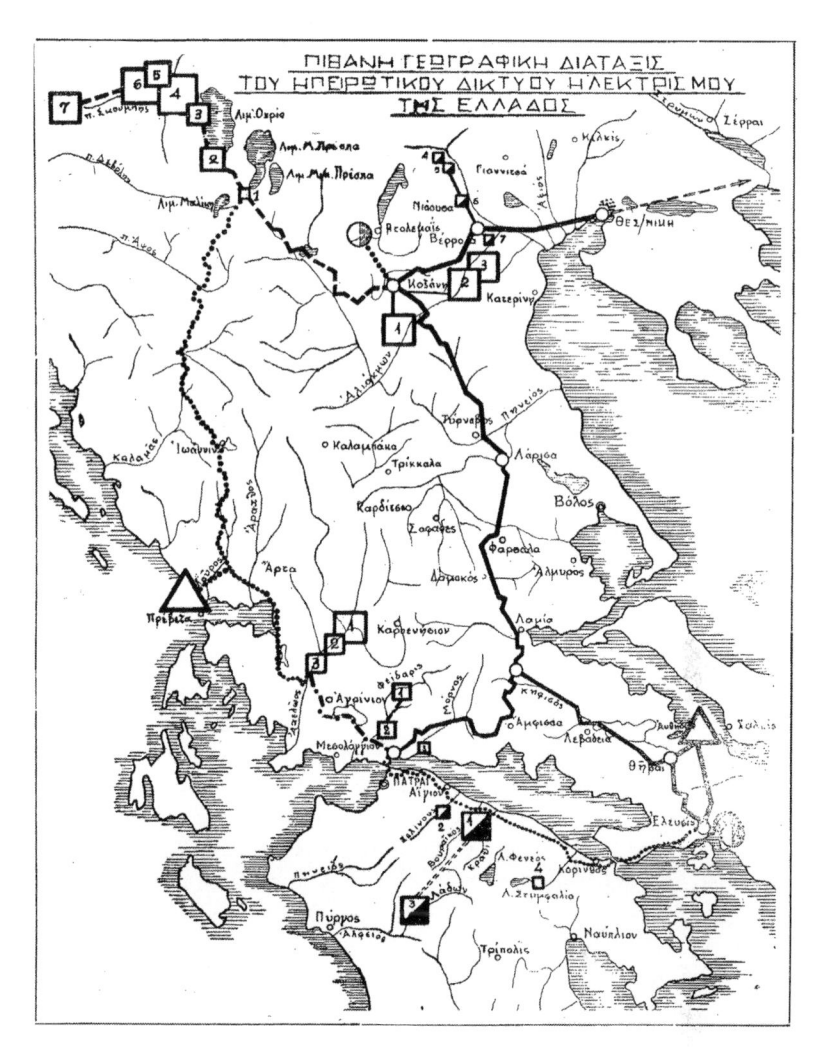

Figure 6.1 Projected Electricity Network of Greece, 1943/4.
Source: Raftopoulos 1943 and 1944, appendage map.

interests that had invested in local (city-wide) networks that were burning coal and oil, which was imported under British control.[12] By the end of the interwar period, Germany and Greece had developed strong economic relations, but the overall course of industrialization remained under British influence, through this country's control of Greek political life and financial institutions. The German occupation of Greece only amplified the engineering voices that warned against dependence on imported coal and oil, and in favor of the indigenous hydraulic resources and lignite ores.[13]

The National Bank of Greece, the institution that employed Raftopoulos as a technical consultant, was rather cautious about publicly endorsing his preliminary study on the Desaretian lakes. For a forum that was prepared to help with the public promotion of the Desaretian lakes plan, Raftopoulos could rely on the Greek Society for the Scientific Organization of Work (Elliniki Eteria Epistimonikis Organoseos tis Ergasias). Instituted during the last years of the interwar period, this society was formed by "rationalist" engineers, economists, and other proponents of a European version of a Taylorist-type economy. The appeal of Taylorism and related technocratic movements in interwar Greece is just beginning to receive historiographical attention. What we know so far seems to suggest that these movements were not very influential in changing the way in which Greek industry worked. At the same time, during the interwar period these movements were attractive to scientific and engineering environments as well as to some intellectual circles. Over the course of this decade, Raftopoulos and other members of the prewar Greek Society for the Scientific Organization of Work moved on to endorse a version of "technological nationalism" that served well the postwar purposes of the ultra-right.[14] As we shall see, the left-leaning Science-Reconstruction Society, an institution formed right after the end of the war, was all but immune to technological determinism.[15]

The 1943 map of Raftopoulos (Figure 6.1) showed no national borders in the area of the Balkans covered. As he saw it, all of the area that he showed in this map could become Greek after the end of the war. Raftopoulos had presented his Desaretian lakes plan at a 1943 meeting of the Greek Society for the Scientific Organization of Work. In 1944, Alexander Sinos, former dean of the polytechnic school, published a book on "the geographical unity of the Greek Mediterranean space," which portrayed the Desaretian lakes as indispensable for the proper development of the Greek national electric power network.[16] From 1944 to 1946, Greek engineers followed up this argument by arguing repeatedly that all of the Balkan area shown in the Raftopoulos map of 1943 ought to be placed within the borders of Greece.

The Raftopoulos plan was also endorsed by engineer Petros Kouvelis, who moved on to publish his "thoughts on our national claims and reparations" in the March 1946 issue of *Industrial Review*, as his contribution to the upcoming Paris Peace Conference, at which the Balkan borders were to be settled. For him, Greece ought to ask for the extension of its borders to the north, because its historical right to do so had been reinforced by "modern technical views."[17] Kouvelis reproduced the geographical determinism of Sinos but moved on to elaborate on the integration of technological determinism into geographical determinism. From the western to the eastern part of the lower Balkan Peninsula, several rivers reached the Aegean after crossing the Greek region of Macedonia. But only the lower part of these rivers was within Greece, which meant that the country had inadequate control of the flow of these rivers. As a result, the agriculture of northern Greece was vulnerable to floods and droughts that could be amplified by careless or intentional acts of those in charge of the upper stretches of the river. But Kouvelis was not simply interested in irrigation infrastructure. In anticipation of the negotiations at the upcoming Paris Peace Conference, he argued to place within the same national borders all of the water resources (lakes and rivers) that could form a natural unity of hydroelectric generation infrastructure.[18]

"In regards to our territorial claims," Kouvelis wrote, "we must emphasize the need to draw the future borders of our country in harmony with the necessity to follow the geographic, and, more importantly, the geo-economic limits that separate us naturally from neighboring territories." He moved on to invite technology to assist history. "In addition to defining the minimum limits of the area where Hellenism has lived and developed in past centuries," he argued, "this way of delineating borders satisfies modern technical and economic views, which, as we shall see, dictate the clear demarcation of the river basins in which neighboring countries are interested."

While Kouvelis sketched a more general scheme of hydroelectric infrastructure development on the northern borders of Greece, he relied on Raftopoulos' calculations to concur that the Desaretian lakes were of critical importance. Raftopoulos had estimated that the hydroelectric exploitation of the Desaretian lakes could generate no less than 3 billion kWh/year. The "energy wealth contained in these lakes," wrote Kouvelis,

> is important for the development of our country, especially considering the comparatively low importance of these lakes for neighboring Yugoslavia, which is exceptionally rich in energy resources, and Albania, which can find energy sufficiency in the rest of its hydraulic resources... This certainly strengthens the view that these lakes ought to become Greek territory.

As a matter of fact, Kouvelis had already submitted an earlier version of his article in November 1944 in the form of a memo to George Papandreou, the first Greek prime minister after liberation.[19] In late 1944, however, Papandreou was too busy to focus on the Kouvelis memo. The very placement of the whole of Greece within the borders of the Western camp was an open issue at the time. When Kouvelis submitted his memo, Athens was about to become a battleground between British troops that supported the establishment of the Papandreou government and communist-led forces that opposed them.[20]

Due to its leadership during the Greek resistance against German occupation, the Communist Party emerged from the war as a key actor of the postwar period. The chain of national events and international deals that took place between the 1944 end of a devastating war and the formal 1946 beginning of the tragic Greek Civil War (1946–9) holds a commanding position in the historiography of modern Greece. By the end of the civil war (1949) the Greek left was defeated, but back in 1944 the hegemony of the political left over the center allowed it to appear as a leading modernizing force. It was convincing enough to make several Greek engineers join the Communist Party and the center-left parties that allied around it.[21]

Nikos Kitsikis, dean of the polytechnic school during the beginning of the German occupation, was one of those who had joined the left. A protagonist in the establishment of the Technical Chamber of Greece and its journal *Technical Chronicles* in the early 1930s, he became secretary general of the Science-Reconstruction Society, which was instituted by a group of communist and other left-leaning intellectuals and scientists in November 1945. Members of this society were key contributors to the journal *Antaeus* (*Antaios*, 1945–51), which featured a wealth of articles on issues relevant to Greek industrialization. It was this publication by the Greek left that hosted the bulk of the debate over the merits of the plan to turn the Desaretian lakes into the critical electricity-generation site of postwar Greece.[22]

In 1945, Raftopoulos attacked the Communist Party for leaving the hydroelectric plans for the Desaretian lakes out of its plans for industrialization in general and electrification in particular. Engineer Stavros Stavropoulos replied, thereby starting a series of exchanges of long letters through *Antaeus*. Raftopoulos, by now an ultra-right political conservative rather than a politically neutral "rationalist," argued that the industrialization plan proposed by the Communist Party required the Desaretian lakes and that Greece ought to request their inclusion in its national territory, because it was the country with the fewest energy resources in the area. "Regarding the distribution of natural resources," Raftopoulos wrote in one of his letters that "Greece is the grand energy proletariat of the Balkans, a group of nations with very rich energy resources available."[23] Stavropoulos argued that incorporating the Desaretian lakes into Greece's national electric power network was immoral and politically dangerous.[24]

The left had not disagreed on the importance of large-scale electrification and on the significance of the lakes as an electricity-generation site. It disagreed only about the control of this site. "There will, of course, come a moment when, in the context of Balkan cooperation, the potential of the Desaretian lakes will be utilized," *Antaeus* editor Dimitris Batsis wrote in 1947. "But," he added, "it will not be the political party of Mr. Raftopoulos that will lead the people to this cooperation."[25] In other words, the left was in favor of a future transnational utilization of the lakes. At the same time, it agreed on the urgency of large-scale electrification. This made it vulnerable to attack by the ultra-right, which argued for the more immediate utilization of the lakes by a Greek state that ought to expand so as to place the lakes within its territory. The left argued that this would be a nationalistic, and therefore immoral, move. As we saw, Raftopoulos countered by arguing that such a move would be only natural, because it was technically rational. Based on the left's agreement on the importance of large-scale electrification and the significance of the lakes as an electricity-generation site, the ultra-right could then accuse the left of being unpatriotic. This was no small accusation against a party that had just gathered considerable support by presenting itself as the leader of a patriotic resistance movement.

By the end of 1946, with the Balkan borders settled in Paris and the formal start of the Civil War, the "rationalist" plan for the Desaretian lakes was quietly withdrawn. The international reconstruction plans of the Civil War years aimed at securing the 1946 Greek borders against neighboring nation-states that were rapidly turning into Cold War enemies. In this context the pragmatist American engineers who eventually assumed the lead in planning the Greek electric power network saw no reason to start from the Desaretian lakes. On the contrary, they had every reason to avoid them because the mountainous area of Greece just below them was a stronghold of the communist army. When the Civil War was over, the borders on the Desaretian lakes were clearly Cold War borders. The winners of the Civil War had every reason to avoid any investment in an infrastructure to be built right at these borders. This, ironically, was what eventually turned them into a critical transnational European resource – a unique habitat of rare birds.[26]

"No payment in currency": Negotiating lines of electric power infrastructure interconnection during the Cold War

In November 1957, the PPC president, Petros Gounarakis, was quoted in the daily *Vima* as announcing plans for a connection between the Yugoslavian and Greek electricity networks.[27] It was emphasized that the use of this connection was not to involve any currency payment, but merely an exchange of electricity through balancing and returns.[28] *Makedonia*, which covered the PPC president announcement extensively, stated this principle in the title of a relevant two-column piece: "The conditions to achieve an exchange of energy between Greece and Yugoslavia: There will be no payment in currency."[29] The press were tireless in stressing this principle. "Those in charge clarify that this interconnection is not aiming at the import or export of energy but simply at the exchange of electric current in case of urgent and unpredictable need," repeated *Makedonia*. At the same time, the press emphasized the Greek interpretation of the Greek-Yugoslavian connection as a means to a broader interconnection with other parts of Europe. "Through this interconnection the Greek national network of PPC is connected to the network of all of the rest of Western European countries," added *Makedonia*, "because these countries are connected to Yugoslavia."[30]

Between 1957 and 1959, the PPC general manager and professor of the polytechnic school, Georgios Pezopoulos, had to give several rounds of clarifications on the capacity of the interconnecting line and reaffirm the principle of the planned exchange of electricity. On June 15, 1959, the Ministry of Industry announced the cost of the PPC part of the Greek-Yugoslavian interconnection, which was estimated at 13.5 million drachmas. Through an 80 km, 150 kV line from Ptolemaida near Kozani to the Yugoslavian border, the amount of electricity to be exchanged was expected to reach around 50 GWh annually, and be limited to a maximum of 250 GWh annually. The PPC president had initially talked about 17 GWh of annual exchange, which required productive capacity of 60 MW (in Greece) to be reserved in order to accommodate the exchange.

Pezopoulos explained that the two systems "could complement each other in periods of seasonal shortages due to the lowering of the rivers because of the lack of rainfall." The Ministry of Industry confirmed that "through their interconnection, these two systems – one hydroelectric [Yugoslavian] and one thermoelectric [Greek] – will complement each other in periods of seasonal shortages due to the lowering of the rivers because of the lack of rainfall."

Kleonymos Stylianidis, a Greek engineer with a long and distinguished career, was the only opponent to this planned power line. He was reading the newspapers and reproduced the above quotes from the PPC officials in his 1959 article, just as he was keeping notes on the *Technical Chronicles*. According to his notes of what was written in the prime engineering publication of Greece,

> the agreement was based on the assumption that one of the two countries has a surplus during various seasons and hours whereas the other has a shortage during the same hours, so that the exchange allows both to get rid of the need to construct a power to serve peak demand, since the supply of energy from Yugoslavia would come from waterfall and from Greece from the strengthening of the Ptolemaida installation.[31]

For Stylianidis, this assumption was "irrational" because a thermoelectric installation (like the one in Ptolemaida) is not subject to seasonal variations that affect a hydroelectric installation (like the one in Yugoslavia). "This agreement," wrote the engineering veteran with irony, "re-baptizes Ptolemaida into a river whose lignite richness depends on meteorological drops." Stylianidis saw additional irrational elements in this agreement. Knowing that the cost of the construction of 1 km of this 150 kV line was estimated to be $40–45,000, he raised the cost estimate of the 80 km Greek section to 100 million drachmas (as opposed to 13.5 million). This was about $3 million. In his estimate, more than $15 million would be wasted in building a station to guarantee the 60 MW of generating capacity. Stylianidis moved on to argue that this installation would operate inefficiently because it would shut down frequently in order to enable the exchange of the projected amounts. Finally, the veteran Greek engineer expressed his disagreement with the expectation of a sharp increase in demand for electricity in Greece, which underlined the drive towards the connection to Yugoslavia.[32] "We have been misled into copying systems of other countries, where the conditions are more or less different," he argued. "It is logical that, for example, Germany and Italy exchange electricity with Switzerland because Switzerland has an energy surplus in the summer from its waterfalls, due to the melting of glaciers, whereas Germany and Italy have surplus of energy in the winter, when Switzerland has a shortage."[33] These "50 or 250 kWh [from Yugoslavia] per year is neither necessary nor usable to us," he added. "The agreement with Yugoslavia is an error against logic," he continued: "We have succumbed to the fashionable obsession with 150 kV lines and, after lighting up our poor lodges with them, we now want to help our neighboring country by spending tens of millions of dollars." In his opinion there was "only one excuse" for this waste of resources. There was capital available through US funds, and this line was a way to use it up.[34]

The US interest in tying the Yugoslavian electric power network to that of the West is now well documented.[35] It surely created a climate that favored some economic cooperation between capitalist Greece and the relatively independent (from the Soviet Union) socialist Yugoslavia, despite the permanent disagreement between the two countries over the naming of the most southern of the

Yugoslavia republics (Macedonia) and the complications that could stem from conflating the Macedonian issue with the Cold War divide.[36] The Greek press welcomed the news about the agreement and paid close attention to the course of its implementation. A 1957 *Makedonia* article offered some further details about the motivation behind the interconnection, citing a Yugoslavian officer who stressed the opportunities arising from the link beyond pure economics, putting it in the context of "the friendship and the good neighborly relations between Greece and Yugoslavia."[37]

The Greek-Yugoslavian connection that started to be negotiated in 1957 could eventually be realized and was inaugurated in 1961. The link was followed by a continued interest in expanding connections with Yugoslavia and in connecting to further countries beyond the Iron Curtain. Tellingly, the pursuit of interconnections between Greece and the Eastern bloc countries was not interrupted during the years of the military dictatorship (1967–74), when scores of Greek communists and other left-leaning citizens were exiled and tortured. Interestingly, in 1971, at the peak of the dictatorship, PPC was even signing agreements with Soviet and Italian institutions for the construction of ultra-high-voltage transmission lines in Greece.[38] Plans were announced to upgrade the connection to Yugoslavia, start a connection to Bulgaria, and undertake shared hydroelectric initiatives with both Bulgaria and Albania. Efforts to connect to Italy and Turkey were also reported.[39]

An agreement with Yugoslavia for a second line was signed between PPC and the Yugoslavian utility YUGEL in the fall of 1973. PPC director Demopoulos explained that "experience of over 13 years has shown the necessity of broadening the exchange so as to improve the economic exploitation of the networks of the two countries, under the broader context of interconnecting the European electric power networks."[40] This was followed in 1974 by the inauguration of a link between Greece and Albania.

At about the same time, Bulgaria also proposed an interconnection with Greece. In contrast to the Greek-Yugoslavian link, the Greek connections with Albania and Bulgaria foresaw a substantial import of electricity, not only an exchange on the margins. In the Bulgarian case, the plans for an interconnection began to be discussed in conjunction with a proposal for the construction of an aluminum production unit in Greece. All of the output of this factory was to be absorbed by Bulgaria. The Greek minister of industry thought that a prerequisite for moving ahead with such a project was the availability of cheap electricity, so that the electricity-intensive production of aluminum would be competitive.

However, the Greek government also expressed its preference for a broader network of Balkan interconnections rather than a mere Greek-Bulgarian line.[41] It pursued this strategy at the mid-1970s meetings of the Coordinating Committee for the Development of the Interconnection of the Electric Network of the Balkans countries. This was an interstate organ of the five Balkan countries (Bulgaria, Greece, Rumania, Turkey, and Yugoslavia) that had been instituted in response to a Greek proposal to the UN Economic Commission for Europe. The interconnection committee was part of a broader expert committee made up of economic and technical representatives of the Balkan countries, which was set up on the initiative of the Greek prime minister, Konstantinos Karamanlis.[42]

There were thus both bilateral and multilateral elaborations of transnational system-building in the Balkans. An agreement on the interconnection of the Greek and the Bulgarian network was eventually announced in the press in 1977, following a Greek-Bulgarian interministerial committee meeting. Several other high-ranking officials were present at the ceremony planned for the signing of the agreement, including PPC director Angelopoulos.[43] However, the implementation of agreements was not automatic. The Greek minister of trade, George Panagiotopoulos would soon state that "the implementation of the agreement is stuck." The reason was the need to organize electricity imports from Bulgaria as part of a broader trade agreement with the same country. The Greek side was clearly concerned with the lack of an increase in Greek agricultural exports to Bulgaria, and this was affecting the interconnection agreement.[44]

Another reason for the hesitancy of the Greek side to conclude an agreement on electricity imports from Bulgaria was the simultaneous Greek effort to advance plans for a Greek nuclear plant (see next section). But the general government strategy was to experiment with the development of economic relationships with the socialist camp, because these relationships were also a diplomatic tool in the hands of the Greek government in its attempt to exert further pressure on the West regarding the Cyprus issue.[45] Energy cooperation was included in economic agreements that the Greek prime minister signed in Moscow during his historic visit to the Soviet Union in February 1979. In 1981, the prime minister, George Rallis, visited the Greek region of Macedonia to lay the foundations of a new PPC electricity-generation unit, devoting a special section of his speech to Greece's "relationship with the neighbors":

> Special mention should be made of the growth of the electric cooperation with our neighboring countries. More precisely: Since 1979, following the interconnection of our network with that of Yugoslavia at 400 kV, the capacity for exchanging electric currents has been doubled. Through Yugoslavia, Greece is now interconnected to the systems of all Western European countries, that is, from Spain and Portugal to Denmark to the north. With Bulgaria, interconnection work is progressing in two stages. During the first stage, there will be a 200 kV interconnection up to the Greek city of Serres by the end of 1981. By 1985, during the second stage, there will be a 400 kV line, which will then reach Thessaloniki. At the end of the second stage, Greece will be connected to the whole of the system of the eastern countries, which includes most of the European part of the Soviet Union, Russia and Turkey: According to the present agreements, we can import electric energy from the Soviet Union through Yugoslavia. The prospects for cooperation will be empowered by the connections to the Bulgarian system. As it is known, our network is connected to that of Albania and there is an agreement in place. We hope to expand this interconnection in the future. Finally, we are in negotiations with Turkey.[46]

Albania, Yugoslavia, and Bulgaria represented very different versions of statist-socialist societies. For a good part of the Cold War, Albania was isolated from both the Soviet Union and the West. Bulgaria remained a faithful Soviet ally throughout the Cold War. The relationship between the Soviet Union and Yugoslavia was notoriously difficult. Yugoslavia tried to remain independent from both the East and the West while maintaining contact with both. As a result, these three countries never formed anything resembling a socialist alliance against Greece. The relationships between Albania, Yugoslavia, and Bulgaria were at times no better (if not worse) than the relationships between Greece and each of these three countries.[47] Greece took advantage of this situation during negotiations over possible electric power interconnections. As the above quote suggests, in order to gain additional negotiating power, Greek politicians and PPC managers habitually coupled references to interconnections with references to new Greek plans (e.g. new plants or connections to Europe through Italy) that would make Greece independent.[48] The quote also points to the broader transnational context of European interconnections that was taken into account during these negotiations.

Generally speaking, in the post-Cold War period, electricity flows in the form of a commodity could be purchased at the market. As we saw, according to the initial rhetoric of the Cold War period, there was not supposed to be any purchase of electricity (only an exchange of equal amounts). Yet, as we can see from Table 6.2, Greece was constantly importing electricity from its Cold War neighbors. To pay for this it gave other goods and even currency.[49]

In Tables 6.3 and 6.4 we have synthesized various sources to present the complete series of electricity flows into Greece from the various connections. Apart from importing electricity from immediate neighbors, Greece used its transnational links to import electricity from third countries. For example, in

Table 6.2 Share of oil and coal imported to Greece from the Eastern bloc, 1953–66 (percentages)

	Oil and oil products	Coal
1953	1	12
1954	6	25
1955	11	10
1956	15	19
1957	18	16
1958	22	32
1959	23	54
1960	41	75
1961	36	79
1962	33	56
1963	34	40
1964	41	29
1965	32	40
1966	26	58

Source: Valden 1991, vol. 2, p. 323.

Table 6.3 Electricity production in Greece and electricity exchanges between Greece and other countries, 1953–2009.

	Electricity production in Greece (GWh)				Exchanges between Greece and other countries (GWh)				
	Thermal	Hydro	Unconnected	Total	Imports	Exports	Net imports	Monthly import maximum	Monthly import minimum
1953	996	31		1027					
1954	1128	43		1171					
1955	1059	332		1391					
1956									
1957	1316	349		1665					
1958	1353	449		1802					
1959	1664	431		2095					
1960	1825	465		2290			9		
1961	1802	541	109	2452	8	22	−14		
1962	1985	599	128	2712	7	8	−1		
1963	2151	789	155	3095	3	9	−6		
1964	2785	738	183	3706	8	23	−15		
1965	3194	745	194	4133	25	3	22		
1966	3520	1699	229	5488	26	18	15		
1967	4406	1635	260	6301	126	18	108		
1968	5324	1352	273	6949	46	26	20		
1969	5705	2030	276	8011	0	47	−47		
1970	6048	2630	313	8991	41	19	22	8.3 (AUG)	≈0 (JAN)
1971	7604	2645	362	10611	10	15	−5	2.2 (AUG)	≈0 (JAN)
1972	8935	2669	429	12034	25	31	−6	2.9 (AUG)	≈0 (FEB)
1973	10830	2215	501	13546	78	33	45	6.4 (AUG)	2.8 (FEB)
1974	10888	2339	497	13724	79	36	43	9.1 (OCT)	5.3 (AUG)
1975	12056	2005	557	14618	18	97	−78	9.9 (FEB)	5.6 (AUG)
1976	13810	1868	645	16323	60	72	−11	7.9 (FEB)	0.9 (AUG)
1977	14773	1912	716	17401	66	25	41	7.1 (FEB)	1.0 (AUG)
1978	15674	2976	808	19458	142	16	126	9.8 (FEB)	3.5 (AUG)
1979	16022	3554	879	20455	105	23	82		

Year							
1980	16678	3395		21045	654	38	616
1981	17210	3396	972	21657	397	89	308
1982	16877	3549	1050	21554	1071	48	1022
1983	18491	2329	1128	22049	1915	32	1882
1984	18652	2853	1229	22820	2592	30	2562
1985	21124	2791	1317	25344	948	209	739
1986	20745	3333	1429	25617	1563	297	1267
1987	25738	2948	1539	27334	826	283	543
1988	28226	2578	1596	29970	464	392	72
1989	28975	2132	1744	30861	825	637	188
1990	29278	1982	1886	31285	1328	619	709
1991	29806	3150	2007	31946	1370	613	757
1992	35015	2389	2141	37410	967	362	605
1993	35807	2541		38396	1093	284	809
1994	37744	2842		40623	816	433	383
1995	37735	3782		41551	1390	593	797
1996	38015	4504		42555	2664	1315	1349
1997	39374	4096		43507	2895	590	2275
1998	42393	3866		46332	2409	889	1520
1999	44640	5058		49860	1813	1652	161
2000	49281	4111		53843	1731	1546	185
2001	50223	2725		53704	3556	1055	2501
2002	50494	3463		54608	4270	1213	3057
2003	52118	5332		58471	4219	2063	2156
2004	53019	5205		59346	4862	2043	2819
2005	53143	5610		60020	5532	1838	3694
2006	42653	6449		49102	6151	1936	4215
2007	47577	3367		50944	6425	1976	4449
2008	46714	2973		51257	7574	1961	5613
2009	41616	4955		48455	7601	3233	4368

64.5 (FEB)	12.8 (AUG)
44.5 (FEB)	17.4 (JUL)
135 (FEB)	16 (DEC)
269 (JUN)	8 (JAN)
440 (FEB)	133 (APR)
734 (JUL)	402 (JAN)
358 (JUL)	44 (DEC)
313 (JUL)	22 (JAN)
289 (JUL)	8 (JAN)
408 (AUG)	143 (APR)
223 (JUL)	42 (DEC)
289 (JUL)	8 (JAN)
526 (JUL)	281 (APR)
585 (JUL)	307 (JUL)
614 (DEC)	342 (MAI)
735 (JUL)	402 (JAN)
853 (MAR)	469 (OCT)
933 (JUL)	454 (OCT)

Note: "Unconnected" refers to power plants operating without connections to the main grid.
Source: Data reconstructed through archival research at the PPC Monthly Report (Minieo Deltio Ergasion, DEI). Data for exchanges from 1994 to 2009 are based on figures provided by the European Network of Transmission System Operators for Electricity.

Table 6.4 Annual electricity exchanges between Greece and its neighbors, by country (GWh), 1961–2009.

	Yugoslavia		Albania		Bulgaria		FYROM		Italy		Turkey	
	Import	Export	Import	Export	Import	Export	Import	Export	Import	Export	Import	Export
1961	8	22										
1962	7	8										
1963	3	9										
1964	8	23										
1965	25	3										
1966	26	18										
1967	126	18										
1968	46	26										
1969	0	47										
1970	41	19										
1971	10	15										
1972	25	31										
1973	78	33										
1974	5	36	73	0								
1975	18	97	0	0								
1976	13	72	47	0								
1977	21	25	45	0								
1978	72	16	71	0								
1979	105	23	0	0								
1980	503	38	151	0								
1981	255	89	143	0								
1982	561	48	113	0	379	0						
1983	1262	32	65	0	588	0						
1984	2368	30	98	0	126	0						
1985	703	84	83	125	163	0						
1986	193	174	763	122	607	0						
1987	267	117	350	166	209	0						
1988	237	250	183	101	45	42						

1989	337								
1990	785	231							
1991	1081	95							
1992		25	265	191	223	216			
1993			181	457	361	67			
1994			289	407	0	180			
1995			571	104	70	104	154		
1996			246	82	684	2	326	200	
1997			340	157	324	0	163	276	
1998			198	391	652	9	153	193	
1999			152	382	647	75	540	858	
2000			82	537	1452	26	1865	27	
2001			11	815	1086	51	1361	23	
2002			126	960	1128	244	1312	448	
2003			49	922	1065	202	559	422	
2004			9.	997	2408	3	617	55	14	6
2005			9	1163	3312	0	1139	50	352	495
2006			51	776	3302	1	949	153	28	1133
2007			205	516	3633	1	838	102	191	1424
2008			15	1056	4553	0	833	71	268	711
2009			26	979	4468	0	796	12	455	945
			1	1802	4297	0	1202	111	1133	163	89	0
			1	1656	4628	0	906	95	1759	179	0	30
			61	1035	3418	0	1188		311	2192	0	0
							3810					

Source: Data reconstructed through archival research at the PPC Monthly Report (Minieo Deltio Ergasion, DEI). Data for exchanges from 1994 to 2009 based on figures provided by the European Network of Transmission System Operators for Electricity.

his *Technical Chronicles* article of 1999, engineer N. Vovos mentioned that the 1961–95 imports through Yugoslavia and FYROM also brought in electricity from Russia and Italy; through Albania, from Switzerland; and through Bulgaria, from Austria.[50]

In her pioneering classification of electricity links, Jane Summerton differentiated between one-link connections of otherwise independent electricity networks and the multilink integration of networks.[51] The Greek national network relied on one-line connections to other national networks, but some of these national networks were further linked to multilinked transnational integrals. This point was stressed by engineer A. Marinakis, PPC transmission department officer, and representative to the Union for the Coordination of Production and Transmission of Electricity (UCPTE),[52] who also presented a paper on the interconnected electric power network of the Balkans at a special Technical Chamber's of Greece meeting.[53]

We have so far focused on interconnections without mentioning the technology involved in implementing them. The technical choice between, for example, synchronous or asynchronous interconnection transmission corresponds to critical yet complex social trade-offs regarding the capital involved in setting up network links, the distribution of vulnerabilities among linked networks, and the flexibility available when it comes to delinking and relinking networks. Choices regarding the voltage of an interconnection (high or low) and the form of the flowing current (direct or alternating) involve similar trade-offs. The high-voltage direct-current (HVDC) asynchronous interconnection is an expensive solution that keeps the connected networks further apart and prevents the spread of blackouts. It was used often, for example, for power lines that cross the Iron Curtain.[54] The initial connections between the Greek network and the networks of Yugoslavia, Albania, and Bulgaria were alternating current (AC) lines, and when these were upgraded the connections remained of the AC type.[55] Only the recent submarine interconnection between the Greek and the Italian network is an HVDC one.

Based on the quantities of electricity imported to Greece from the various interconnections, we could conclude that the second of the Yugoslavian interconnections has been the most important, with the Bulgarian one placed second. But other criteria may change this hierarchy. The importance of a particular connection may also be measured in terms of the bargaining power that a country wielded during a critical political conjuncture or an international energy crisis. Moreover, the same connection could be more important during some years and less important during others (see Table 6.4). While the relative importance of each interconnection may have varied, the importance of the whole of these interconnections becomes apparent through three observations. First, in most years the net flow to Greece was clearly non-negligible. Second, the import of electricity to Greece was actually even more important, occasionally reaching or even exceeding percentages in the order of 10 per cent. Third, this import was even more important because there was substantive seasonal variation, which means that there were months when the transnational flow of electricity to Greece became even more critical.

Transnational infrastructures and "nuclearity": How to have nuclear energy without a nuclear power plant

To better understand the historical importance of the interconnections introduced in the previous section, and, also, the historiographical significance of a transnational history of technology that retrieves and interprets such interconnections, in this concluding section we move on to compare the amount of electricity imported to Greece through interconnections with the amount of the flow expected from a prospective Greek nuclear reactor. To arrive at a quantitative comparison of these flows, we rely on an engineering debate about the merits of a Greek nuclear plant that took place in 1977. Following this comparison, we conclude by revisiting this chapter's opening theme, namely the contribution of transnational links in moments of actual crises (blackouts).

References to the promising future of nuclear energy in Greece go back to the 1950s, but plans for a Greek nuclear plant were given serious consideration only after the fall of the military dictatorship (1967–74), during the conservative governments of Konstantinos Karamanlis and Georgios Rallis. In 1976 a plan for a nuclear reactor was proposed in a PPC ten-year development program. The nuclear facility was planned to be operational at the end of this program, by 1986. In 1976 the southern Evoia town of Karystos – about 90 km away from Athens – was chosen as the most likely site for the installation of a 1000 MW nuclear plant.[56] However, the Evoians strongly opposed its construction in their region.[57]

The engineering community, for its part, was ambivalent about the plans for a Greek nuclear plant. In late May 1977 the Technical Chamber of Greece organized a conference on "The Present Energy Problem of Greece," which became an important forum for debate. The discussion about the nuclear plant started with two presentations, one by a group of physicists called "Physics in the Service of Man" and one by the chamber's Permanent Committee on the Environment. In the context of the discussion, the pro-nuclear Union of Greek Nuclear Scientists and a PPC group also presented their views.[58] Arguing against the construction of a nuclear power plant in Greece, the "Physics in the Service of Man" group mentioned a 1976 study that estimated the cost of a nuclear plant in Greece to be 30 per cent higher than earlier anticipated, due to extra costs for the transfer of technology and education of technical personnel. This cost could increase further if the plant was built underground. The group thought that this was necessary for a nuclear plant that was to be built in a politically volatile region such as Greece.

The group's arguments concerning the percentage of the national energy to be supplied by a nuclear plant deserve special attention. It started by assuming that the capacity of a nuclear reactor could not be less than 1000 MW, because international experience had allegedly shown that even a 600 MW reactor was economically unsustainable. The installed capacity of the Greek electricity system at the time totaled 3000 MW. By 1985, the earliest possible year of the installation of a nuclear plant, this capacity would not be higher than 6800 MW, according to the most optimistic scenario. This meant that the nuclear plant would account for at least 15 per cent of the country's total generation capacity. The group stated that "no other country in the world has made such a huge jump" by relying on nuclear energy for such a large percentage of the overall electricity-generating capacity. While it may not be correct that no other country had made such a jump, this certainly was a huge jump for Greece. Providing reserve generating facilities for such a large percentage during periods when the nuclear plant would close down for regular or other maintenance – at least once a year – would require considerable investment.[59]

R. Papadopoulou, one of the participants in the discussion, elaborated on this point. To her, the introduction of nuclear energy in the energy mix had made energy planning unpredictable. In her opinion, this explained "why Britain, after 25 years of practical experience with nuclear energy, covers only 10 per cent of its electricity generation through nuclear energy." "Other Western countries," elaborated Papadopoulou, "rely even less on nuclear energy."[60]

Interestingly, those in favor of the construction of a nuclear power plant in Greece proposed a smaller percentage of nuclear power. One of the recommendations by the Union of Greek Nuclear Scientists was to "study and properly plan the expansion of the Greek electric network based on conventional sources, so as to keep the percentage from the integration of nuclear energy below 8 per cent."[61] Taking this percentage as the maximum to be contributed by nuclear energy, we may note that it was within the reach of what Greece could obtain through the available international interconnections of its network. In 1984, when the electricity supplied to Greece from abroad reached a peak, the country imported about 10 per cent of the total of the electricity it consumed (see Table 6.3). Domestic nuclear power and imports of electricity from abroad could thus be seen as alternative solutions to the same problem.

PPC's K. Kasapoglou argued that "introducing a nuclear unit in Greece by 1986 would not be premature, because countries in the region that have richer energy resources than Greece or have comparable resources already have or will have nuclear plants by 1986: Yugoslavia, Bulgaria and Turkey." The Bulgarian case was of particular interest. Kasapoglou referred to a PPC study that confirmed that the Greek grid could remain stable and absorb a surplus of 7 per cent following the integration of a nuclear plant. For him, this meant that a 600 MW plant could be safely integrated into the Greek network. Taking the floor after Kasapoglou, energy specialist K. Mihalakis agreed that a 7 per cent surplus could be absorbed by the Greek network. But he noted that, first, the Bulgarian network was successfully absorbing surpluses of this order because it was twice the size of the Greek network and, second, it was fed by 440 MW reactors. By contrast, the projected Greek nuclear reactor was larger – 600 MW – while at the same time the Greek network was smaller than the Bulgarian. The implication was that the tested stability of the Bulgarian network could not be a safe guide for Greece's nuclear visions.[62]

The nuclear future of Greece continued to be an open issue for a few more years. The Three Mile Island accident in 1979 does not appear to have decisively influenced the debate. A greater concern appears to have been local geology. The chamber's Permanent Committee on the Environment again raised "the difficulty to find in Greece an area that is non-seismic and is adequately distant from the urban centers."[63] This difficulty became clear to all in February 1981, when a 6.7 Richter earthquake hit Athens, with its epicenter located about 77 km to the west. This was about as far as Athens was from Karystos to the east. By then, Athens was already home to more than a third of the Greek population. The conservative government came under massive fire from the leaders of the opposition parties, who took the parliament floor one after the other to argue that the state had been proved totally incapable of preparing for a strong earthquake.[64]

The combination of the earthquake shock, engineering doubts, and political resistance eventually led to the abandonment of plans for a nuclear plant.[65] In the PPC plans, a nuclear unit of 600 MW appeared in the future planning up to 1981, but from 1982 there was no longer any mention of a nuclear unit in the PPC platform.[66] As we saw in the introductory section of this chapter, during the 1995 earthquake, Greeks did not have to worry about the possible impact on a domestic nuclear power plant. They could rely on nuclear plants located outside their seismogenic country for an adequate inflow of electricity during this critical event.

Having adopted a "transregionalist" perspective on the history of nuclear energy, Gabrielle Hecht has recently challenged canonical definitions of "nuclearity" as having to do only with countries that have a nuclear reactor. As her argument goes, it was this definition that has resulted in the problematic exclusion of poor uranium-producing areas from (nuclear policy) decision-making centers of the first world.[67] Based on the history of a critical transnational flow of electricity into the Greek national electric power network, we further argue that areas and countries may actually be nuclear without having either nuclear plants or nuclear fuel. Hecht has shown that the definition of nuclearity and the location of "nuclear things" ought to take into account a uranium-producing country such as South Africa, in addition to a country such as France or the Soviet Union. We have here introduced the argument that a country such as Greece – neither a producer of nuclear energy nor a producer of nuclear fuel – should also be included in proper definitions of nuclearity. A critical share of the electrons imported to this country through the lines that connected it to the electric power networks of other countries were nuclear electrons.

After taking into account the concept of nuclearity introduced above, the notion of the "shared vulnerability" of critical transnational infrastructures discussed in our introduction can be elaborated. Let us conclude by spotlighting a set of "critical events" analyzed by the PPC Greek engineering team that authored the 1999 *Technical Chronicles* article analyzing the impact of the 1995 Kozani earthquake. The first concerned two "anomalies" at the Bulgarian nuclear power station at Kozloduy, and the second two anomalies at the Rumanian nuclear power station at Cerna Voda.

On May 24 and June 1, 1996, there was a loss of a 1000 MW unit at the Kozloduy nuclear power plant. "Despite their large size, these losses did not cause problems in the interconnected system," which at the time included Greece, Bulgaria, Romania, Albania, and parts of the former Yugoslavia (FYROM/Macedonia, Montenegro, Serbian Republic of Bosnia).

On September 9 and October 1, 1997, there were "anomalies" at the Cerna Voda nuclear power plant in Romania. These, too, were handled without a problem by the interconnected Balkan system. According to the PPC engineers, the 1995 event (Greece), the 1996 event (Bulgaria), and the 1997 event (Romania) were handled by the interconnected Balkan network in a manner that "made clear the importance of interconnection and its great contribution to the secure operation of the power stations of the Balkan peninsula."[68]

The importance of this contribution is captured in the technical diagrams that the PPC engineers included in their article. In the case of the anomaly at Kozloduy, these diagrams show how power imported from the transnational interconnections filled the critical gap between the minute when national power production dropped abruptly at the Bulgarian nuclear power plants and the minute when national production at the Bulgarian hydroelectric plants was raised to offset this drop (Figure 6.2). The diagrams given in reference to the 710 MW drop in nuclear production at Cerna Voda (Romania) in September and October 1997 offer additional visualization of the successful transnational handling of critical events (Figure 6.3). Figure 6.2 visualizes the story of how a transnational network came to the rescue of a national one. In this case the engineers show the transnational story from the perspective of the country where the critical event took place (in this case Bulgaria). By contrast, Figure 6.3 shows a similar transnational story from the perspective of a country (in this case Greece)

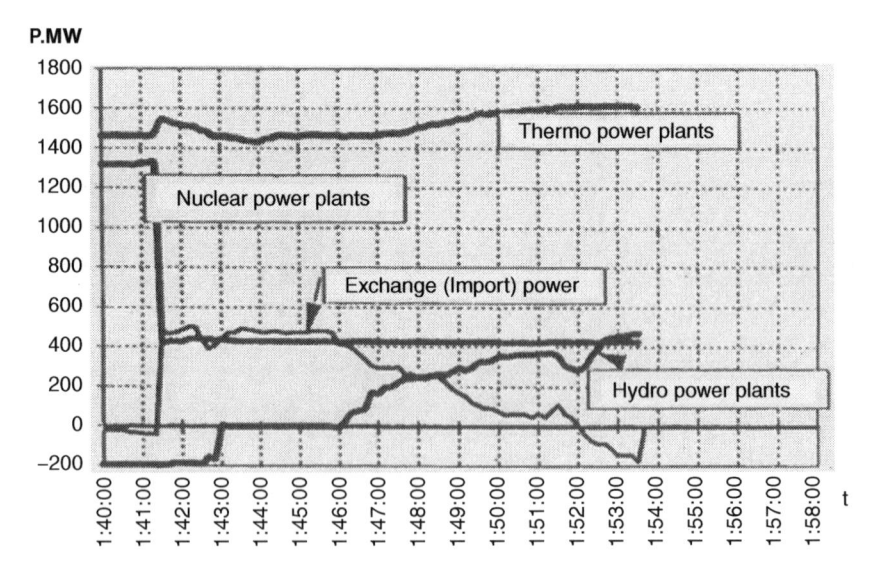

Figure 6.2 Diagram showing how a 1996 import of electricity from the interconnected Balkan system saved the Bulgarian network during anomalies at Kozloduy, which resulted in a drop in nuclear power generation. It took some time before hydroelectric plants could be set in operation to replace the drop in nuclear power. This is when imported electricity was proven crucial.
Source: Kampouris et al. 1999, p. 80.

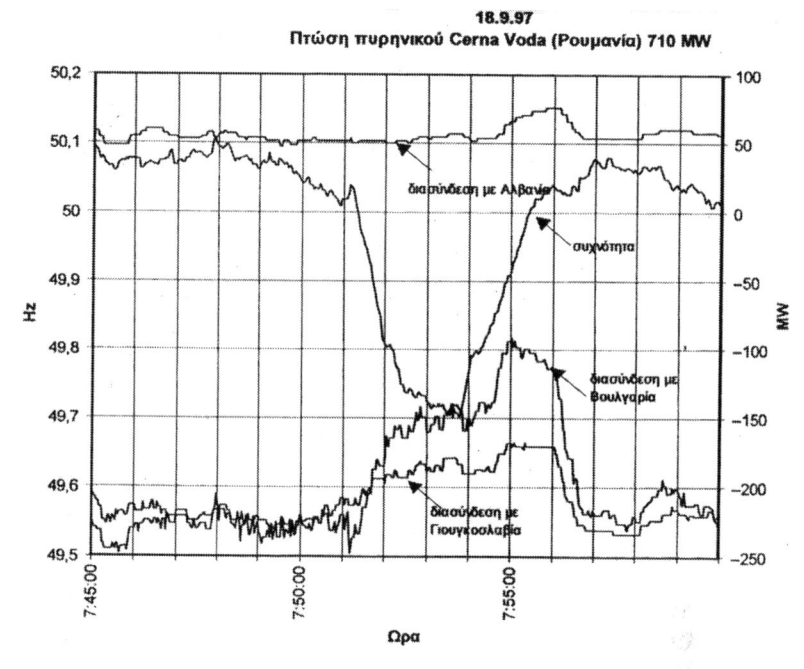

18.9.97
Πτώση πυρηνικού Cerna Voda (Ρουμανία) 710 MW

Figure 6.3 Diagram showing how a 1997 drop in frequency at the Greek power network due to anomalies at the Cerna Voda nuclear plant of Romania was offset by positive changes in the exchange of electricity between Greece, Albania, Bulgaria, and Yugoslavia.
Source: Kampouris et al. 1999, p. 81.

that experienced a critical event that took place in another country (in this case Romania). More specifically, Figure 6.3 shows how each of the lines connecting Greece to Albania, FYROM/Macedonia, and Bulgaria was affected by the September 1997 anomaly at the Rumanian nuclear station, by illustrating how each line was used to channel electricity from Greece to Rumania amid a loss of 710 MW in Rumania that caused a drop in the frequency of the Greek network.[69]

Steven Lubar has offered a convincing interpretation of engineering representations (diagrammatic and otherwise) as normative means to social power.[70] In the case of the diagrams just discussed, we find not simply representations of social power but, more specifically, representations of the social power of the transnational interconnections of national power networks. As such, these diagrams offer maps of southeastern Europe that are especially suitable for accommodating a transnational historiographical perspective. They help us to realize that the distance between, for example, Kozloduy or Cerna Nova and Kozani has been radically shortened through the transnational interconnections of national technological networks. It has always been rather clear that a nuclear accident at Kozloduy could mean catastrophe for a good part of Greece. This much has been well known to Greek society – its journalists, its politicians and its citizens – who have been arguing in favor of shutting down Kozloduy.[71] By contrast, the transnational history we sought to advance in this chapter focuses on something that is much less well known: a history of technological integration that shows how the Greek non-nuclear power plants were actually linked to the Bulgarian Kozloduy nuclear plant.

Notes

1. See "Horos apo Richter [Dance of the Richters]," *Nea*, May 12, 1995, p. 9.
2. "Geologiki Ekpliksi itan i Sismi Kozanis Grevenon [The earthquakes of Kozani, Grevena were a geological surprise]," *Makedonia*, September 26, 1995.
3. "6,6R Horis Parelthon [6.6R without a Past]," *Nea*, May 15, 1995, p. 21, and "Sismologi: Giati den to Perimename [Seismologist: Why we did not expect this]," *Nea*, May 15, 1995, pp. 22–23. In the context of Greek debates over the validity of earthquake prediction techniques from a history of technology perspective, see Katsaloulis 2006.
4. "6,6R Horis Parelthon [6.6R without a Past]," *Nea*, May 15, 1995, p. 21.
5. "I Etoimotita tis DEI Apetrepse to Blackout [The Readiness of PPC Prevented the Blackout]," *Nea*, May 15, 1995, p. 27.
6. Kampouris et al. 1999, p. 78.
7. For a historiographical argument that invites us to investigate the transnational history of technology in order to reveal the "hidden integration" of Europe, see Misa and Schot 2005. For a historiographical introduction to the benefits of a transnational history of technology, see Van der Vleuten 2008. For something special on energy infrastructures, see Van der Vleuten and Kaijser 2005. For a sample of the first crop of histories of electric infrastructures that were written in response to this invitation, see Van der Vleuten and Kaijser 2006, Van der Vleuten et al. 2007, and Lagendijk 2008. All the above works refer explicitly to the historiographical orientation of the "Tensions of Europe" network of historians of technology. For a historical work that paved the way to this historiographical orientation, see Kaijser 1997. For an edited volume that covers representative cases from around the globe, see Lynch and Trischler 2003.
8. For an introduction to such periodization written from the perspective of the transnational history of electric power infrastructures, see Van der Vleuten and Lagendijk 2010a and 2010b.
9. The history of the inability to move beyond the interest of local companies to form a national electric power network around indigenous resources offers an appropriate conclusion to the pioneering economic and business history of early Greek electrification by Nikos Pantelakis. It also offers an appropriate entry point to Stathis Tsotsoros, who has written an informative economic and business history of more recent Greek electrification, which focuses on the establishment and development of the Greek Power Company. See Pantelakis 1991 and Tsotsoros 1995.
10. On the transnational Prespa Park, see http://www.prespapark.org/ (retrieved on March 1, 2011).
11. Raftopoulos 1943, p. 13. For a masterful study of what engineers see when they look at natural resources, see Sinclair 2009. For League of Nations' attempts to institute conventions regarding the transformation of transborder water resources into electricity infrastructures, see Lagendijk 2008, Chapter 2.
12. This story is covered by Nikos Pantelakis in his history of the early Greek electrification. See Pantelakis 1991, Part 2.
13. For an excellent interpretation of the Greek economy during German occupation as a battleground between the pursuits of the occupying forces and their local allies and, on the other hand, the spontaneous and organized resistance of the Greek population, see Hatziiosif 2007a. This chapter includes special references to Germany's efforts to bring Greek industrial production under its own influence from very early on during the occupation.

14. On nationalism and internationalism in relation to technology and technocracy, see Fridlund and Maier 1996, and Schot and Lagendijk 2008.
15. For revealing references to the interwar history of these movements in Greece, see Agriantoni 2002, Antoniou 2004, 2006, and 2009, and Noutsos 1988 and 1994.
16. Sinos 1946, p. 37.
17. Kouvelis 1946, p. 73.
18. On the Sinos book, see Sinos 1946.
19. Kouvelis 1946, pp. 72–73. Kouvelis reported on his 1944 memo to Papandreou at the end of his 1946 article (p. 76).
20. Ibid., p. 73. On the situation by the end of the war in connection with the December 1944 battle over Athens between a British force that had a clear objective, and a leftist resistance-born army that was strong but committed only a limited force because it was operating under an unclear mandate, see Hatziiosif 2007b.
21. On the tremendous empowerment of the Communist Party during the resistance and how this brought it to the front of "the challenge of history" by the end of the war, see Papathanasiou 2007. For references in English to the turbulent postwar history of Greece in the 1940s, see Koliopoulos et al. 2007, pp. 68–98, Iatrides 1987, Woodhouse 2002, and Mazower 2000.
22. On Kitsikis, see Antoniou 2004 and 2006. On *Antaeus*, see Pappa 2000.
23. Raftopoulos 1946, pp. 371–372.
24. See Stavropoulos 1945a, 1945b and 1946a and Raftopoulos 1945 and 1946.
25. Batsis 1977, p. 292.
26. "For many years," we read in *Wikipedia*,

> the Greek part of the Prespa Lakes region was an underpopulated, military sensitive area which required special permission for outsiders to visit. It saw fierce fighting during the Greek Civil War and much of the local population subsequently emigrated to escape endemic poverty and political strife. The region remained little developed until the 1970s, when it began to be promoted as a tourist destination.

See http://en.wikipedia.org/wiki/Lake_Prespa (retrieved on March 1, 2011).
27. Gounarakis 1959.
28. Ibid.
29. "Ipo poias synthikas tha epitefchthi I antalagi ilektrikis energias metaksy Ellados ke Yiougoslavias: I pliromi the tha ginete is nomisma [Under which conditions the exchange of electric power between Greece and Yugoslavia will be achieved: the compensation will not be in currency]," *Makedonia*, November 7, 1957, p. 3.
30. "I syndesis ton energiakon diktyon tis Elladas ke tis Yiougoslavias tha epitefchei mehri tou fthinoporou [The connection of the power networks of Greece and Yugoslavia will be realized by autumn]," *Makedonia*, July 28, 1960, p. 3.
31. Stylianidis 1959, p. 763.
32. Ibid., pp. 763–764.
33. Ibid., p. 764.
34. Ibid., p. 765.
35. See Lagendijk 2008.
36. For a history of Greek-Yugoslavian diplomatic relations during this period, see Valden 1991a and Hatzivassiliou 2006. The first Greek-Yugoslavian line and the Greek-Turkish line were constructed amidst favorable political conditions. The Greek-Yugoslavian line was built while collaboration with Tito's Yugoslavia matched well with Greece's foreign affairs and its continuously dynamic and problematic relations with Bulgaria. The postdictatorship political setting was much different. After 1974 the Cyprus problem and the tensions with Turkey redefined the relations with Bulgaria as well as the Soviet Union. For a sample from the rich literature, see Kassimeris 2008 and Voskopoulos 2008.

37. "Ipo poias synthikas tha epitefchthi I antalagi ilektrikis energias metaksy Ellados ke Yiougoslavias; I pliromi the tha ginete is nomisma [Under which conditions the exchange of electric power between Greece and Yugoslavia will be achieved; the compensation will not be in currency]," *Makedonia*, November 7, 1957, p. 3.
38. "Symvasis metaksy DEI, Sovietikou Organismou ke megalon italikon viomihanion pros kataskevin tis grammis iperipsilis tasis [Contract between PPC, Soviet Organization, and big Italian industries to construct the HV power line]," *Makedonia*, December 21, 1971, p. 11.
39. "Poson 36,000.000 drachmon tha diatethi ipo tis DEI di ektelesin megalon ergon [The amount of 36,000,000 drachmas will be appropriated by the PPC for carrying out construction works]," *Makedonia*, June 16, 1972, p. 9.
40. "Grammi ipsilis tasieos di antallagis revmatos me tin Yiougoslavia [HV power line for the exchange of electricity with Yugoslavia]," *Makedonia*, September 19, 1973, p. 7.
41. "Sindesis di antallagin ilektrikis energias eprotinen i Voulgaria [Bulgaria proposed a connection for the exchange of electric power]," *Makedonia*, January 23, 1975, p. 10.
42. "Eliksan e ergasie tis Syntonistikis Epitropis dia tin Anaptyksin ton Diasyndeseon ton Ilektrikon Diktyon ton Valkanikon Horon [The work of the Coordinating Committee for Development of the Electric Power Networks' Interconnections of the Balkan Countries has come to an end]," *Makedonia*, April 2, 1976, p. 6.
43. "Sta 150.000.000 dollaria tha fthasoun i antallages Ellados ke Voulgarias [The exchange between Greece and Bulgaria will reach $15,000,000]," *Makedonia*, June 4, 1977, p. 10.
44. "Iparhoun dysheries stis eksagoges mas pros ti Voulgaria [There are difficulties regarding our exports to Bulgaria]," *Makedonia*, June 1, 1978, p. 9.
45. See, for example, "I Sovitiki Enosi theli na efarmostoun i apofasis tou OIE gia tin Kypro tonise o k. Kosygin stin prosfonisi tou [Mr Kosygin stressed in his address that the Soviet Union wants the UN decisions regarding Cyprus to be implemented]," *Makedonia*, January 2, 1979, p. 7.
46. "Nea erga sti Makedonia gia tin paragogi ilektrikis energies [New projects in Macedonia for the production of electricity]," *Makedonia*, April 14, 1981, p. 9.
47. See Hatzivassiliou 2006.
48. See also "I themeliosi tou neou idroilektrikou ergou [The founding of the new hydropower plant]," *Makedonia*, September 13, 1975, p. 4.
49. See Valden 1985, 1991b, and 2009.
50. See Vovos et al. 1999.
51. See Summerton 1999.
52. On the history of UCPTE, see Lagendijk 2008.
53. Vovos et al. 1999.
54. For a periodization of interconnections that connects HVDC to a specific period, see Morton 2000. For an insightful sketch of such trade-offs from a transnational history of technology perspective, see the discussion of HDVC in Lagendijk 2008. On the social meaning of HDVC, see also Fridlund 1998. The Greek engineers were following and reporting on the potential of HDVC connections. See, for example, Vovos 1989.
55. See Vovos et al. 1999.
56. Kouloumpis 2008. As far as the EBASCO involvement goes, the September 15 issue of *Makedonia* had reported that the minister of industry, M. Evert, announced that EBASCO was hired because it had won a relevant competition. See "Tous 1000 Tonous Fthani to Katharo Ouranio stin Periohi Serron [The pure uranium at the area of Serres is estimated to be as much as 1000 tons]," *Makedonia*, September 15, 1979, p. 6.
57. On the history of resistance to the building of a nuclear plant in Greece, see Arapostathis et al. 2010.
58. See Group of Physicists "Physics in the Service of Man" 1978, and Permanent Committee on the Environment of the Technical Chamber of Greece 1978. For information on other fora for scientific and engineering debates over the plans for a nuclear plant, see Nikolinakos 1978, especially pp. 55–67.

59. Group of Physicists "Physics in the Service of Man" 1978, p. 264.
60. The entire discussion was presented in *Tehnika Hronika* [Technical Chronicles], March–April 1978, pp. 261–298. The quote here is from p. 288.
61. Ibid., p. 289. This was about the percentage used in all other contemporaneous estimations. For example, in a 1976 estimate published in the journal *Economic Postman* (no. 1166, September 9, 1976, p. 5), it was expected that the contribution of nuclear energy in the decade between 1980 and 1990 could not exceed 9 per cent. This estimation was endorsed by subsequent studies, including the ones that Marios Nikolinakos and Nikos Filias wrote for the special energy issue of the journal *Economy and Society* (Οικονομία και Κοινωνία) published in February 1980. See Nikolinakos 1980, table on p. 21, and Filias 1980, table on p. 58. According to another article in the same *Economy and Society* issue, the Ministry of Industry and Energy had announced that it planned to operate a 900 MW nuclear unit by the end of 1988. The reactor would be chosen between two types: (a) the Light Water Reactor (Pressurized), which used enriched uranium U-235 (3.5 per cent) as fuel and water under pressure as decelerator, or (b) the Heavy Water Reactor (Candu-type), which burned natural uranium and used heavy water as decelerator. See Group of Special Contributors 1980, p. 31. This percentage compares well with the percentages of electricity exchanged in areas of Europe with a tradition of transnational electric power integration. See, for example, Verbong 2006.
62. *Tehnika Hronika*, March–April 1978, p. 293.
63. Ibid., p. 279.
64. "I Vouli gia ta metra meta tous sismous [The Parliament on the measures to be taken after the earthquakes]," *Makedonia*, March 14, 1981, p. 6.
65. The earthquake had exposed the vulnerability of the country's existing infrastructure. For an engineering report written by an international team that included an EBASCO employee, see Karydis et al. 1982.
66. On the place of the first (1968–72) and last (1981–90) PPC programs to include plans for nuclear plans in the history of PPC, see Tsotsoros 1995, pp. 95 and 102, respectively.
67. See Hecht 2007. For an explicit historiographical reference to a transnationalist history of technology that supports such a view on "nuclearity", see Hecht et al. 2007.
68. Kampouris et al. 1999, p. 79.
69. Ibid.
70. Ibid., pp. 81–82, and Lubar 1995.
71. Several studies by engineers and others sought to calculate the devastating effects that a Chernobyl-type accident at Kozloduy would have in Greece. See, especially, Kollias 1993, Antonopoulos 1993, Papastefanou 1993, and Mousiopoulos 1993. In the context of the 1990s, after the plans to build such plants in Greece were decisively abandoned, all Greek newspapers frequently hosted extensive reports that reproduced the calculations of such studies. For an example, see "Piriniki Apili gia tin Ellada [Nuclear Threat Against Greece]," *Nea*, April 24, 1996, pp. 16–17.

Part III
Coping with Complexity

Introduction

In Parts I and II we investigated energy vulnerabilities illustrative of Europe's critical infrastructure interdependencies across political and geographical borders. In Part III we turn to the second major category of critical infrastructure, namely information and communication technology (ICT). Here we find an additional form of cross-boundary vulnerability: growing interdependencies between Europe's transnational infrastructure systems implied that failures could transcend system boundaries and spill over from one infrastructure system to another. To investigate such cross-infrastructure vulnerabilities, the following chapters discuss the introduction of ICT infrastructure in electric power, air-traffic control, and emergency communication systems. Although Part III focuses mainly on developments in northwestern Europe, it brings into play such different geographical scales as mesoregional collaboration, bilateral linkages, national policy priorities, and microregional dynamics. All were important units of building, experiencing, and governing infrastructure vulnerabilities.

The chapters in Part III tackle three aspects of the cybernation of critical infrastructure and its vulnerabilities. In Chapter 8, Lars Thue addresses the increasing complexity of critical infrastructure entanglements and their implications for vulnerability. Apart from integrating infrastructure networks of different countries, ICT also allowed for an enhanced integration of different systems and subsystems, thus creating what Paul Edwards calls "networks of networks" or "internetworks." Did such increased complexity make systems more or less controllable, and more or less vulnerable? Thue analyzes the relations between the Norwegian power industry and the European mainland. He tries to make sense of the notion of technological complexity by introducing the notions of horizontal and vertical integration. Horizontal integration refers to the geographical extension from local or regional to national and international networks. Vertical integration often takes place in parallel with this process and refers to the increasing dependence of power systems on ICT-based control systems. Thue describes the developments in Norwegian electricity as a process leading from local electricity systems in the 1930s to transnational systems in the 1970s. In the transition from local to regional electricity systems in the 1930s, control functions were rather tightly connected to high-voltage installations. From the

1930s, regional high-voltage systems developed into national systems. Control systems were then based on telecommunications and relays. In the 1970s the number of transnational connections grew, and so did computers, digital data communication, and automation. The system became more complex and thus an intriguing paradox emerged: the ICT-based control systems installed to reduce vulnerability may themselves cause more complexity and vulnerability. Although key actors in the history of Norwegian power argued that increased control is related to increased vulnerability (e.g. the engineer Brochmann in the 1930s), Thue shows that vulnerability judgments are socially constructed and infused with a high degree of uncertainty. Despite all potential sources of vulnerability, he argues that "given the huge and increasing flows of electricity generated, transmitted and distributed, the stability of technological complexes are impressive." In the end, complexity did not result in greater chances of blackouts, but rather in more uncertainty about vulnerability, reliability, and risk distribution in the power industry. Thue concludes his chapter with the warning that this uncertainty and the resulting knowledge vacuum can be exploited by strong economic interests, and that this constitutes a challenge for present-day vulnerability governance.

The second theme that Part III scrutinizes is the relation between standardization, harmonization, and vulnerability. In order to create networks across borders, historical agents worked hard to make different systems, based on different standards, compatible and interoperable. The same applies to the rules and agreements in international organizations, such as Eurocontrol. It is often assumed that standardization reduces vulnerability, but does standardization indeed lead to a better international coordination and, hence, to a decrease in vulnerability? This problem is addressed by Anique Hommels and Eefje Cleophas in Chapter 9. They analyze the history of emergency communication from a vulnerability perspective. For a long time, emergency communication between police, ambulance, and fire-fighter services took place in separate local analog radio networks. Since the late 1970s more efforts have been made to establish cross-border cooperation between local emergency services and to harmonize agreements about technical equipment and radio frequencies. In the 1990s the European Telecommunication Standards Institute began to develop a digital technical standard for emergency communication called Tetra. Initially the idea was that Tetra would be the technical backbone of a pan-European network for public safety, but this dream never became a reality. To analyze the dreams and dilemmas of a common standard for emergency communication, this chapter studies two cases of microregional cross-border emergency communication. The first case focuses on the explosion in May 2000 of a fireworks storage facility in the Dutch city of Enschede, close to the German border in one of the EU's first Euroregions. German fire-fighters came to the rescue of their Dutch neighbors, and their collaboration required ample improvisation. The second case studies a major test of a standardized, Tetra-based network for emergency communication in the Euroregion of Maastricht (the Netherlands), Aachen (Germany), and Liège (Belgium). The chapter shows how vulnerabilities did not disappear, as intended, but merely changed character.

In Chapter 7, Lars Heide, too, discusses the dilemmas involved in the standardization and harmonization of air traffic control when Eurocontrol was established. The standardization of infrastructure and software allowed a number of geographically dispersed civilian and military air-traffic control centers to be combined into one center in Maastricht. The idea was that this single center would provide far better possibilities for coordination between the controllers operating in different sectors. An expansion of Eurocontrol across Europe would require a further standardization of air-traffic control across the member countries, so that all centers would operate on the same system. However, Eurocontrol's member states were unwilling to surrender national sovereignty to their supranational organization. The basic idea that technological standardization reduces vulnerability was adopted, as in the case of Tetra, but in the case of air-traffic, the advantages offered by the rigid Maastricht air-traffic control system based on mainframe computer technology of the 1960s and 1970s offered limited advantages compared to separate national air-traffic control operations, which most West European countries preferred. In contrast, the central flight flow management system of the 1990s and 2000s, built upon a more flexible technology and was accepted by all countries in Western Europe (except Iceland), most countries in Eastern Europe, and several countries in adjacent parts of Asia. It contributed to Western Europe's extension to the east after the fall of the Soviet Union in 1991.

A third concern of Part III is the scale and character of cooperation and vulnerability management. Cooperation took place at the local, regional, national, and international levels in the case of emergency communication, at the national, and international levels in the case of the Norwegian power industry, and at the national, international, and supranational levels in the case of European air-traffic control. For instance, while national air-traffic control remained operational, Eurocontrol advocates assumed that "once a technical air traffic control infrastructure was established, national governments would accept its advantages and implement a unified, advanced air traffic control system across Western Europe." However, originally a mere six Western European countries joined the Eurocontrol project. Meanwhile another organization, the International Civil Aviation Organization, added a third layer. Advocating the full integration of national air-traffic control systems for the upper airspace to reduce risks, this organization worked to introduce analog radar, computer-mediated radar pictures, and a flight-plan database from the 1950s. Similar to Thue in the case of electricity sector cybernization, Heide concludes that this multilayered European air-traffic system has turned out to be surprisingly safe. If facilities such as the Central Air Traffic Flow Management Unit break down, air traffic can still continue, but less efficiently. Air-traffic controllers are trained to operate when one or more of their technical facilities fail, but this kind of operation always imposes limits on traffic density. Still, Heide shows, like Hommels and Cleophas in their case of emergency communication, that the dream of a pan-European system also turned out to be hard to realize in air-traffic control because its operational advantages were limited.

Together, the chapters in Part III show how the integration of ICT into systems for air-traffic control, electricity, and emergency communication took shape in a context of attempts at European integration, vulnerability reduction, and infrastructure control. Ironically, in many cases these attempts triggered new dilemmas for the governance of such geographically dispersed, complex, vulnerable, hard-to-control, ICT-based internetworks.

7
Eurocontrol: Negotiating Transnational Air Transportation in Europe

Lars Heide

Introduction

On June 30, 1956, a United Airlines DC-7 collided with a Trans World Airlines (TWA) Lockheed "Connie" over the Grand Canyon in the United States. The accident killed all 128 people on the two craft, the greatest loss of life in any air crash of the time.[1] This major mid-air collision had an enormous impact on the public debate about air-traffic safety in the United States as well as in Europe, as press coverage shows.[2] The crash was caused by the inability of pilots of modern propeller-driven aircraft of the 1950s to stay clear of each other using the "see and be seen" principle. This rule was the basis for air traffic in vast areas outside clearly demarcated airways and zones around major airports of the United States, including the Grand Canyon area. Although this principle worked at slower speeds and lower altitudes, it was inadequate to separate the two four-propeller aircraft in the Grand Canyon crash. They flew at greater speed and higher altitudes where the atmosphere could be hazy on even an apparently clear day. Higher speed and increased traffic made air transportation more vulnerable.

The Grand Canyon crash and several smaller midair collisions in the mid-1950s were a major reason for the passage of the Federal Aviation Act of 1958.[3] By this means, Congress created the Federal Aviation Agency (renamed the Federal Aviation Administration in 1967), which established a nationwide air-traffic control network for civilian flights. Though Western European countries did not experience any mid-air collisions of large craft at that time, the Grand Canyon disaster was widely reported in the media and became a prominent reason for improving air-traffic control in Europe.[4]

Air-traffic control is a transnational infrastructure that directs aircraft in traveling between airports as well as in landing and taking off. Its objective is to reduce risk in air transportation or, in industry terms, to "improve safety." Ground-based air-traffic control centers are responsible for controlling and monitoring movement between origin and destination airports. Each center is responsible for a defined geographic area; as an aircraft continues to fly beyond its area of origin, the responsibility for monitoring and directing the craft is transferred to the next air-route center. The flight continues to be transferred until it reaches the control area at its destination. At this point the air-traffic control function is turned over to a controller in an airport traffic-control tower, and the craft is guided to land and taxi to a parking place. It is based on the English language as a standard of communication, in contrast with the situation in, for example, railroads and emergency communication systems (see Chapter 9, this volume).

The pressure to improve air-traffic control in the late 1950s was given a further impulse by growing air traffic and the introduction of faster civilian aircraft – turbopropeller and jet-propulsion craft – that were even faster than the four-engine propeller-driven craft of the Grand Canyon disaster. Growing air transportation meant that more frequent flights and faster speeds gave air-traffic controllers less time to react. At the same time, they had to handle a more complex picture of aircraft operating over a greater range of speeds. In the late 1950s, these challenges were answered through improved air-traffic control in the United States and Western Europe. The United States established a country-wide air-traffic control system based on radar monitoring. The Western European endeavor to improve the safety of civilian air traffic was hampered by the division of the airspace into many separate national airspaces along national borders. Western European efforts, therefore, in addition to seeking improved national technical air-traffic control systems, called for a reduction in vulnerability through transnational governance or coordination.

Very few scholars have studied the air-traffic control infrastructure.[5] Johan M. Sanne's detailed analyses of air-traffic control operations focused on the essential interaction between pilots and air-traffic controllers, and their mutual trust in an institutional context. Ralf Resch studied the general development of the air-traffic system from an organizational perspective. This chapter complements Resch's study by focusing on the interplay between the technical and organizational aspects of establishing and operating air-traffic systems.

At the heart of Western European air-traffic control was an organization known as Eurocontrol. Between 1960 and 1981, it existed as a transnational air-traffic control project intended to reduce vulnerability in air transportation through supranational governance and improved monitoring technology. Six Western European countries worked together to establish a unified air-traffic control system for their upper airspace (above about 20,000 feet [6000 m]) based on monitoring air traffic with a computer-mediated radar system, which significantly improved monitoring over analog radar systems, and on the renegotiation of governance. Eurocontrol successfully established this information technology platform for transnational air-traffic control but proved unable to extend operations beyond four of its six Western European member countries.

In this chapter I first analyze the discrepancy between technical progress and organizational restraint between 1960 and 1981, showing how the opportunities for a more sophisticated technical infrastructure came to prevail over its technical limitations and how the problems of establishing transnational governance reduced the scope of the project. The supranational Eurocontrol project was based upon an assumption that once a technical air-traffic control infrastructure was established, national governments would accept the inevitable and implement a unified, advanced air-traffic control system across Western Europe. This assumption did not materialize, which changed the organizational scope of the project from supranational to international.

The reduced organizational scope became the basis for establishing a register of scheduled flights, in the late 1980s, stored in a large computer in the Eurocontrol headquarters in Brussels. I discuss how this new system emerged and how it was related to Eurocontrol's transformation in 1997 into a more open international governance institution with airlines and national air-traffic organizations as stakeholders in the extension of the Western Europe system to the East after the fall of the Soviet Union in 1991.

Air-traffic control in the 1950s

In the earliest phase of civil air-traffic developments, only the immediate airport environment had traffic control to safeguard approach, landing, taxiing, parking, and takeoff, which has always been the most vulnerable part of a flight. Beginning in the 1930s, en route control was established to manage the traffic between airports based on audio radio communication between pilots and ground-control units.[6] Every country controlled its own airspace as an integrated part of its claim to sovereignty. They had separate control of civilian and military aircraft, and the military was responsible for rejecting unauthorized craft entering the national airspace. The mode of cooperation between civilian and military air-traffic control varied between countries.[7]

Every country's civilian airspace was divided into sectors, each of which was controlled by an operator. The operator monitored the progress of airplanes through pilots' reports of passing distinct "way points" on the ground along the route. Safe separation by this rough means required a wide distance between craft and was feasible only when traffic density was low. The air-traffic control unit could cover several sectors, receiving flight plans filed by the aircraft captain at the air-traffic facility in the departing airport via the international aeronautical telecommunication network. At the beginning of the flight, flight information (aircraft identification: e.g., aircraft registration or a flight number, aircraft type, assigned altitude, departure and destination, and at least one time) was written on "flight control strips" and distributed to a controller in each sector. Each controller organized their flight control strips on a "flight strip board" displaying information about all craft in their sector. The controllers monitored aircraft progress based upon radio communication with the pilots and kept track of their movements by the use of flight strip boards and maps in the air-control unit of the airways marked on the ground.[8] The Grand Canyon crash sadly demonstrated the limitations of this mode of control. Because air-traffic control did not have the means to control the two craft's exact positions along the route, the reduction in vulnerability would have required sufficient separation between the two craft to avoid the faster United Airlines craft overtaking the slower TWA craft before their flight directions separated. This would have delayed the departure of the United Airlines flight and reduced the scale of air transportation.

In the 1950s, growing air transportation and faster craft challenged the established air-traffic control mode. Growing air traffic could be coped with by introducing more air-traffic controllers. Safety concerns required that controllers be able to monitor the separation between planes, which limited traffic density. This could be overcome by dividing a control sector into several smaller sectors, each operated by an air-traffic controller. However, this way of coping with vulnerability had limited effects because increasing the number of controllers caused more communication between controllers, making it difficult to monitor a sector through which flights passed very quickly. Instead, air-traffic control organizations chose to improve their operations through technical means, enabling one controller to monitor more aircraft and operate safely with less separation, first by monitoring with analog radar, later with computer-mediated radar systems, and finally through a register of scheduled flights. However, each time a new technical device was introduced to reduce vulnerability it also added a new reason for failure.

Analog radar systems were introduced in the 1950s and facilitated closer control of air traffic. The air-traffic controller was now located in front of panoramic radar displays of their sector and still used flight-control strips to keep track of aircraft. Each air-traffic control unit was based on radar screens with rotating antennas that recreated the radar image at each turn of the antenna; for each rotation, the controller had to identify the dots with dots in earlier radar images and flights in their sector. The well-established flight-control strips were kept at the screen as a quick way of annotating flights, to allow others to see instantly what was happening, and to pass on information to other controllers in the same control unit who would subsequently control the flight. Analog radar was accepted by air-traffic control despite several technical shortcomings that subsequently surfaced. The new technical device was a complement to a smoothly functioning system, and the controllers were always trained to operate a system with any element failing, be it radar, radio, or another tool.

Controllers matched the flight information with the spots on their analog radar screens made by aircraft in their sector, some of which might have been military craft beyond their control. They compared data on position and planned course for all craft in their sector and identified upcoming problems, like changes in airway or collision risk, and formulated instructions to control the aircraft. An aircraft entering a new sector would report to the new controller on their specific radio channel, and in most cases no contact was needed between the old and new controller, as they both referred to the same flight plan from the departing airport. Significant deviation from the original flight plan or, for example, a detected collision risk in a neighboring sector would cause one sector's controller to contact the neighboring sector's controller, whether located at the same or in another control unit. If they belonged to the same unit, two controllers could easily make contact directly. If they belonged to different units, however, telephone contact was necessary. In the latter case there was a risk that the line was busy.

To summarize, in the 1950s, national air-traffic control networks were based on several separate infrastructures – telephone, telex, and various radio communication networks, civil and military, for communication with aircraft pilots. Further, the control units were located at the radar installations because of technical problems related to transmitting analog radar images. The work of air-traffic controllers had two sets of limiting factors. First, controllers had to use their power of judgment to distinguish the various altitudes of the aircraft in order to be able to foresee potential collision risk based upon their identification of dots on radar screens with flight data on flight strips. Identification was further complicated by irrelevant radar signals on the screen and weak representation of planes located immediately above the radar antenna. These technical limitations reduced the number of flights they were able to monitor and guide. Second, air-traffic control services were organized separately in every country. In Europe, this complicated the task of handing over a flight from one country to a neighboring air-traffic control service.

Establishing Eurocontrol

The introduction of civilian turbo-propeller and jet-propulsion aircraft in the mid- and late 1950s challenged the established air-traffic control structure in two ways. In contrast with traditional propeller craft, these craft reached their optimal cruising altitude in upper airspace, where only military craft had operated so far. In addition, their faster speed challenged the established organization and practices for air-traffic control in Western Europe, particularly the division into separate national airspaces. High-speed craft quickly passed through the airspace of small countries. This was expected to become a serious problem particularly after the introduction of even faster, supersonic aircraft, which producers started considering in the late-1950s (though the first Concorde test flight did not take place until 1969). Growing volumes of civilian air transport and several midair crashes in the United States provided additional motivation for the endeavor to improve Western European air-traffic control.[9] These problems were regularly discussed by government representatives at meetings of the International Civil Aviation Organization (ICAO). ICAO had been established in 1944 as an agency of the United Nations, superseding an organization founded in the interwar period. The establishment of ICAO was based on the perception that airspace was a sphere of national sovereignty regulated by national governments. Airlines were organized in the International Air Transport Association (IATA), established in 1945, and only influenced air-traffic control issues through national governments.

In the mid-1950s, ICAO responded to the problems of air-traffic control for the upper airspace by advocating the full integration of air-traffic control systems for the upper airspace across existing national borders. In 1958 the Federal Republic of Germany (FRG, West Germany), Belgium, Luxemburg, and the Netherlands agreed to establish a common civilian air-traffic service for their upper airspace and invited adjacent countries to join the project. France and the United Kingdom accepted.[10] These countries devised a plan for a common organization and established the International Convention relating to Cooperation for the Safety of Air Navigation, Eurocontrol, which they signed in 1960.[11]

The original convention's objective was to reduce risk in civilian air transportation by creating an international control organization that would be entirely responsible for civilian traffic in the upper airspace in Western Europe with due regard to the requirements of national defense, which was essential in the midst of the Cold War. National governments were to keep control of lower airspace used for landing and takeoff. All founding members were also NATO members. The air-traffic service agency, later called Eurocontrol like the convention, was to provide civilian air-traffic services for the upper airspace. Eurocontrol was to be managed through a board of directors (the Permanent Commission) with representatives from all member governments, and a management committee, also made up of representatives from all member governments. Member states had votes weighted according to gross national product (GNP), assigning large votes to large states. It was decided that Eurocontrol's headquarters were to be located in Brussels.

Eurocontrol had an implicit objective of advancing the integration of Western Europe, constituting one of many attempts to integrate Western Europe after the Second World War. Eurocontrol can be said to have belonged to a group of less conspicuous public and private projects that contributed to a "hidden integration" of Europe and thereby complemented the more well-known (and more well-researched) high-profile projects, such as the European Economic Community (EEC).[12] Eurocontrol's founding members were different from the original EEC members: five countries were both founding members of EEC and Eurocontrol, while the United Kingdom joined the EEC only in 1973. Eurocontrol's founders appropriated "Europe" for the new organization's name. Like the founders of the EEC, the European Coal and Steel Community and the European Atomic Energy Community, the founders of Eurocontrol tended to equate "Europe" with Western Europe.

Computer-mediated radar and the Maastricht Upper Area Control Centre

An original objective of Eurocontrol was to establish common control of the upper airspace of Belgium, Luxembourg, the Netherlands, and West Germany. This linked Eurocontrol to ongoing efforts to attain better radar images through technical innovation. In 1964, Eurocontrol decided to establish an upper airspace control unit at Maastricht for the northern part of this area, covering Belgium, Luxembourg, the Netherlands, and the northern part of West Germany (the border was drawn between Cologne and Frankfurt). The Maastricht Centre would replace nation-oriented upper airspace control units in Brussels (Belgium and Luxembourg), Amsterdam (Netherlands), and Hanover (northern Germany).[13]

When discussing the situation in the 1960s, it is essential to note that the FRG was a completely different country than the Germany of today. As a result of the Second World War, West Germany had been occupied by the United States, Britain, and France, and this occupation lasted until 1992. The allied occupation powers imposed extensive restrictions on air-traffic operations over the FRG, and Allied air-traffic organizations operated the three air-traffic corridors linking West Berlin with West Germany.

In the late 1950s, the German air force funded R&D projects at Siemens & Halske, AEG-Telefunken, and other companies to sharpen the quality of analog radar images, with the aim of improving the monitoring of air traffic. The air force discarded analog amplification and "extracted" aircraft images from irrelevant radar signals ("shadows"), such as clouds and unwarranted responses to different interrogation signals. This extraction and amplification was achieved by converting the analog images into digital images, which were manipulated by computer. Software for this task was developed in the years around 1960, based upon predictions of each aircraft's movements, which constituted an early example of software-based image processing.[14] When the Eurocontrol project started, Germany provided this expertise.[15] This contributed to the project's sophisticated mathematical and computer ambitions, merging radar signals from several radar units into a virtual model of the area's complete airspace. The project included extensive studies to formalize information about the behavior of all civilian aircraft types in order to facilitate the virtual model's prediction of their behavior while airborne, which enabled the linking of several radar dots and the projection of their flight during the next several minutes.[16]

Eurocontrol used this as a basis for designing the Maastricht Centre's computer system, incorporating results from projects at the US Federal Aviation Agency.[17] A complex and advanced information technology system should facilitate the air-traffic controllers' willingness to rely on the computer-enhanced radar images. The computer generation of radar images and all other elements was tested and the system was designed at great expense to enhance reliability, in many instances through the duplication of facilities. The system processed flight-plan data from departing airports and radar data from several radar stations to produce digital radar images that were distributed to air-traffic controllers' working positions.[18] The flight-plan data from departing airports, particularly the time, were checked and updated, and a computer printed flight-control strips for use by the controllers. This produced strips that included the time of the craft's entry into each control sector, which reduced the risk of the erroneous identification of craft.

At the five radar installations at Bremen (FRG), Hannover (FRG), Frankfurt (FRG), Leerdam (Netherlands), and Brussels (Belgium), computers converted the analog radar signals into digital signals and "extracted" aircraft radar signals from irrelevant signals. The digital signals were then transmitted to the main computer in Maastricht via two independent telephone lines to enhance reliability. The main computer at Maastricht merged the radar data from the five sites, correlated the outcome with flight plans, and produced a virtual model of the area's airspace that was continually updated. All aircraft in the upper airspace were detected by several radar systems, and the computer system integrated all incoming radar data so that the controller only had a single aircraft presented on their display. The flight plans facilitated safe identification of planes with radar images. Originally there were errors in the representation of the data on the screens, which were replaced, but the clarity and legibility of the data on the new screens was significantly preferable to traditional displays of analog radar data. Even if one radar system failed, the controller still got their information.

Air-traffic control was based upon primary and secondary radar systems. Primary radar systems detected only electromagnetic reflection from the aircraft, while secondary radar identified an aircraft based on information from a transponder in the plane. Normally the transponder would inform a secondary radar of the aircraft's identity and altitude, which the Maastricht system used to display as a label at the aircraft's current position showing its identity and altitude. To enhance safety, the Maastricht system also identified craft based on data from primary radar or from aircraft equipped with defective transponders. This substantially improved controllers' identification of craft and safety. Though secondary radar imparted much more information to air-traffic control, it was considered less reliable than the original – primary – radar. Air-traffic control systems therefore continued to use primary radar in addition to the new secondary radar.

To increase reliability, the Maastricht system had three identical main computers. The first ran the system for air-traffic control; the second was on standby, copying the operations on the first computer, and was immediately available at any failure. The third computer was off, and went on standby as soon as the second computer started actively operating air-traffic control. This double backup ensured non-interrupted operation, and allowed the third computer to be used for test work, as long as the first two computers were operating correctly. This provided a unique opportunity for the operational testing of programs developed by Eurocontrol's experimental center in France. The Maastricht Centre had a test sector with up to four working positions available for program tests. Controllers were assigned to these positions and tested the new programs without interfering with live air-traffic operations. The virtual upper airspace model in the main computer covered the Maastricht Centre's complete area; the parts of the model needed for the display of specific sectors were selected and distributed to the controller displays by several secondary computers.

Since industry did not yet have much experience with such advanced systems, Eurocontrol designed its own computer system for the Maastricht Centre. Various companies were able to build computers and displays, but Eurocontrol wanted a homogenous system and developed it through a dialog with industry.[19] In 1966, Eurocontrol invited tenders from a large number of European companies and received five tenders in 1968. Eurocontrol's technical experts preferred the tender from a consortium of Companie Générale de Télégraphie Sans Fil (CSF, France), Plessey Radar (UK), and Telefunken (FRG). However, they requested the Telefunken main computers to be replaced by IBM 360/50 computers, because only IBM was able to supply a large, proven computer and could provide reliable after-sales service, including the immediate availability of spare parts. Eurocontrol needed a 24-hour service guarantee to ensure prompt repairs in case of technical problems.[20] The British and French members of Eurocontrol's management committee objected to acquiring IBM computers because similar computers from European producers were, or would be, completed by the planned opening of the Maastricht Centre in 1972. But Eurocontrol's staff experts insisted on the use of

IBM computers. If other computers were to be used, the experts declared that they would not be able to guarantee safe and secure operation of the Maastricht air-traffic control system from its opening. The management committee with representatives from all members finally gave in, with the decisive stipulation that the IBM computers be built in England.[21] But the computers were designed in the United States and high reliability had thus won out over nationalistic concerns.[22]

During the development of the software in the period from 1969 to 1971, demand emerged for greater capacity on the central computers, for three reasons. First, Eurocontrol's experts originally underestimated the size of the software complex. Second, real-time programming languages became accepted tools for writing real-time applications, replacing assembler language. Assembler language varied from one computer model to another and maintenance was easier for real-time programming language programs, but these were larger than equivalent assembler programs. Third, air traffic grew at an unprecedented rate. For these reasons, Eurocontrol extended the main computers of the Maastricht Centre through several phases. Originally it was based on IBM 360/50 central computers, and first upgraded to IBM 360/65 computers (1970), and then to IBM 370/155 computers (1971).[23] Ironically, the original choice of the small IBM 360/50 computer had probably simplified the acceptance of acquiring IBM equipment because it cost less than its larger successors.

The design of the wide-area Maastricht Centre also entailed a new organization of air-traffic control operations in order to reduce control transfers between sectors and controllers and associated coordination tasks. The plan was for one large sector of Belgium, one large sector of northern West Germany, and one large Dutch sector. The large sectors would be controlled by teams, in contrast to the traditional mode of an individual controller operating each small sector. The team design was based upon results from air-traffic control research in the United States, where sector teams of one planning controller were assisted by up to four radar controllers.[24] The planning controller would plan the passage of an aircraft through the larger sector. A planning controller applied the fairly wide separation standards applicable at the planning stage and found a route for many flights before the plane entered their sector, which enabled the radar controller to control the flight with little interference. For some flights, the planning controller might be unable to resolve all potential conflicts, and he would then decide to allow the aircraft to continue with less separation than used for planning purposes, but under close surveillance by the radar controller. The radar controller maintained radio contact with the craft and monitored its progress along the cleared path, observed minimum radar separation, and watched out for the emergence of unanticipated conflicts. The arguments for this new organization were that it combined safety and efficiency, particularly by reducing the number of transfer of craft from one control sector to another. Finally, data processing at the Maastricht Centre included automatic updating of flight-plan data, computer printing of flight-control strips, and use of flight data to facilitate safe identification of planes on radar images.[25]

The complexity and cost of the sophisticated computer-mediated Maastricht Centre made its introduction different than the start of analog radar in air-traffic control in the 1950s. Once the Maastricht Centre assumed control of an area, upper airspace control operations moved significant distances from Brussels, Frankfurt am Main, and Amsterdam, which rendered it difficult to return to analog radar-based control at the original locations. This made it essential to reduce the Maastricht Centre's vulnerability. The system received several alternative computers, as we have seen, as well as other equipment to reduce vulnerability. This was costly but proved successful. In addition, Eurocontrol assumed phased control of its vast upper airspace, which made it possible to return operations to the former analog computer-based localities. The transfers were, eventually, successful. The success of the computer-mediated air-traffic control system was based on English as a well-established standard of communication upon which the Maastricht system was built from the outset. It established reliable standards of technical communication that remain valid even today.

In 1964, Eurocontrol had already assumed control of the upper airspace of Belgium and Luxembourg, operating at Brussels Airport based upon existing technology. In 1972 it successfully moved this operation to the new control center at Maastricht and inaugurated the core of the center's computer system. Two years later, Maastricht took control of the upper airspace of the northern part of the FRG. In 1975, Maastricht planned to take over control of the Dutch upper airspace. However, the Dutch national controllers rejected this transition, which included relocation of operations from Amsterdam to the far south of the country. This was a conflict between the organization's supranational objectives and national concerns, and the Dutch government proved to be unwilling to honor its commitment to transfer operations to Maastricht. This situation lasted until 1986, when control of the upper Dutch airspace was transferred to Eurocontrol in Maastricht.

West Germany decided to move its military air-traffic control of the northern part of the country to Maastricht in 1970. There it occupied the vacant Dutch control positions.[26] The military had two air-traffic control systems: air defense, which was linked with many other countries through NATO, and operational air-traffic control, which dealt with training fights and transport flights. Belgium, West Germany, and the Netherlands had separate operational air-traffic control operations, a trade-off between air-traffic vulnerability and the military perception of sovereignty. Luxembourg had no air force. At that time, operational military air-traffic control for the northern part of the FRG was negotiating with industry to acquire a new air-traffic control system. Eurocontrol invited the Germans to join Eurocontrol in the Maastricht Centre and assume the advanced, completed, and reliable working positions built for Dutch civilian control. The German military accepted the invitation, probably for cost reasons. Since then, Eurocontrol has colocated military and civilian air-traffic control, facilitating an informal reduction of vulnerability according to Eurocontrol air-traffic controllers in 2008,[27] but this does not appears to have been a reason for the German military to accept the invitation. The military air-traffic controllers were stationed at certain workplaces in Eurocontrol's control room, with their own military supervisor sitting next to Eurocontrol's civilian supervisor. They had practically the same information on their radar screens, including flight-plan data. The advantage for both military and civilian controllers was that they could work from the same system using the same information. This enabled them to make efficient common use of the airspace even though the cooperation remained informal and the civilian and military controllers remained part of separate organizational spheres. The Dutch and Belgian military refused to move to Maastricht.

The hesitation of the Dutch to move the air-traffic control of its upper airspace to Maastricht caused modifications to the planned introduction of larger sectors for organizing the control operations.[28] The absence of the planned large Dutch sector between 1975 and 1986 from the area controlled by Eurocontrol implied that there was little advantage in carrying out the initially envisaged cross-country planning. Instead, Eurocontrol established a planning control and radar control as a functional distinction in every sector. The planner did his work prior to the aircraft's arrival. When it arrived in the sector, the radar controller took responsibility, both when the flight followed the plan and when deviation occurred, such as with a late aircraft or an aircraft at a different altitude. When the Dutch eventually joined Eurocontrol in Maastricht in 1986, these established procedures remained.

The Maastricht air-traffic control system was a complex and integrated system for the four national air-traffic control units that it replaced (civilian and military). The computer-based facilities enabled improved safety, higher traffic density, and greater controller productivity. With traditional radar control, a controller had originally been able to control up to six or seven aircraft at the same time, while the Maastricht system allowed them to control 12–15 aircraft. The replacement of three civilian and one military air-traffic control centers linked through several separate infrastructures with an integrated system in one unit provided far better opportunities for coordination between the controllers operating in various sectors, to a large extent based informally upon colocation. The main limitation to the advanced Eurocontrol system was that gaining full advantage required standardizing air-traffic control across Eurocontrol member countries, as in the United States, which has 21 air-traffic control centers operating on the same system. However, Eurocontrol's member states were not willing to relinquish sovereignty.

Negotiations between technical success and organizational constraints

The momentum of creating the original convention in 1960 remained until it had been ratified by the member states and began operation in 1964. Subsequently, Eurocontrol experienced two divergent sets of developments: the successful establishment of an advanced, computer-mediated air-traffic control infrastructure, and the members' resistance to ceding sovereignty to establish joint control of the upper airspace of its member states.

Eurocontrol had been successful in building sophisticated, reliable air-traffic control units for the upper airspace at Maastricht in the Netherlands (opened in

1972), Shannon in Ireland (1975), and Karlsruhe in West Germany (1977) that could act as nodes of a united Western European air-traffic control infrastructure network. Eurocontrol had clearly advanced the technical capability of air-traffic control.

Simultaneously, the Eurocontrol project reached a deadlock with regard to establishing supranational control of the upper airspace, or even a common structure for this airspace. Originally, Belgium, the FRG, Luxembourg, and the Netherlands had divergent lower limits for their upper airspace, with the lower limits at 25,000 feet (7600 m) in Germany and 20,000 feet (6100 m) in Belgium, Luxembourg, and the Netherlands. They did not accomplish a complete harmonization by the opening of the Maastricht Centre, though Germany reduced the lower limit of its upper airspace to 24,500 feet (7500 m).[29]

While Belgium, the FRG, Luxembourg, and the Netherlands pursued a common supranational solution from the outset, the United Kingdom and France went for national solutions and kept complete control of their upper airspace. The official reason was national sovereignty, but requests from national unions of air-traffic control personnel to keep control of these jobs were a reason as well.[30] The British government opened the discussion as early as 1963 by stating that it would keep full control of the civil air-traffic control in its upper airspace in order to facilitate coordination with its military air operations, referring to the vague reference to military operations in the original Eurocontrol convention. The action seems not to have been coordinated with France, which jumped to support it.[31]

This split between countries pursuing the supranational agenda and the nationalists surfaced in the yearly discussion of Eurocontrol's funding. Originally the organization was funded exclusively by annual contributions from the member countries, which the members apportioned based upon GNP. In 1965, France complained about what it considered to be Eurocontrol's drain of the French national budget. The other members countered this with a suggestion to introduce a centralized system, in which a single charge per flight would be collected on behalf of the Eurocontrol member states and then reimbursed to the member states, which would also enhance Eurocontrol's organizational position. The member states decided to establish this system in 1969, locating the Central Route Charges Office at the Eurocontrol headquarters in Brussels. The system was introduced in 1971. It charged aircraft operators for each aircraft that used a given airspace, the exact cost being dependent on the distance flown and the weight of the aircraft. The proceeds financed nationally operated and Eurocontrol navigation aids, air-traffic control facilities, radar systems, and associated support and safety services.[32]

Subsequently, the existence of Eurocontrol-operated air-traffic control facilities was also questioned, this time because the plans for new Eurocontrol centers in Karlsruhe and Shannon encountered organizational problems. In 1967, Eurocontrol decided to establish a second upper airspace control center in Karlsruhe for the southern parts of the FRG. However, the German government conditioned its approval on the premise that the Karlsruhe center would control more than German territory, like the Maastricht Centre, which implied control of the upper airspace of eastern France. As this area was controlled from Paris, France rejected the suggestion. In response, the West German air-traffic authority argued that it should operate the Karlsruhe center if it only covered German territory. Accordingly, when Karlsruhe was key-ready in 1976, the German government requested the transfer of responsibility of the Karlsruhe center to its authority, which Eurocontrol accepted. The Karlsruhe center launched operations in 1977 and operated according to Eurocontrol standards, which remain in force today.[33]

Ireland applied for membership of Eurocontrol and joined the convention in 1965. It needed help to establish an air-traffic control unit at Shannon that would be able to manage the increasing transatlantic traffic leaving and entering European airspace over Ireland. Flights departing for North America were put onto specific tracks and instructed to reach certain levels and speeds before they began their ocean crossing. Accession was complicated, as Ireland was not a member of NATO like the six founding members.[34]

From 1969 on, Ireland received financial compensation from Eurocontrol due to services for transatlantic traffic, and in 1971 the organization decided to build a new air-traffic control center at Shannon to cope with the increasing traffic. Just as it did for the center at Karlsruhe, Eurocontrol drew on its experiences in building the Maastricht Upper Airspace Control Centre. The Shannon Centre was inaugurated in 1975. However, it was operated by the Irish national air-traffic authority, which was distinct from the Eurocontrol operations of the Maastricht Centre and the planned Karlsruhe Centre. Eurocontrol planned to take over the Shannon center in 1976, but right beforehand the Irish government notified Eurocontrol that it wanted to keep operation of the center. It argued that Eurocontrol personnel would replace Irish air-traffic controllers, which might cause conflict in Ireland. In addition, the Eurocontrol personnel would introduce higher salaries, which could cause demand for higher salaries in the Irish civil service.[35] The Irish government might also have been inspired by France and the United Kingdom's continuing operation of their own national upper airspace control. Indeed, Irish accession to the nationalistic line within Eurocontrol might have inspired the FRG's request the following year for national control of the Karlsruhe center.

In 1970, Belgium had already raised the issue of Eurocontrol's inability to implement the original convention's objective of common air-traffic control in the upper airspace. Subsequently, this issue was entrusted to a committee of the deputy members of the Permanent Commission, which discussed the problems for several years and by 1976 came to the conclusion that the main direction of Eurocontrol activity should be the less ambitious objective of promoting cooperation and coordination between the member states. In an era when air traffic grew slowly due to the economic crisis, it found that this would provide a sufficient framework for improving air-traffic control for the next ten years.[36] Eurocontrol as a facilitator of cooperation was accepted as realistic objective, and this became the objective for the second Eurocontrol convention, which was signed in 1981 and came into power in 1986 after ratification by the members.[37]

Growth through cooperation and central flight flow management

Eurocontrol's new function as a facilitator of cooperation changed the role of its technical infrastructure. Technical infrastructure remained essential for its member states' ability to handle the extensive growth in air traffic and, eventually, the expansion of Eurocontrol membership from seven members in 1980 to 38 in 2010. Eurocontrol's successful development of the software complex for the Maastricht Centre and the expertise gained from its implementation was essential for improving the capability of air-traffic control units across Western Europe, though Eurocontrol failed to implement its original objectives of operating all air traffic in the upper airspace of all member states. It established the successful new computer-mediated technology through cooperation with industry, which diffused the new technology beyond Eurocontrol's authority and established a new industry standard. When you compare the current system at Maastricht (which I had the pleasure to see in operation during a visit in October 2008) with the two previous systems at Maastricht since 1972, two observations are worth remarking on. Technical improvements to the computer-mediated air-traffic system have significantly enhanced the "radar" images and produced better functionality for the air-traffic control operations of the controller through features like touch screens and mice.[38] However, these latter improvements remained incremental compared with the revolutionary improvements achieved by the first Maastricht system over analog radar-based control. Further, two kinds of expertise remained with Eurocontrol after the Maastricht center opened in 1972: the expertise of operating advanced computer-mediated radar monitoring, and the ability to act as a forum for international negotiations on air-traffic control in Western Europe.

Back in the early 1970s, several members had perceived the implications of Eurocontrol's failure to establish a unified air-traffic control system across the member states for resolving safety and capacity problems in European air transportation. Further, these problems were aggravated by Eurocontrol's lack of success in attracting new members. Ireland was the only state that acceded to the convention between 1960 and 1988.

In 1979, Belgium, Luxembourg, the Netherlands, and the FRG – the Maastricht Centre countries – found that the key problem in Western European air-traffic control was the lack of coordination in its flight planning, which caused excessive delays in airports and in the air, resulting simultaneously in unused capacity in other locations. By then, these countries had developed extensive expertise in separating planning control and radar control operations at the Maastricht center, and they realized the limitations on controlling traffic in separate sectors. They attained low vulnerability but at a cost of an inefficient use of airspace, because the busiest sector determined traffic density. The four countries suggested solving this problem by introducing institutionalized planning of air traffic across Western Europe, beyond the existing Eurocontrol member countries, which they called "Air Traffic Flow Management." The purpose was to make the best possible use of the airspace for civil aviation by balancing demand and capacity, and centralizing air-traffic flow management in one location. Therefore the four countries suggested a technical solution to reduce delays without increasing vulnerability, which implied the surrender of limited national sovereignty by participating governments. The various national air-traffic control organizations were to surrender their flight-planning operations prior to departure, but the operational air-traffic control was to remain in the diverse countries and at the Maastricht Centre.[39]

This new conceptualization of efficient transnational air-traffic management was based upon the emergence of relational databases and software products that facilitated improved interactive computer operations by allowing several people to access the same data simultaneously.[40] The Maastricht software concept had been based upon a one-way process of digitally improving radar images and the establishment of a mathematical model of the airspace of the area controlled by the Maastricht Centre. The air-traffic controllers at Maastricht had successfully used the dynamic information in this mathematical model of the airspace as a basis for controlling the flights in their air sector. In contrast the proposed Air Traffic Flow Management system was based upon a simple structure. Like analog radar, it would add a new layer of infrastructure across the operational network of national air-traffic control organizations, which controlled live aircraft. The new layer would control the planned use of the airspace by future flights. Air Traffic Flow Management would be based upon data from all national air-traffic organizations in the area and a large number of air transportation companies (100 aircraft operators in 1995[41]), which they would enter into the system's database. In any case it proved more complex to collect and harmonize these limited sets of data than all of the data from a small number of radar sites, which slowed implementation.

In the late 1970s and the 1980s the growth of air traffic strained the system of national air-traffic control, particularly in the summer vacation months. This caused extensive and increasing delays for tens of thousands of vacationers. The key problem was the congestion of air traffic in central parts of continental Western Europe, which would probably have been a problem even for a unified system of closely linked air-traffic control units that covered half of the continent. The Maastricht Centre design handled flight planning separately for each sector. Against this backdrop, Eurocontrol discussed the suggestion by the Maastricht Centre consortium for centralized air-traffic flow management.[42] In 1987, Eurocontrol began preparing for the establishment of a Central Flow Management Unit by establishing a computer-based register of flight-plan information on every flight. In order to manage the flow of transnational air traffic, the central flow managers needed access to flight-plan information on every aircraft that was planning to fly in their airspace. Collection of data from several national air-traffic organizations and many airlines proved complicated, so that it took several years to establish coherent information in the database.[43]

In any case, a decision to establish a Central Flow Management Unit for a wider area than Eurocontrol's members required a wider forum. The Western European governments used the European Civil Aviation Conference (ECAC) for this purpose. They had established ECAC as an association of Western European governments in 1955 to promote the improvement of safe and efficient Western European air transportation.[44] (ECAC acts as suborganization to ICAO.)

In October 1988, ECAC decided to establish the Central Flow Management Unit for the airspace of Western Europe and asked Eurocontrol to establish the unit, because Eurocontrol was the only civil organization with operational experience of air-traffic control across several countries.[45]

By 1988, flight plans in Western Europe were coordinated through 12 databases that were located at selected national air-traffic authorities and covered separate airspaces across Western Europe. In the early 1990s, Eurocontrol consolidated these databases in five locations (Frankfurt am Main, London, Madrid, Paris, and Rome). In 1995, Eurocontrol established the operational Central Flow Management Unit at Eurocontrol headquarters in Brussels. Apparently to please France, Eurocontrol decided to keep the flight information database in two locations – Brussels and Paris – both of which had a copy of the full database for safety. Brussels managed the database containing flight-plan information on every aircraft that was planning to fly in the northern European airspace. Paris managed the database containing flight-plan information on every aircraft that was planning to fly in the southern European airspace. In 1995–1996 the flight databases in Frankfurt am Main and London moved to Brussels, and the flight databases in Rome and Madrid moved to Paris.[46] The Central Flow Management Unit in Brussels began providing flight-plan service and air-traffic flow management in a limited area (France and Switzerland) in April 1995. It progressively extended its area of operations, which by 1996 encompassed all of the 22 Eurocontrol members at the time (including three former Warsaw Pact countries). It handled an average of 20,982 flights per day in the summer of 1996.[47]

Based on the central database of planned flights at the Eurocontrol headquarters in Brussels, the Central Flow Management Unit issued a calculated take-off time for every flight, also known as "slot time" or simply "slot." Actually, the slot is a period of time within which take-off has to take place, defined between five minutes before and ten minutes after the calculated take-off time. When an aircraft cannot comply with its slot, a new slot must be requested from Brussels. The slot and any revisions are communicated to the aircraft operator as well as the air-traffic control unit at the departure airport through the Aeronautical Fixed Telecommunication Network. This was an already established worldwide system of aeronautical fixed circuits which comprised air navigation service providers, aviation service providers, airport authorities, and government agencies.

The success of the Central Flow Management project provided Eurocontrol with technical and organizational expertise that reached beyond the original Eurocontrol members. In 1987, Eurocontrol began developing the Central Flow Management Unit. That year, Eurocontrol only had seven members: the six founding members and Ireland, which had joined in 1965. Four Western European countries joined in the late 1980s, and today the convention has 38 members. It extends from the Atlantic as far east as Ukraine and Armenia, and covers all of Europe, except for Russia, Belorussia, and Iceland. Eurocontrol became a major agent in integrating the countries of Central and Eastern Europe into Europe.

The Central Flow Management project also implied new stakeholders. Originally, only the member state governments were stakeholders. In the Central Flow

Management project, aircraft operators and national air-traffic control organizations also became stakeholders. This reflected the liberalization of air traffic, which weakened the governments' links to aircraft operators and air-traffic organizations. In addition, new privately owned airlines emerged. The new stakeholders were included in the (new) third Eurocontrol Convention, which was negotiated between 1992 and its signature in 1997.[48] It also allowed for the expansion of Eurocontrol's authority to include airport taxiways and runways. This was a response to the changes in air-traffic management due to the reduction of the state's operational role in air transportation. Further, Eurocontrol explicitly became a civil-military intergovernmental organization, which reflected its new role after the end of the Cold War. The military had pre-eminence in the Cold War. Many military people contributed to the work in the board of member states and the management committee, but most discussions of issues of the relations between civil military flights took place outside Eurocontrol.

The third convention addressed the broader sphere of activities and additional stakeholders by advocating the establishment of a number of advisory bodies to facilitate the transparency of Eurocontrol's work beyond national governments, which included a Civil-Military Interface Standing Committee.

Dynamics of infrastructure control technology and governance

The development of European air-traffic control since the 1950s was based upon three revolutionary new technologies for monitoring air traffic and the establishment of trust among several Western European countries that these technologies could be used for air-traffic control to reduce vulnerability: analog radar, computer-mediated radar images, and the flight-plan database. Analog radar was developed in several countries by research institutions and the military in the interwar years and was applied to detect enemy aircraft during the Second World War. This technology was applied to civilian air-traffic control from the 1950s on.

The second revolutionary new technology was computer-mediated radar images, which constituted a significant improvement on analog radar technology. The West German air force began developing this technology in the late 1950s and Eurocontrol completed its development for civilian air-traffic control in the 1960s and 1970s. It was based on a digital approach that resembled the choice of digital technology for designing programmable machine tools in the United States in the 1950s.[49] Because the machine-tool industry was not willing to risk developing such a revolutionary technology, the US government financed the design of programmable machine tools through research and development projects at Massachusetts Institute of Technology and other non-government research institutions. Similarly, the producers of analog radar systems were not willing to risk developing a revolutionary technology, so the West German air force began developing computer-mediated radar technology in the late 1950s and chose to ensure its advancement along a similar path through contracts with Siemens & Halske, AEG-Telefunken, and other companies. Then Eurocontrol emerged in 1960 as a rich, government-funded institution and chose to continue developing this new technology itself. Dr Hansjürgen von Villiez, the first director of the Maastricht Centre, explains this choice with Eurocontrol's unwillingness to wait until industry could develop a similar technology.[50] In the 1960s the momentum of the project of a computer-mediated radar technology was essential for the Eurocontrol dynamic. We can speculate about what would have happed if the decision-makers had opted to wait until industry developed a technology. This would have delayed the completion of the technology by five or more years, causing the Eurocontrol project to fizzle out.

Eurocontrol succeeded in developing and operating significantly better technical monitoring of the airspace in the Maastricht Centre, opened in 1972. Its functionality and software were copied by industry and became standard. This improved the operation of the international air-traffic control network, but the overall assumption, namely that this would lead to an integrated, transnational air-traffic control infrastructure, proved false. Eurocontrol's role as an integration project lost momentum as a consequence of the organization's inability to extend its operations of air-traffic control beyond Belgium, Luxembourg, and West Germany by 1976. Eurocontrol's member countries were not willing to cede control of their national airspace. This caused Eurocontrol's objective to be reconceptualized from supranational to international, codified in the second Eurocontrol convention of 1981.

While the member states negotiated the new convention, new ideas for the international governance of air-traffic control emerged. The focus shifted from the efficiency of separate operational control in one sector to introducing the control of flight planning for all sectors, called Air Traffic Flow Management, which facilitated locating bottlenecks and rescheduling flights, or redirecting them through less crowded airways. When Air Traffic Flow Management was introduced, crafts no longer had to queue in circular waiting positions for tens of minutes before landing in crowded airports. Instead they were kept in their departing airport until clearance was granted for a straight flight to their destination, which made the sky safer and saved fuel. The Air Traffic Flow Management function added a new layer of infrastructure across the operational network of national air-traffic control organizations that controlled live aircraft. The new layer was concerned with future flights through a central database of planned flights in Eurocontrol's headquarters in Brussels. It added a new network, feeding the flight data computers in Brussels and Paris, and used the established Aeronautical Fixed Telecommunication Network for operational communications. Like the Internet, this telecommunication network was completely decentralized in the sense that a failure in one node did not cause the network to break down. The failed node was cut off, but communication between other nodes found alternative routes to its destination.

The Aeronautical Fixed Telecommunication Network was extremely reliable, but it was based upon the conception of each air-traffic control unit operating separately. The original Eurocontrol concept – implemented in the centers at Maastricht, Karlsruhe, and Shannon – focused on improving efficiency in every center, which reduced vulnerability. In contrast, the Air Traffic Flow Management concept was more vulnerable as it was based on a single computer in Brussels (and its always updated duplicate in Paris), but centralized information and the allocation of slots were seen as prerequisites for the efficient use of European airspace. However, a breakdown or failure of the Central Air Traffic Flow Management Unit would not cause the cessation of air traffic. Traffic would continue – though less efficiently because of the absence of overall planning – in the same way as air traffic might bypass the airspace of troubled air control units, as illustrated by the closing of the airspace of Yugoslavia during its wars of secession in the 1990s. Air-traffic controllers are always trained to operate when any of their technical facilities does not work, but this kind of operation always imposes limits on traffic density.

Since the 1950s, air-traffic control organizations repeatedly introduced new and improved technical systems – analog radar, computer-mediated radar, and a flight-plan database – to improve the monitoring and management of flights. Each new system was introduced because air-traffic control experts convinced the politicians on Eurocontrol's board of directors that it would improve the efficiency of the air-traffic control network without increasing the risk of failure. Analog radar and the flight-plan database were complements to a well-functioning system, and the controllers were always trained to operate a system with failing elements. The introduction of computer-mediated air-traffic control became Eurocontrol's first technical objective. Eurocontrol's staff experts saw it as more vulnerable because they saw its success as essential for the Eurocontrol organization's survival and growth. Computer-mediated radar was a success as a basis for the exemplary Maastricht Centre's air-traffic control infrastructure, but it did not suffice to realize Eurocontrol's supranational ambitions, so the organizational scope changed from supranational to international.

The new, reduced organizational scope became the basis for establishing central flight flow management in the late 1980s, built upon a register of scheduled flights stored in a large computer. A new, more flexible technology with different technical and organizational potentials transformed Eurocontrol from a technical agent of integration to become a facilitator of cooperation between air-traffic organizations. Before, integration had been based on the cession of air-traffic control in national airspace to a Western European authority, Eurocontrol. Now, flight-flow management had decentralized control. Slots were distributed by national authorities, while Eurocontrol controlled their implementation, which often comprised airspace in several countries. This was a more attractive model of integration in Western Europe of the 1980s and also appealed to the former Warsaw Pact countries in the 1990s because of less perceived infringement of national sovereignty across Europe. This made central flight-flow management a facilitator in expanding Eurocontrol's operations to encompass all of Western Europe (except for Iceland) and to contribute to Western Europe's extension to the east after the fall of the Soviet Union in 1991.

Notes

1. Accident Investigation Report, Civil Aeronautics Board: Trans World Airlines, Inc., Lockheed 1049A, N 6902C, and United Air Lines, Inc., Douglass DC-7, N 6324C,

Grand Canyon, Arizona, June 30, 1956, released April 17, 1957, http://www.doney.net/aroundaz/CAB_report_TWA-UAL_1956-06-30.pdf (accessed on February 7, 2010); "Grand Canyon Collision: The Greatest Commercial Air Tragedy of its Day!" http://www.doney.net/aroundaz/grandcanyoncrash.htm (accessed on February 7, 2010).
2. "Crash Victims Insured," *New York Times*, July 4, 1956, p. 13; "U.S. Official Lays Responsibility Before Collision to T.W.A. Pilot," *New York Times*, July 8, 1956, p. 1, "C.A.B. Chief Denies Fault was T.W.A.'s," *New York Times*, July 9, 1956, p. 8; "C.A.A. Defends Airway Control," *New York Times*, July 10, 1956, p. 21; "Air Rules Change Called Difficult," *New York Times*, July 14, 1956, p. 32; "Air Agency Bill Signed," *New York Times*, August 24, 1956, p. 48.
3. Public Law 85-726.
4. Rapport de présentation d'un avant-projet de conversation internationale de coopération pour la sécurité de la navigation aérienne "Eurocontrol", February 23, 1960, pp. 1-2, non-registered material on pre-1963 endeavors, Eurocontrol Archives.
5. LaPorte 1988, pp. 215-244; Resch 1994. Johan M. Sanne studied individuals in the systems: pilots and controllers. See Sanne 1999 and 2003.
6. Resch 1994, pp. 88-97; Hermansen and Strøjberg 1986.
7. I was unable to locate accessible material on the relation between civilian and military air-traffic control in several countries. Such archives seem to be classified.
8. "Chronological Survey of Eurocontrol," 1978, p. 67, box 1, series 21, Eurocontrol Archives; Resch 1994, pp. 92-97; author's interview with Hansjürgen von Villiez, June 2008.
9. See note 4.
10. Originally Italy participated in the project but left in 1960. Minutes of diverse meetings 1958-60 in box 166, archive series on Eurocontrol agency organization, Eurocontrol Archives.
11. Eurocontrol convention of 1960, box 1, series 24, Eurocontrol Archives; "Chronological Survey of Eurocontrol," pp. 1ff.
12. For example, Urwin 1995; Milward 1992; Eichengreen 2007; cf. Misa and Schot 2005.
13. "Chronological Survey of Eurocontrol," pp. 57-72; Permanent Commission meeting minutes, February 28, 1964, June 16, 1964, and October 19, 1964; "Report on the Results of Negotiations with the MADAP Consortium," WP/CE/57/10 of February 8, 1971; box 3-5, series 1, Eurocontrol Archives; author's interview with Villiez, June 2008.
14. Jessen 1963.
15. Working Group Automatic Transmission, "Draft report of 1st meeting at Amsterdam," March 11, 1959, Bundesarchiv (BArch) B108-24361; "Versuchsgerät für die Digitalisierung, Verarbeitung und Darstellung von Sekundär radar-Informationen," March 7, 1964, BArch B108-24358.
16. Folders "Meetings of the program, 1966," and "General, 1964," Box 144, Eurocontrol Agency organization archive series, Eurcontrol Archives.
17. Management Committee minutes, 50th Session, June 1970, pp. 9-10, box 74, series 45, Eurocontrol Archives; "Chronological Survey of Eurocontrol," p. 59.
18. Hansjürgen von Villiez, "Der Schritt in die europäische Flugsicherung," box 23, archives series on Euorocontrol agency organization; "Chronological Survey of Eurocontrol," pp. 57-72.
19. Folders CSF-Decca-Telefunken; Hollandse Signaalapparaten; both in box 38, Eurocontrol Agency organization archive series, Eurcontrol Archives.
20. WP/CE/36/11 (1968), series 45, Eurocontrol Archives.
21. Minutes of Management Committee, 36th session, July 24-26, 1968, pp. 11-29, box 52, series 45, Eurocontrol Archives.
22. Pugh et al. 1991, pp. 50, 167, and 213-214.
23. Minutes of Management Committee, 51st session, July 8, 1970, agenda item 2, box 74, archive series 45; minutes of Management Committee, 57th Session, March 8-9, 1971, agenda item 5, box 80, archive series 45, Eurocontrol Archives.

24. Minutes of Management Committee, 14th session, April 6–7, 1965, agenda item 3, box 16, archive series 45, Eurocontrol Archives; "Chronological Survey of Eurocontrol," p. 59; author's interview with Hansjürgen von Villiez, 2008.
25. Hansjürgen von Villiez, "Der Schritt in die europäische Flugsicherung", p. 4, box 23, archives series on Eurocontrol agency organization; author's interview with Villiez, 2008.
26. Minutes of Permanent Commission session 28, June 25, 1970, agenda item 3, and session 29, November 18, 1970, agenda item 3, boxes 26–27, archive series 1; minutes of Management Committee, 54th Session, October 20, 1970, agenda item 5, box 77, archive series 45, Eurocontrol Archives.
27. Heide, discussion with three civilian air-traffic controllers during visit to the Maastricht Control Center in October 2008.
28. Author's interview with Villiez, 2008.
29. Reduction of the lower limit of the upper airspace of the Federal Republic of Germany, WP/CN/24/11 of June 4, 1969, box 21; Permanent Commission meeting minutes, July 3, 1969, box 21; series 1, Eurcontrol Archives.
30. Author's interview with Villiez, 2008.
31. Permanent Commission meeting minutes, December 13, 1963 and February 28, 1964; "The Scope of the Eurocontrol Organisation and the Operational Activities of the Agency," paper presented by UK, WP/CN/V/2, September 23, 1963; boxes 2–3, series 1, Eurocontrol Archives.
32. Permanent Commission meeting minutes, October 26, 1965, January 28, 1966, June 27, 1966, April 4, 1967, December 7, 1967, April 3, 1969, July 3, 1969, October 1, 1969, November 18, 1969, March 19, 1970, and July 8, 1971; Examination of the general financial policy of the organization with regard to the positions of the Member States, working paper presented by France, WP/CN/12/2 of October 5, 1965; boxes 9–11, 13, 16, 19, 21, 23, 24, 26, 30, series 1, Eurocontrol Archives.
33. Permanent Commission meeting minutes, November 25, 1976; "Report on the operational, technical, financial and social implications of the proposal put forward by the Federal Republic of Germany for the future of the Karlsruhe UAC," WP/CN/48/19 of October 28, 1976; box 48, series 1, Eurocontrol Archives.
34. Permanent Commission meeting minutes, June 26, 1964, October 19, 1964, and December 15, 1964; boxes 4–6, series 1, Eurocontrol Archives.
35. Permanent Commission meeting minutes, June 26, 1968, July 3, 1989, November 18, 1971, June 22, 1972, May 15, 1975, and November 20, 1975; Letter from the Irish Minister for Transport and Power concerning the exercise of air-traffic control at Shannon UAC, WP/CN/45/6 of April 16, 1975; "Report by the Committee of Management concerning the financial implications of the operations of the Shannon Centre by Ireland," WP/CN/46/14 of October 22, 1975; boxes 18, 21, 31, 34, 43, 44, 45, series 1, Eurocontrol Archives.
36. Permanent Commission meeting minutes, March 19, 1970, November 8, 1973, November 21, 1974, May 15, 1975, and May 15, 1976; "Note presented by the Belgian Government," WP/CN/27/3 of February 17, 1970; "Third report by the study group of civil and military alternates to the members of the Permanent Commission on future activities of Eurocontrol," WP/CN/47/2, 3 of May 28, 1976; boxes 24, 38, 42, 43, 44, 47, series 1, Eurocontrol Archives.
37. Permanent Commission meeting minutes, July 8, 1980 and November 20, 1980, boxes 67, 72, series 1; Protocol amending the Eurocontrol International Convention relating to co-operation for the safety of air navigation of December 13, 1960, folder 26, box 3, series 24, Eurocontrol Archives.
38. Author's interview with Villiez, 2008; Villiez 1987; Paylor 2003.
39. Papers for agenda item 19: "Proposal for an international Air Traffic Flow Management System in Europe," WP/CN/54/4, October 29, 1979, box 64, Series 1, Eurocontrol Archives.

40. Bergin and Haigh 2009, pp. 38–39.
41. "Report on the 1996 Air Traffic Situation and on CFMU Implementation," p. 3, WP/CE/84/13, November 29, 1996, box 179, Series 1, Eurocontrol Archives.
42. Minutes of the 54th Session of the Permanent Commission, November 22, 1979, agenda item 19, box 64; Minutes of the 60th Session of the Permanent Commission, June 29, 1982, agenda item 3, box 86, Series 1, Eurocontrol Archives.
43. Minutes of the 73rd Session of the Permanent Commission, July 5, 1988, agenda item 5; "Weiterentwicklung der zentralen Datenbank für de europäische Verkehrsflussregelung," WP/CE/73/13, June 13, 1988; box 128, Series 1, Eurocontrol Archives.
44. http://www.ecac-ceac.org/index.php?content=presentation&idMenu=1 (accessed on July 1, 2010). After the fall of the Soviet Empire, the former Soviet satellite countries acceded to ECAC, which now encompasses all European states to the west of Russia and Belorussia.
45. Minutes of the 74th Session of the Permanent Commission, November 22, 1988, agenda item 4b, box 132; Minutes of the 79th Session of the Permanent Commission, December 3, 1991, agenda item 5, box 149; Series 1, Eurocontrol Archives.
46. Minutes of the 74th Session of the Permanent Commission, July 4, 1989, agenda item, 4, box 135, Minutes of the 79th Session of the Permanent Commission, December 3, 1991, agenda item 4, box 149; Minutes of the 82rd Session of the Permanent Commission, December 5, 1995, agenda item 3, box 171; Minutes of the 84th Session of the Permanent Commission, December 10, 1996, agenda item 3, box 179; Report on the 1996 Air Traffic Situation and on CFMU Implementation, WP/CE/84/13, November 29, 1996, box 179; Series 1, Eurocontrol Archives.
47. Minutes of the 176th Session of the Management Committee, June 28–29, 1994, agenda item A.2, box 329; Minutes of the 179th Session of the Management Committee, March 21–23, 1995, agenda items A.1 and A.2, box 334; "Evolution of costs of the CFMU since its inception," WP: CE 97/186/62, October 16, 1997, box 356; Archive Series 45, Eurocontrol Archives.
48. Eurocontrol revised convention, September 1997, folder, box 2, Archives Series 21, Eurocontrol Archive. This convention is not yet in force, because it has not been signed by all original signatory member countries.
49. Noble 1984; Reintjes 1991.
50. Author's interview with Villiez, 2008.

8
Connections, Criticality, and Complexity: Norwegian Electricity in Its European Context

Lars Thue

Introduction

> The NorNed cable has been out of operation since Saturday April 11, due to a fire at the Eemshaven converter station. No interruption in the supply of electricity occurred when the cable, which has a capacity of 700 MW and was transporting electricity from Norway to the Netherlands at the time of the fire, was taken out of operation. However, it is expected that no electricity transmission will be possible for several weeks at least.[1]

These were the words of a press release from the transmission system operator (TSO) of the Dutch electricity system, TenneT, on April 15, 2009. One month later a new press release stated that "from the 16th of May the cable will be back in operation."[2] Was this a critical event?

Complex information and communications technology (ICT) systems of computers, sensors, electronic devices, relays, and communication lines protected consumers from being blacked out. The control systems' flows of information controlled the flows of electricity. However, the concept of "critical event" refers to events with serious or critical consequences, and certainly there can be serious consequences without any blackouts. Norwegian power companies and the Norwegian and Dutch TSOs lost millions of euros during this single month. Dutch consumers had to pay higher electricity prices because of a smaller supply of electricity on the market. The winners were Norwegian households and companies, who took advantage of the reduced market price following the increase in national supply. Economically, the NorNed disruption certainly was a critical event (Figure 8.1).[3]

That the breakdown of NorNed cable had not only losers, but also winners, demonstrates that the construction of infrastructure, as well as its breakdown, involves multifaceted risks. The perceived social and geographical distribution of these risks influences the discourses on transnational infrastructure connections. The purpose of this chapter is to analyze such discourses in the history of Norway's transnational electricity connections. To actualize this historical narrative, I first give a short overview of recent discussions of such connections. These indicate what may be a new formative period in the development of a European grid.

Figure 8.1 At 580 km the NorNed link between Kvinesdal, Norway, and Eemshaven, the Netherlands, is the world's longest high-voltage submarine power cable. Here the cable is loaded onto the cable-laying vessel.
Source: Statnett. Used by permission.

Norway as the "battery of Europe"

"Super-grid gets super-serious, but does it rely too much on Norway?" read a headline in the *Financial Times* on March 8, 2010.[4] The article referred to the ambitions to create an offshore European "supergrid." Among the different proposals for such a high-voltage direct current grid, the *Financial Times* referred to ideas supported by an association known as the Friends of the Supergrid. This organization was supported by ten companies, among them Siemens of Germany and France's Areva. What was characterized as the most realistic phase 1 of the association's proposal was to connect Britain and other offshore wind farms and Norway's hydropower stations to the land-based European electricity network. Several positive effects of the joint offshore grid were identified:

> If such a super-grid existed, it would indeed have a dramatic effect on European power supplies, unlocking the potential of offshore wind, improving the resilience of the grid and reducing electricity prices by allowing much more international trading.[5]

In December 2009, then, nine European states signed a declaration referred to as the "North Sea Countries Offshore Grid Initiative." In February 2010, Norway joined this regional cooperation for the development of the supergrid.[6] The mentioned advantages of the supergrid all referred to the European level, and the initiative was supported by the European Commission. But what was in it for Norway?

The discussions of Norway's transnational electricity connection have always been closely related to the country's vast hydropower resources. Norway is the largest producer of hydropower in Europe. Currently its annual output is about 124 TWh, corresponding to 99 per cent of the country's total electricity production.[7] Together with Iceland, Norway has the highest per capita production of electricity in the world. A further 35 TWh could be added in the future by exploiting additional waterfalls, but environmentalists are critical of such initiatives and have often showed themselves willing to take action against new installations. Waterfalls and watercourses that could produced 45.5 TWh are already permanently protected.

Three main characteristics of the Norwegian hydropower system are of special interest for its transnational connections. First, hydropower stations are easily regulated. Second, Norway has a storage capacity for water corresponding to 82 TWh of electricity, or about half of Europe's total storage capacity. Third, the big variation in precipitation means that Norway sometimes has more and sometimes less than the average, resulting in a varied need for export or import. If seen in relation to the supergrid discussion, the large storage capacity combined with the easy regulation of production is highly significant. During peak load periods and in times with little wind blowing in the North Sea, Norwegian power companies are able to increase production and sell it at high prices. In windy periods with low consumption, the companies are able to buy electricity cheaply and store their own water. Bård Mikkelsen, chief executive of Norway's state-owned power company, Statkraft, explained that "hydropower in Norway should be valuable for compensating for the irregularity of wind power. That position – being a swing producer to the European market – is a very important role for us." Or as one commentator in the *Financial Times* article put it more bluntly,

> When the wind farms are running near to capacity, the system marginal price will fall to near zero – Statkraft will be able to buy power for storage at next to nothing. When the wind farms are not running, we'll see huge price spikes – during the cold snap we saw half-hours when power was being bought at £300/MWh. And, that's when Statkraft can sell the stored power. Nice business if you can get it.

For the same reason, pumped storage power stations are also regarded as an interesting option for the future of Norwegian electricity in its European context, and several are under discussion. The country's extraordinarily large capacity for storage has resulted in the concept of Norway as the "battery of Europe." In addition to the supergrid, several planned cables to the Continent are planned by Statnett, the Norwegian TSO, and some of the bigger power producers. These cables are also important for a planned net export of electricity. Norway's implementation of the European Union's (EU's) renewable energy directive is expected to increase the country's power production. Transnational connections will be necessary in this context to prevent a substantial fall in domestic electricity prices.

The differing points of view on the construction of a supergrid and transnational cable connections are very much connected to the perceived distribution of risks. The big Norwegian power producers naturally support the supergrid project, but also transnational connections in general, given the possibility for the huge profits promised by operating as "swing producers." In addition, the prospect is that Norway and the Nordic countries will have a large power surplus in the years to come. Without new cables or a supergrid, this increased supply will substantially reduce the market prices in Norway and in the Nordic market. Power exports will prevent this from happening. As the most probable co-owner of international cables or a North Sea supergrid, Statnett would have substantial income from the use of the transnational connections. The electrotechnical industry, like the Norwegian cable producer Nexans, also supports the construction of new transnational connections. All supporters use the climate threat and the need for new renewable energy as a main argument.

However, representatives of power-intensive Norwegian industry are critical. These companies export semimanufactured goods, such as aluminum bars and ferroalloys. The industry consumes about a third of the country's electricity, traditionally at very low and partly subsidized prices.[8] Many local communities are crucially dependent on it and therefore vulnerable to change. Certainly, other companies operating in other businesses and many ordinary consumers are also critical of the ongoing integration with the broader European electricity market, fearing continental electricity prices in their cold and dark environment. Many households are worried by possible price increases linked to the subsidization of offshore wind power through taxation. However, such consumers are neither organized nor well informed, and some are receptive to the powerful rhetoric of new transnational connections as saviors of the world's climate.

I will come back to these contemporary policy questions at the end of this chapter. In the next section I present some concepts and approaches to help us understand the discussions about Norway's transnational electricity connections.

A conceptual model of transnational connections

In the following, Norway's transnational electricity connections are discussed with reference to three structural aspects: one technological, one institutional, and one concerning the perceived place of the country in its European context. These structural aspects are closely linked to each other. All three aspects have consequences for both the vulnerability of power supply and the distribution of risks. Together with the fluctuating economy, they constitute the central environment for the development of what might be called the connection discourses and the connection regimes. By a "connection discourse" I mean the structure and logic of policy discussions and negotiations, while a connection regime refers to the governance structure – or the rules of the game – surrounding an actual transnational connection. Figure 8.2 tries to make some of these relations explicit. In the following they will gradually be historicized, exemplified, and further elaborated.

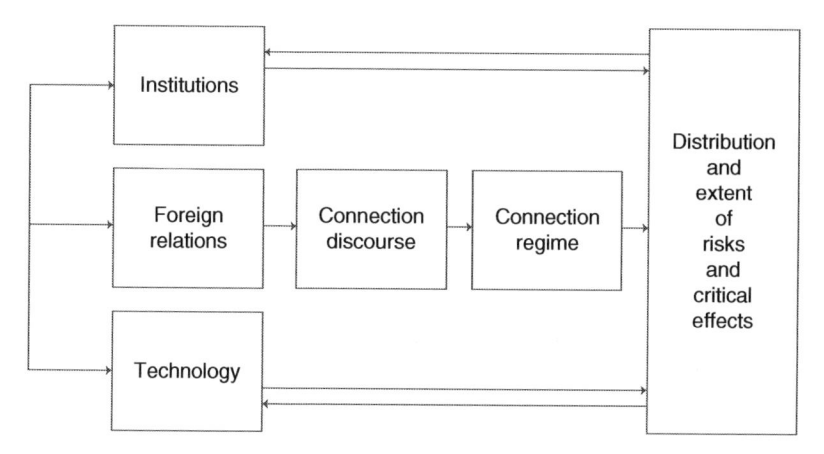

Figure 8.2 A model of influences in transnational connections.

Table 8.1 Periods of technological and institutional development

Approximate period	Horizontal integration	Vertical integration of control technology	Institutionalized political economy	Foreign relations as expressed in connection discourses
1880–1930	transition from local to regional systems	mostly integrated with the high-voltage system	unstable classical liberalism	from international to Nordic to European
1930–70	transition from regional to national systems	expansion of a low-voltage analog telecom system	expansion of a more coordinated market economy	from European to Nordic
1970–	transition from national to transnational systems	expansion of digital ICT systems	expansion of the neoliberal economy	from Nordic to European

In Table 8.1 I have tried to periodize the three structural aspects of the model. The technological part is divided into two columns: the electricity networks' horizontal and vertical integration. By horizontal integration I mean the geographical extension from local and regional to national and international networks. The main focus of this chapter will be on the parallel vertical integration, which refers to the power systems' increasing dependence on ICT-based control systems. Until the 1930s, during the transition from local to regional electricity systems, the control functions were rather tightly connected to high-voltage installations.

From the 1930s, when regional high-voltage systems gradually developed into national systems, the control systems were based very much on the extensive use of relays and analog telecommunications. The telephone became particularly important. Relays are cybernetic devices and introduce automation on a broad scale in the electricity supply. In the 1970s, in parallel with an increasing number of transnational connections, computers and digital data communication were transforming the control systems – with further and extensive automation.

The technological developments have interacted with broad institutional changes, both in the European capitalist economies generally and in infrastructural sectors more specifically. Roughly, we may divide the development of the Western European business systems, or political economies, into three main periods, from classical liberalism, over a more coordinated economy, toward versions of neoliberal economic organization.[9] Each period saw different styles of economic governance of the power-supply industry, with shifting balances between public involvement and monopoly on the one hand, and private involvement and market coordination on the other.

In the table, foreign relations, or the perceived place of Norway in the European context, are described by the changing orientations of Norway towards the Nordic and the broader European communities. The discourses on transnational electricity connections have been affected by the changing centers of gravity in Norway's foreign relations. A strong Nordic focus after the First World War replaced the more global orientation of the free-trade regime during the previous era. In the 1920s, this Nordic focus was followed by a broader European orientation. After the Second World War, this shifted again, and Norway worked actively for further Nordic integration. Even though Norway stayed out of the EU, from the late 1980s the country once again developed a strong orientation toward the rest of Europe.

Until 1960, when Norway's first transnational connection was established with Sweden, the three structural aspects in the model affected only connection discourses. Since then they have also influenced the connection regime and the governance of the actual transnational connections.

A history of increasing complexity

The complexity of both the technological and the institutional structures increased during the three periods in the table. Complexity is a relevant variable when considering risks, not least in the electricity industry. The most common hypothesis is that increased complexity leads to increased risks.[10] But there is not any simple, linear or direct relation between increased complexity and increased risks or vulnerability.

The main stages in the historical growth of this complexity can be described as changes in paradigmatic technologies. Simplified, the First Industrial Revolution, from the late eighteenth to the late nineteenth centuries, represented a shift in the paradigmatic technology from tools to machines. The second, from the late nineteenth century to about 1970, represented a change from machines to large technical systems. These large-scale systems posed complex problems of coordination and regulation that were met by institutional and organizational changes, such as "the visible hand" of big private organizations, direct state ownership, and extensive regulatory and standardization measures taken by national and international associations.[11] Complex institutions and organizations were deemed necessary to cope with complex technology. Perhaps "technological complexes" could be an appropriate term for the changes that we experience during the ongoing Third Industrial Revolution. Not only do we find larger systems but we also see vertical layers of systems on systems, with increasing interdependencies between them. A fundamental role is played by ICT-based control systems. These have the same large-scale extension as the operational systems that they control, but in addition they are characterized by an extremely high density of functional elements and microstructures, such as computers, microprocessors, sensors, optical fibers and other communication technologies.

James Beniger sees the Information Revolution as a "response to problems arising out of advanced industrialization – an ever-mounting crisis of control."[12] Industrialization resulted in "larger and more complex systems – systems characterized by increasing differentiation and interdependence at all levels."[13] Infrastructures, such as the power-supply industry, are among the core examples of such complex systems. The coevolution of these infrastructures' operational systems and control systems contributes to further complexity. Better protective devices, control and supervising technology, and routines make it possible to more safely expand the scope of the operational technology, increasing the complexity even further. For this reason the technological control systems installed to reduce risks have the side-effect of increasing risks. It is not easy to assess the resulting sum of risks.

We find the same ambiguity in institutional development. Organizational departments and offices for control and supervision make infrastructural firms more complex. Through their steady production of rules, the state and regulatory authorities further add to institutional complexity. Regulatory authorities normally also demand protective devices, control routines, the hiring of supervisory employees, and extensive reporting for the operation of infrastructures. To sum up, technological, organizational, and institutional complexities related to the core operational functions on the one hand and the control functions on the other evolve in parallel and in interaction. In the next part of the chapter, I analyze in greater depth the ways in which these structural aspects have interacted historically, and how they have affected the connection discourses and connection regimes – and the perceived distribution of risks and critical effects.

A hydropower nation and its transnational connections

Norway's many beneficial waterfalls provided strong incentives for electrification of the country early on, and the abundance of cheap hydropower also stimulated power-export projects. What seems to be the first Norwegian plan for a transnational electricity connection was made public in 1913. It was a private, purely commercial business project headed by two engineers, Hans Abel and

Fritjof Heyerdahl, who wanted to build a sea cable from Norway to Jutland in Denmark. Seeing no technical obstacles to such a project, they told the press that they held appropriate waterfalls in southern Norway.[14] We do not know why they did not succeed. At the time, the envisaged cable would have been technically very demanding. But even if had been technically and economically possible and the First World War had not broken out, the project would have been difficult to fulfill since it was launched during a period of institutional transformation. This transformation concerned both the Norwegian business system in general and the hydropower regime in particular.

The electricity sector was born during the era of classical economic liberalism. In 1885, the first Norwegian electrical utility, private company Laugstol Brug, was established in the city of Skien. In accordance with the liberal principles of the era, the Norwegian parliament (Stortinget) in 1887 approved a Watercourse Act that confirmed private ownership of watercourses and waterfalls. This was contrary to the situation in most of Europe, where the main rivers were used for transport and therefore subject to public ownership. The liberalist regime made Norway's hydroresources an object of private and international speculation. The possibilities for excellent reservoirs in the mountains combined with very high waterfalls, often close to the seashore, were attractive for establishing large-scale, science-based, and power-intensive industries.

The word *fossespekulanter*, or "waterfall speculators," became widely used. The speculators bought waterfalls at low cost from peasants and local people. Some buyers were also industrial entrepreneurs. They allied themselves with German, Swedish, or other foreign capital, and contributed to the building of electrochemical and electromechanical companies and power stations. They launched Norway's export-oriented, power-intensive industry, which is still important today for indirectly exporting a substantial part of Norway's power production.

The first electricity law from 1891 was called the "Law on measures to protect against dangers related to electrical installations," and it was meant to protect people and property from electrical shocks and fire. After a short period when the police were tasked with monitoring compliance with the rules, professional public inspectors were employed in 1898.[15] Norway's insurance companies had already agreed on common rules for electrical installations back in 1882, however. To have the installation properly insured, the companies had to follow these rules. This dual structure of risk management, relying on both market-based insurance and governmental regulations, is still in place today.

Both the regulatory authorities and the insurance companies prescribed quality requirements for electrical installations that influenced utility managers' choices and contributed to increased technological sophistication and complexity. Rules and institutions for approving types of electrical equipment developed through the mutual effort of the utilities and the authorities. In 1909 the Control Department in the Watercourse Administration was set up to supervise the quality and safety of the dams being built.

During the first decades of the twentieth century, state regulation and involvement gradually increased. In 1905, Norway declared its independence from

Sweden. In the following years, strong nationalism went hand in hand with a policy inspired by American politician, journalist, and political economist Henry George. His main message was that natural resources belonged to society.[16] Economic rent of land and natural resources should not be expropriated by private property owners.[17] George's economics inspired a "water movement" in Europe, and in many European countries the use of hydropower became strictly regulated and subject to duties. Influential Norwegian politicians, such as radical lawyer Johan Castberg, were members of the Henry George Association. As a reaction to the invasion of foreign capital, the parliament passed several bills to secure national and public ownership of hydropower resources. The majority of legislators wanted to limit the influence of big business and encouraged a widespread electrification of households, agriculture, crafts, and small industry. It was a battle between a "small-scale" and a more "large-scale" modernization strategy.

This policy was headed by the Liberal Party, especially by Prime Minister Gunnar Knudsen and Johan Castberg, who served as minister of justice in Knudsen's government during a critical period. The main opposition was found within the Conservative Party. In 1909 and 1917 the so-called concessions laws were passed by parliament, strictly regulating the ownership, development, and allocation of resource rent. In 1920/1 the government established the Norwegian Watercourses and Electricity Administration (NVE), both as a strong regulatory authority and as a major state-owned utility. With some important amendments, the concession laws from 1917 are still in place at the time of writing, and both their words and their "spirit" continue to influence current energy policy.[18] But they have often been contested.

Changing discourses on transnational connections[19]

1906–20 was a formative period for the institutions of the Norwegian electricity sector, but also for the transnational connection regime. From 1909 the early concession laws explicitly forbade export of electricity without government permission. This rule was restated in 1917.

The discussions about the concession laws took place during a long period of economic expansion. Good times created a strong national demand and need for electricity, and even if there was no ban on exporting electricity there were few economic incentives for Norwegian power producers or authorities to prioritize such projects. In 1918, however, the Norwegian government received an official request from neighboring Denmark to import electricity. The war had made it difficult and expensive for Denmark to import coal, and the country saw the possible import of cheap Norwegian hydropower as a solution.

The request from Denmark came in a period of growing "Nordism" – that is, strong loyalties between the Nordic countries. In 1919 the Nordic Association was established, working to promote friendship and cooperation among Sweden, Denmark, and Norway.[20] The most enthusiastic supporter of the power trade scheme was the chairman of the Confederation of Danish Industries, Alexander Foss, a prominent proponent of Nordic cooperation. Foss and the Norwegian premier, Knudsen, knew each other well, but Knudsen appeared to be in no hurry to comply with the Danish request. However, the quest for Nordic cooperation made it difficult to give a blunt refusal.

In 1920 the Conservative Party came to power. The party had argued for more liberal hydropower legislation and, although the war had now ended, the new government gave a positive response to Denmark's request. A joint commission was appointed to report on the matter. Sweden was represented, too, as a possible transit country for the transmission lines that would be necessary. In 1922 the commission's technical committee proposed transporting 42 MW through Sweden. However, the economic downturn after 1920, with decreasing domestic demand for energy and decreasing coal prices, dampened the Danes' interest in the project. In 1925 they finally decided to refrain from further negotiations. At this time many Norwegian utilities had surplus power and there was an outspoken interest in power export.

The course of events following the Danish initiative shows an economic logic that also manifested itself during later discussions about power export and transnational power connections. To establish a transnational link for power export, Norway and the importing nation should preferably be economically "out of sync," with an economic downturn in Norway and an upturn abroad. If both Norway and the potential importer experienced a boom, Norway wanted to keep the power to serve its own industry and consumers. And if both Norway and the potential importer experienced an economic downturn, Norway was interested in export but there was a lack of foreign demand. Given international economic interdependencies, however, the ups and downs in the economy rarely followed a different course in Norway than elsewhere in the Western world, and this posed a significant obstacle to export projects.

Technocratic hubris

Until the 1920s the operational and control technology for electricity installations evolved gradually, without any radical systemic leaps. At the start, the voltage of generators, the gate openings of turbines, and the circuit breakers and switches were operated directly and manually on the machines and in the powerhouse in accordance with fluctuating loads. In this way the control technology was directly integrated with the electrotechnical and mechanical equipment. That was also the case with the relays and circuit breakers, which were gradually introduced a decade after the turn of the century. The relays, by the way, may also be seen as a first generation of cybernetic devices in electricity networks.

To cope with the evolving complexities of growing numbers of generators, higher voltage, and more outgoing lines and cables, the controlling instruments – like the voltmeters and ampmeters and switches – were removed from the machines. They were centralized on an easily monitored control panel and sometimes in a special control room. At this time the control technology did not impress the general public. This changed during the 1920s through the regionalization of electricity production, transmission, and distribution. In the Norwegian

1930 edition of *The Great Inventions*, engineer Georg Brochmann wrote about the Norwegian Power Pool's newly established dispatch center:

> The load dispatching engineers at the Norwegian Power Pool stand as a symbol of technology in its highest stage: Networks of control centers connected with one another and to a main center, where one will direct the whole... The Power Pool is the glorious ideal for all organization... Such technological and centralized societal machinery will be able to work more efficiently, more economically and in all directions more satisfying than anything else – but is at the same time correspondingly more vulnerable.[21]

Interestingly, Brochmann related increased control to increased vulnerability. Increased control implied more interconnections and dependency on telecommunications. In the 1930s, a blackout occurred in Oslo because the machine operator at a power station was using the station's telephone to discuss affairs with comrades in the local Labor Party. The dispatcher at the Power Pool in Oslo observed the overload in the system but was unable to get through with his order to halt a generator. In such cases the impression of order and control at the Power Pool's headquarter vanished for a moment: "When an accident happened, telephones chimed, lamps flashed, the alarm boomed out, and if the lights went out the reserve generator started with an extremely loud sound." (Figure 8.3)[22]

The growing telephone network was at the start separated from the electricity system, since the dispatcher and the power station operators normally used

Figure 8.3 The dispatchers at the control center of the Norwegian Power Pool in the 1930s.
Source: Statnett. Used by permission.

the public telephone system. In the early 1920s, however, Norwegian power stations started to use carrier-wave telephony over the power lines, for telephones, teleprotection and telemetry.[23] The Power Pool began utilizing telemetry in 1932, and 15 telemetry installations with automatic measurements were in operation in 1940.[24] In this way a separate low-voltage, analog, electronic control system emerged with its own technological logic, its own professional competence and gradually its own organizational units. It was connected to the electricity system through several technical, organizational, and professional interfaces. The division of labor represented an advantage, but also a challenge for the reliability of the system. It depended on the capacity for professional and organizational "border crossings."

In James Beniger's terminology, the Power Pool's control room can be seen as a functional answer to a control crisis – a mismatch between the regional growth of the electricity system and the old, distributed control technology. According to Thomas Hughes, engineers in the 1920s increasingly used concepts such as "coordination," "integration," "control," "flow," "concentration," "centralization," and "rationalization."[25] There was a hubris linked to the belief in the benefits of large-scale projects – and in the prospects of controlling such projects. This hubris manifested itself not only on the national but also on the European level.

Initiatives for Norwegian power export

In his thesis "Electrifying Europe," Vincent Lagendijk examines the discussion about the integration of European electricity systems since the late nineteenth century.[26] The political idea of European unity gained momentum in the 1920s, and was related to both engineers' and businessmen's interest in creating a more integrated European electricity network.[27] In one way or another, the question of Norwegian power export was involved in most plans and visions. Norway itself also started to take an interest in a broader European electrification, and the far-reaching restrictions and strong governmental involvement in the hydropower sector were relaxed by the mid-1930s in a temporary liberalist turn.

In the discussions of the Nordic commission's work on a possible power export to Denmark in 1923, entrepreneur, engineer, and industrialist Sigurd Kloumann underlined that an export project had to be large-scale, taking especially Germany into consideration as a market. Kloumann further held that "we have to look at ourselves as members of the European polity... Let us look at the electrification of Europe as a dream."[28] In 1924, at the first World Power Conference in London, Kloumann was one of several Norwegian speakers. In his paper entitled "Export of Electrical Power from Norway" he referred to the concept of "super-power" supply that had been introduced in the United States. Kloumann would not at this time "enter into the question as to whether a similar idea would be realizable in Europe," but he thought that several countries could be supplied with power from Norway.[29] The transmission of a more or less constant load to northern Germany would be of special interest. Germany was relatively close and had a large enough demand to justify the cost of transmission. The Netherlands was also mentioned.

Kloumann's vision was well in line with that of other European engineers. At the World Power Conference in Berlin in 1930, the General Address by German engineer Oskar Oliven and the presentation given by a Norwegian delegation were largely in agreement with each other. In Oliven's long-term vision of a vast European electricity network, one of the power lines ran from Norway to Rome.[30]

The Norwegian delegation presented a comprehensive plan for transmitting 750 MW of power from Norway to Germany. Newspapers dubbed Norway "the powerhouse of Central Europe." At about the same time, a large consortium was formed to study the export question. State-controlled German power company Elektrowerke Aktiengesellschaft procured the most capital certificates and thereby had the greatest influence in the consortium. German subcontractors were also well represented. In addition to representatives of Norway's hydropower community, the most important Norwegian suppliers of hydropower machinery and electrotechnical materials were members of the consortium – along with Sweden's state-owned power company, Vattenfall, and the Danish state.

In the late 1920s the German economy experienced a boom, while the Norwegian economy stagnated. The two economies were out of sync and the prospects for export were ideal. In the early 1930s, however, the German economy stagnated as well and the export plans were put aside.

During the German occupation of Norway (1940–5), plans to exploit Norwegian hydropower for transnational purposes were an integral part of the occupying power's economic policy.[31] Power-intensive industry was to be radically expanded, mainly to supply light metals to the German armaments industry. One of the first tasks of the organization "Working Group for Expansion of Norway's Electricity" (Arbeitsgemeinschaft für den Elektrizitätsausbau Norwegens), established in 1940, was to encourage the export of electric power to Germany. However, German industry did not have the capacity to manufacture the electrotechnical equipment and materials needed to accomplish the transmission network. The Norwegians themselves were not supportive of these plans. Even some of the Norwegian Nazis referred to the concession laws in an attempt to secure the country's hydropower resources for domestic use. In the end, the German efforts to expand Norwegian power production and power-intensive industry yielded marginal results. In retrospect, however, it is clear that the German installations that started to be built became valuable assets for the Norwegian state after the war.

A "risk-free" and coordinated electricity supply

Just after the war the Danes once more approached the Norwegians hoping to import cheap Norwegian hydropower. This led to hard discussions both within Norway and between Norway and its Nordic and allied friends. The director-general of the Norwegian Water Resources and Electricity Directorate, Fredrik Vogt, fought for a long time against proposed plans for extensive power export to Denmark. Both the Norwegian and the Danish economies were growing fast, and national demand for electricity skyrocketed. Vogt and many others preferred a secure national power supply and to support the expansion of power-intensive industry by supplying low-cost electricity.

In 1954 the Norwegian minister of trade, Erik Brofoss, argued in a letter to Vogt that "if energy export could help bring the Nordic countries closer together, it would be a greater achievement than almost any conceivable application in the smelting industry."[32] Brofoss saw power export as an instrument to create a "consolidation" among the Nordic countries, partly as a response to what he regarded as the "great danger" in Europe – Germany – with its fast-expanding economy. Brofoss considered Vogt's one-sided national orientation to be at odds with his ambition to strengthen the Nordic community.

In 1955 the Norwegian parliament approved an agreement foreseeing an annual electricity export of 330 GWh to Stockholm. An important part of the deal was that the Swedes agreed to help finance the exploitation of the Nea watercourse in the county of Sør-Trøndelag. After several amendments, a 15-year contract was approved by the parliament in 1959. In 1960 the Nea power station and the transmission line to Sweden went into operation. After decades of ambitious plans and discussions, this was the first concrete case of the actual export of so-called firm power and of a transnational high-voltage connection with Norwegian participation. But the agreement did not become the starting point for a new trend. Until 1995, no further Norwegian power export deal was implemented. Instead, national self-sufficiency and the cross-border exchange of "occasional power," in combination with the provision of mutual reserves, became the golden rule for Nordic electricity cooperation. In 1963, an organization called Nordel (see Chapter 3) was created to govern this transnational power exchange on the margin.

In Norway, as in most industrialized countries, efforts to deal with the economic crises of the 1930s had resulted in a movement towards a more coordinated business system. In general, the Norwegian state's involvement in the economy was stronger than in most other European economies. Even though Brofoss and Vogt differed somewhat in their views on power export, they were both working in line with the Labor Party's ambitions to electrify the country and to build a strong power-intensive industry. "Power socialism" has been used as a term to characterize the party's industrialization strategy. Except for the period of German occupation and some minor interruptions, the Norwegian Labor Party was in power from 1935 to 1965. It also played a prominent political role in the following decades. The early 1970s represented a culmination of what has been labeled "the social democratic order" in Norway. During this period, most economic problems were understood as market failures that had to be compensated for by political action. Whether in politics, engineering, or the economy, the quests for coordination were an expression of "systemic" thinking and often explicitly inspired by different strands of system theory.[33] Norway also had a strong group of economists oriented toward planned economy, among them two Nobel laureates, Ragnar Frisch and Trygve Haavelmo.[34] Frisch, especially, saw computers and ICT as important instruments for tight control of the economy.[35]

The Labor Party was instrumental in establishing an extensive Norwegian welfare state. The welfare state is about the distribution of risks, and the Nordic welfare states were among the most developed in the world.[36] In 1981, Yair Aharoni published his book *The No-Risk Society* about the expansive welfare states. He starts with the following observation: "In all countries of the developed world, government is being used to reduce or shift the risk borne by individuals."[37] A new social order had evolved that "include[s] pressures on government to mitigate almost every risk any individual might be asked to bear." The welfare state had "turned into an insurance state": "We are insured against a variety of mishaps that range from earthquakes and other natural disasters, to poor health, unemployment, and the infirmities of old age."[38]

Even if the concept "risk-free" rhetorically overstates the phenomenon, there is something in it that also is relevant for our discussion of critical infrastructures and the question of Norwegian power export. First, the whole concept of national self-sufficiency indicates a "risk-free" inclination, a deep concern for the security of electricity supply. Until the 1990s the net electricity trade among the EC countries amounted to only a good 1 per cent of production. In Norway the situation was extreme, partly due to its resource base. The precipitation and thereby the supply of power could vary significantly between years and seasons. Similar variations in temperature created changes in the demand for power. For Norway to attain self-sufficiency in years with very low average precipitation and average temperature, it had to generate a large surplus of power in many other years. A rule developed for energy planning that Norway should be self-sufficient in at least nine out of ten years. Local and regional utilities also had to be self-sufficient; if not, they could join the national Power Pool and participate in the exchange of occasional power.[39] As a result, in the 40 years preceding liberalization in 1991, Norway was a net exporter during all years but two. The State Power Board, later named Statkraft, was given monopoly on the export and import of electricity.

Towards a digital shift...

The increased institutional coordination of the power industry was matched by a parallel and necessary development of the sector's control technology. Until the late 1960s, the expansion of control systems was mostly based on analog telecommunications using electronic vacuum tubes. However, the growing number of power stations, substations with transformers, circuit breakers, lines, and consumers stretched the analog control systems to their limits. There was a similar capacity problem with the transmission of information. The growing number of telephones in the Norwegian Power Pool resulted in the installation of telephone exchanges to reduce the number of communication lines. But the further growth both of telephones and of telemetry connections led to capacity problems, even though to some extent radio transmission both replaced and supplemented transmission over power lines.[40] For good reasons, Norwegian power and the electrotechnical industry started to develop protocols for digital data transmission in the early 1970s.

Gradually, however, remote control and automation made it possible to centralize the operation of several power stations into one local or regional control center. These control centers were connected either directly to the control center of the Norwegian Power Pool or indirectly through the utility's own energy-management center. This multilevel hierarchical organization, a vertical division of technology and labor, represented a new way of coping with complexity.

The operational complexity of the large-scale electricity system also created a need for appropriate complex control systems. At a Nordel conference in 1972, Swedish Vattenfall's chief engineer, Lars Gustafsson, gave a concise summary of the leap forward within the electricity sector:

> At the start, one managed the operations of the power system with paper, pen and a telephone. Since then, the need for information has increased, and today it is natural to base an information system on automatic data capture. We have been forced to do so because of the dimension and the complexity of the system.[41]

The 1972 conference witnessed the early years of the transformation from analog to digital control technology. Norwegian representatives talked about the computerized control of the Tokke power stations in southern Norway. Gustafsson presented Vattenfall's Totally Integrated Data System and Bengt Smith of the Swedish utility Sydkraft talked about a similar information system, DATABUS. The conference was primarily focused, however, on the integration of the Nordic electricity networks. This process obviously posed new problems for control and coordination. Gustafsson explained that if the Nordic system were to be operated as "one system," this would pose a "high demand on well functioning computer-to-computer connections between the separate operational centers."[42] It also demanded people with "huge theoretical knowledge" to develop application programs for production planning, security tests, and the like.[43] Gustafsson even discussed the prospect of a joint control center for the entire Nordic system. The Nordic interconnections put strong demands on control systems, and contributed to the introduction of more complex, digital equipment and computer-based modeling.

Up to the second half of the 1980s, parts of the electricity industry evolved in the direction of more planning. New ICT opened up for a planned operation of the whole hydropower system, a possible realization of the engineer's old dream of large-scale, centralized coordination. What might be called an ICT triangle of men, models, and machines emerged. In the early 1960s, work on an ambitious computer-based decision support model started as a collaborative effort between the Norwegian Power Pool, the Norwegian Electric Power Research Institute (EFI), and the NVE. The model was based on decades of hydrological and meteorological data on precipitation and the inflow of water in Norwegian rivers through the year. The main computer model was "EFI's Multi-area Power-market Simulator" (EMPS), called *Samkjøringsmodellen* in Norwegian. The characteristics of rivers, reservoirs, power stations, and power lines were modeled, today both in Norway and in the other Nordic countries.[44]

In parallel, an extensive system developed for the real-time monitoring of hydrological and meteorological data, like changing depths of snow and its distribution in different catchment areas. A variety of sensors and sensor technology was applied, including remote sensing by satellites. Data from the sensors were automatically sent to the control centers by several forms of telecommunications. The data were processed by the computer-based models to calculate the marginal values of the water in a country's reservoirs, in the short and in the long term. In this way the Power Pool planned the most economically efficient operation of the whole Norwegian system. A project called Norwegian Operation was developed in the second half of the 1980s. Since ICT also revolutionized the monitoring and operation of power stations, it was not that difficult to use the water values processed by the EMPS model for decisions on when and how to operate which power stations. In parallel, Nordel performed extensive studies on the possibilities to plan, develop, and operate the whole Nordic system. Why did these plans not materialize?

...and a neoliberal shift

Certainly, an international trend towards liberalization played a substantial role. From the late 1970s the dominant policy paradigm gradually shifted from the social democratic order toward a new "neoliberal order." Now, economic problems were generally interpreted as political failures that had to be compensated for by market-oriented reforms and "new public management." This political and institutional transformation developed in parallel with the growth of the "digital society," a transition into the Third Industrial Revolution.

As for the place of Norway in the broader process of European integration, there was a gradual shift from a focus on Nordic cooperation to a concern for Norway's relations with the European Commission. A parallel shift in social mentality and cognition both mirrored and influenced the different aspects of the transition. The reduced faith in centralized management and control in politics and business went together with increased sensitivity for the aspects of complexity, risk, and even the phenomenon of chaos in both technology and institutions. All of these more general trends were both reflected in and strengthened by the development in the power sector.

In addition, problems of "overproduction," which were already being debated in other branches of the economy, surfaced in the electricity industry. As we have seen, the Nordic connection regime was based on the principle of national self-sufficiency. However, Norway was the only country exclusively based on hydropower, and the only country that normally would have to export large quantities of power. There was a problem with the price tag for this power. Nordel's rule put the price between the marginal variable cost of the Norwegian hydropower production and the marginal cost of alternative production in the importing country. Since about 90 per cent of production costs in Norway were fixed costs, the price normally became extremely low. Some of the Norwegian power exported to Denmark and Sweden was transited on to Germany and other countries – at higher prices. In 1990, Norway used 99 TWh of electricity domestically, while exporting 16 TWh to Sweden and Denmark. The same year Sweden imported 13 TWh, mostly from Norway, while exporting 15 TWh to Denmark and Finland. Denmark imported 12 TWh while exporting 5 TWh to Germany.

By the late 1980s the "risk-free" strategy, in both Norway and the rest of the Nordel area, had thus generated a huge electricity surplus. Norway, in particular, no longer merely exchanged electricity on the margin with other countries. Nordel's annual report from 1987 stated that in a year with average precipitation there was a surplus of 60 TWh in the area. During a "wet year" the theoretical surplus was 100 TWh, to which Norway contributed heavily.[45] Economists concerned with "efficient allocation of resources" certainly saw the need for either the export of generated power at reasonable prices or, even more preferable, a liberalized market regime. Many environmentalists, having criticized the extensive development of Norwegian hydropower, supported the introduction of a market regime that would make the large power surplus visible. Politicians – not only conservatives but also some of the leaders of the Labor Party – supported the Energy Act passed by the parliament in 1990, making the Norwegian electricity sector the world's most liberalized. In 1992 the former integrated, state-owned power producer and network operator was divided into a transmission and system operator, Statnett, and a power producer, Statkraft (Figure 8.4).[46]

Figure 8.4 Statnett's national control center coordinates the operations of all players involved in the Norwegian main grid with its international connections.
Source: Statnett. Used by permission.

Exporting liberalization

An indispensable part of the successful liberalization of the Norwegian and later the whole Nordic electricity market was the creation of a new marketplace for trading electricity. Its birth was closely linked to the variation in and surplus of hydropower generation. Even back in the late 1960s, the director general of NVE, Vidkunn Hveding, had been critical of how the Norwegian Power Pool administered surplus hydropower. The pool had been given authority to dispose of the surplus in a rather centralized way. Hveding, inspired by economic theory, demanded that the pool establish a market for occasional power based on decentralized decisions of supply and demand. In 1971 the pool's "market for occasional power" was established, using advanced computers and data communication technologies. The various market actors could submit their bids online, and the computers automatically calculated the right market price. In 1993 the power exchange was formed as a wholly owned subsidiary of Statnett. In 1996 the Norwegian power exchange took on a new role as the Swedish electricity market was connected to it. Nord Pool, as the world's first international power exchange was called, expanded during the latter part of the 1990s to include Finland and Denmark.

Until the opening of the NorNed cable in 2008, the Nordic orientation had dominated the flow of Norwegian power. However, the broader European dimension was drawn into the discussion in the early 1990s. A key reason for this was a growing dissatisfaction with the Nordel regime. This resulted not only in a Norwegian interest in direct export to the Continent but also in active support for the liberalization efforts in Sweden, Finland, and Denmark: "From an economic perspective, a free Nordic and European electricity market is the best alternative for Norway."[47] This was the conclusion of a working group appointed by the Ministry of Petroleum and Energy, published in December 1990, only six months after the parliament had passed the Energy Act. The working group recommended that "the Norwegian Government actively go out and give clear signals to the Nordic countries that Norway supports the establishment of a Nordic electricity market as soon as possible, and that the Norwegian Government create favorable conditions for such a development."[48]

Certainly the Norwegian government, along with Statnett and the Norwegian utilities, worked on several levels to support liberalization in the neighboring countries.[49] The combination of these conscious efforts and the seemingly successful Norwegian liberalization was important for the Swedish parliament's decision on liberalization, implemented in 1996.[50] Finland liberalized fully in 1997. As argued by Johan Lilliestam, "the successful Norwegian liberalization and the decreasing price there – which were largely decoupled, however – were of major importance for the liberalization processes in Sweden and Finland."[51]

The most reluctant liberalizer, Denmark, found it impossible not to follow Norway, Sweden, and Finland, given that the EU had also started to take liberalization seriously, liberalizing in 1999 (western Denmark) and 2000 (eastern Denmark).[52] Finland joined Nord Pool in 1998 and Denmark in 1999–2000.

By the early 2000s the Nordic electricity market was seen as a model for further liberalization within the EU.

Liberalization meant both fragmentation and centralization. Competition authorities and sector regulators pushed for the organizational separation of production, transmission, distribution, and trade. Responding to toughening competition, however, a wave of horizontal mergers and acquisitions followed. Some observers considered vertical disintegration – especially the organizational separation between production on the one hand and transmission and distribution on the other – problematic for both reliability and economic efficiency. Competition and privatization resulted in the producers focusing more on profit and savings than on the security of supply and reliability. A lack of supply resulted in high prices, and this was often to the benefit of the power producers.

During the 1990s the investments in the Norwegian grid were historically low. ICT devices were introduced to expand the capacity without building new lines or cables. For instance, the capacity between eastern Norway and Sweden through the Hasle–Borgvik connection increased by 10 per cent for export and 33 per cent for import, primarily because of ICT-based system protection. The integration of the liberalized Nordic markets demanded stronger interconnections to become efficient. Interestingly, from 1989 to 2001 the transmission capacity between the Nordic countries more than doubled, from 4000 MW to 10,000 MW, most of it between Sweden and Norway, and between Sweden and Denmark.[53] This network expansion was closely related to the ambitions to advance the integration of the liberalized Nordic market.

Increasing flows of cross-border electricity within Nordel called for better communication between the Nordic TSOs. In January 1994 a separate telecom network for the Nordic power industry began operation. After several years of discussions and planning, in 2008 the TSOs were able to take advantage of a web-based information system called the Nordic Operational Information System. This served as an aid for balancing regulation with updated information on the Nordic electricity system, with real-time information on the cross-border flows of electricity.[54] In 2006 the Nordic TSOs signed an extensive agreement "regarding operation of the interconnected Nordic power system."[55] Part of this agreement was the use of the ICT-based Nordic Outage Planning System, a software tool for coordinating production outages.

Complexity and climate creating a "turning point"?

Some evolutionary economists and researchers think that the recent economic crisis represents a turning point. To get out of the crisis, they argue, it will be necessary to fully exploit the potential of information and communication technology.[56] There is also much talk about all sorts of smart and intelligent things and infrastructures, such as "smart cars," "smart transportation," "smart water systems," "smart houses," or merely "smart infrastructures" and "smart technologies" or, alternatively, "intelligent" devices.[57] The "smart grid," the "self-healing networks," the "Intelligent Grid" or just "Intelligrid" refer to the electricity complex, and some of the rhetoric seems to promis everything. For example, on the home page of the US Electric Power Research Institute (EPRI) we read: "EPRI's IntelliGridSM initiative is creating the technical foundation for a smart power grid that links electricity with communications and computer control to achieve tremendous gains in reliability, capacity, and customer services."[58] In the EU, the Smart Grids Task Force was established in 2009. Its aim is to advise the commission on policy and it will coordinate the first steps towards the implementation of smart grids.[59]

The work on smart grids and other new ICT-based investments can be seen as a response to a mismatch between operating technology and the existing control system. A significant share of the control crises relates to the combined forces of liberalization on the one hand and energy, environmental and climate policy in the EU on the other. First, these policies have created strong incentives both for more cross-border trade and for the introduction of decentralized production units, such as windmills, solar panels, wave energy, and small hydropower stations. These energy sources often have less stable and predictable output, which increases the demands on the control systems. There is an obvious need for the planned supergrid in the North Sea to be a smart grid as well.

Second, the emergence of several European power exchanges creates substantially more complexity in the industry's operation. The power exchanges, such as the Nordic NordPool, German EEX, and Dutch APX, are all heavy users of ICT. A World Bank report concludes: "The complexity of the energy exchanges, especially arrangements for pricing and settling commercial transactions on a real-time basis, increases dramatically when a competitive market regime is introduced."[60] These ICT-based control systems of the power exchanges include, among other things, trading systems, settlement systems, risk-management systems, and systems for fund transfers. In addition, the software on the computers of buyers and sellers must communicate safely with the power exchanges. The close relations between the functioning of the market system and the balancing activities of TSOs make it decisive not only that the power exchanges function as intended but also that the communication lines between exchanges and the TSOs work properly. Third, consumer rights and the aim to make efficient use of consumers' economic incentives, in combination with the need for energy conservation, have made the installation of smart meters an essential part of the smart grid projects.

The smart grid is closely connected to politics and institutional change. US President Obama has pointed at the smart grid as one of his main concerns.[61] His support for smart-grid development combines policy on climate, security of supply, and energy efficiency, but also a sort of modern "New Deal" politics of employment. We see some of the same tendencies in Europe. Significant parts of the existing grids are old, and would have to be renewed in any case. The implementation of the EU's climate and energy package, with ambitious targets to be met by 2020, certainly requires active public involvement, on both the EU and the national level. Combined with the challenges stemming from the 2008 financial crisis and its aftermath, there is indeed potential for more extensive coordination of the European business system. If so, there are indications that the electricity sector will be an integrated part of such a transformation. What about Norway's transnational electricity connections to the Continent in this context?

Norwegian power producers: Profiting on policy

The turn from the old Nordel to a liberalized connection regime has been very profitable for Norwegian power producers, given their very low costs of production. Out of self-interest, actors within the Norwegian power supply industry have also actively supported the further liberalization of the EU electricity market. Norway was actively involved in establishing the European Network of Transmission System Operators for Electricity and the European Regulators Group for Electricity and Gas in 2003. Norway's participation in these EU organizations is very much connected to the evolving integration of the structures of its electricity sector related to energy, information, and the market. Paradoxically, this institutional adaptation and integration also makes it possible for Norway to formally remain outside the EU without too much inconvenience. In parallel, from the early 1990s, numerous plans to lay cables directly from Norway to continental Europe have been proposed. Such cables would make Norwegian power producers independent of transit through Sweden or Denmark.

Some cable plans of the 1990s were cancelled by continental utilities when the European Commission started to push harder for liberalization of the electricity market, resulting in substantial compensation for its Norwegian contract partners. However, after more than ten years of planning, the auctions of transmission capacity in the NorNed cable began in May 2008. During the rest of that year, Statnett and TenneT earned €113 million each. Norway's electricity producers probably earned even more than Statnett because of the high electricity prices in the Netherlands.[62] Dutch consumers saved about the same amount, whereas Dutch generators were the main losers. The one-month break in transactions meant losses for the two TSOs amounting to €28.5 million, and the losses by the Norwegian producers and the Dutch consumers were also very large.

The perceived success of the NorNed cable has stimulated new cable plans. As of early 2011, at least five new cables between Norway and the Continent were under planning or serious discussion. Statnett has plans for new cables to Denmark and the Netherlands, but also to Germany and Britain. In the EU, such cables are welcomed. They would provide new couplings between the Nordic and the continental electricity market and thereby make the European electricity market more efficient. It was also argued that such cables dovetailed nicely with Europe's climate policy. As mentioned, Norway's large water reservoirs and flexible generation make Norway an excellent supplier of peak power and a "swing producer," contributing to the security of supply when windmills or solar energy are unable to meet demand. Clean energy from hydropower is seen as the perfect complement to the clean energy from the wind and the sun, as envisioned in the plans for the North Sea supergrid.

The largest and most ambitious Norwegian power producer, Statkraft, has a heavy portfolio of hydropower stations in Norway, but also some in Sweden,

Germany, Finland, and other countries. Statkraft praises itself as "Europe's largest renewable energy company."[63] It also has some wind-driven and gas-fired power plants. Both Statkraft and other Norwegian power producers make rhetorical use of climate policy to legitimize their plans for power export and their strategy of being a "battery for Europe." For example, the president and CEO, Bård Mikkelsen, states in Statkraft's 2009 annual report: "We ourselves are driven by the world's need for pure energy."[64] Statkraft is well aware of the large amount of subsidies for green power production within the EU, and the firm tries to take advantage of this. For the Norwegian power-intensive industries and their workers, in contrast, the cable plans represent a considerable risk for higher electricity prices, decreased international competitiveness, and possible factory shutdowns.

Statkraft's European strategy is also very much related to its well-developed ICT systems, not least for market operations. Having operated in a liberalized market since 1991, and being a co-developer of the financial market at Nord Pool, Statkraft is a very experienced market actor and trader. When investing abroad, it most often has a double motivation, both to earn money directly from the project and to get first-hand information about prices and markets. In addition to a tailor-made version of the mentioned EMPS model for the Nordic market, Statkraft has developed a special model for the fossil-fueled continental market. Statkraft sees these models, their extensive systems for information acquisition, and the experienced personnel operating them as a main competitive asset for its European operations.[65]

Transnational connections, risk, and critical effects

One aspect of the recent technological and institutional development of the electricity supply industry seems obvious: complexity is increasing. Horizontal integration in the form of transnational connections is one reason; vertical integration between the high-voltage system and numerous ICT systems is another. At a Nordic utility meeting in 1953, a Norwegian engineer stated: "In a way, the telephone constitutes the nervous system of the electricity network."[66] Since then, electricity supply has had not only a nervous system but something close to a brain, if not an ICT-based consciousness. Not only computers but thousands of intelligent electronic devices fill up the electricity system, like remote terminal units and programmable logic controllers with installed microprocessors networks. From sensors, these electronic devices receive data about voltage, current, temperature, pressure (in dams), frequency, water level, and snow weight. These data are checked by microprocessors, or are forwarded to the control centers. Since microprocessors may have close to 1 billion transistors coping with many billions of instructions per second, there certainly is a substantial amount of "intelligence" distributed in the network. The high-voltage macrocomplexes have many low-voltage microcomplexes inside them. These complexities also affect the transnational connections, and they call for agreements regarding information and communications standards in addition to the standards of the high-voltage system.

If there were an obvious and direct link between complexity and vulnerability, we would have reason to be afraid. In addition, the focus on both efficiency and profit inherent in liberalization and the possibility of cyberterror through hacking ICT systems have been said to increase the vulnerability of the electricity complex.[67] Transnational connections have also created institutional complexity, with several national regulators and TSOs operating on the same system along with a number of EU institutions. In coping with the economic crisis there is also a reasonable chance for a recurrence of the more protectionist and antagonistic Europe of the 1930s.

However, as we have seen, there is no obvious connection between complexity and vulnerability. Given the huge and increasing flows of electricity generated, transmitted, and distributed, the stability of the technological complexes is impressive. Technological and institutional complexity in the operational systems has been coped with rather efficiently so far through the addition of new complexities in the control systems. A report entitled "Vulnerability of the Nordic Power System" to the Nordic Council of Ministers from 2004 concluded:

> With respect to blackouts, the system is in a medium risk state. This is due to the fact that large blackouts in southern Scandinavia cannot be completely ruled out. Such blackouts involve many consumers resulting in major or potentially even critical consequences. However, this is not different from the situation before deregulation.[68]

The report in general saw "no indications that the situation will become worse towards 2010."[69] Norway, with the longest history of deregulation in the Nordic countries, had a sharp decrease in investments in the grid and other parts of the electricity network during the 1990s and early 2000s. Despite this the number of reported outages decreased in the years 1996–2001, and then remained stable in the years 2002–8.[70] The impressive development of control systems and protection devices seem to do their job properly for the most part. However, this relates to the more serious breakdowns of the networks. Local problems and outages have always been part of the game.

The history of electricity supply is certainly a story of increasing technological and institutional complexity. Transnational connections have contributed to this complexity. The result is an ambiguity and uncertainty concerning vulnerability, reliability and the distribution of risks and critical effects in this industry. And it is not possible in any strict sense to calculate the probability of the outcomes. We have genuine uncertainties, which are lately combined with the lack of certain knowledge about the effects and the nature of climate change. This knowledge vacuum gives an opportunity for strong economic interests to exploit the situation. The combined forces of a climate-industrial complex and a security-industrial complex might contribute to a huge redistribution of risks – and resources. The use of CO_2 quotas, enormous subsidies for "green energy," and large investments in grids for connecting renewable energy sources with low efficiency affect the distribution of wealth and tax burdens. The dominating actors in the discourses on critical infrastructures are consultancies, utilities, and researchers working for profit. To what extent are the warnings about and focus on vulnerabilities, criticalities, and risks influenced by their own self-interests? For power-intensive Norwegian industry and for Norwegian households, the cables to the Continent will probably result in higher electricity prices. Perhaps they might also have some favorable effect on our climate.

Notes

1. TenneT, press release, April 15, 2009.
2. TenneT, press release, May 15, 2009.
3. Economic risk is one of the aspects defining "critical infrastructures": "Critical infrastructures were originally considered to be those whose prolonged disruptions could cause significant military and economic dislocation" (Moteff, Copeland, and Fischer 2003).
4. http://blogs.ft.com/energy-source/2010/03/08/super-grid-gets-super-serious-but-does-it-rely-too-much-on-norway/ (retrieved on June 17, 2010).
5. Ibid.
6. Norwegian Ministry of Petroleum and Energy, press release, February 2, 2010.
7. About 19 per cent of power production worldwide is hydropower.
8. Facts 2008, p. 39.
9. Theoretical inspiration for this approach is primarily found in Whitley 1999 and Hall and Soskice 2001, and used on the Norwegian development in Thue 2008.
10. For instance, Perrow 1984/99. Perrow 2007 also deals explicitly with the power grid.
11. Chandler 1977 and 1990 give an account of how significant institutional and organizational changes took place within the private sector during the Second Industrial Revolution.
12. Ibid. p. 10.
13. Ibid. p. 11.
14. *Teknisk Ukeblad* 48/1913.
15. Jensen and Johansen 1993, pp. 38–9.
16. George 1884.
17. Thue 2003.
18. The early and successful regulation of the country's petroleum business from the late 1960s was to a large extent inspired by the model of hydropower regulation in this formative period.
19. Most of the narrative on the discussions of power export until the 1990s is based on Thue 1992 and 1994.
20. Later, Finland, Iceland, Greenland, the Faroe Islands, and Åland became members.
21. Brochmann 1930, p. 385.
22. Samkjøringen 1959, p. 76.
23. Schwartz 2007.
24. Telemetry is measurements combined with telecommunications.
25. Hughes 1983, p. 368.
26. Lagendijk 2008.
27. Ibid., p. 69.
28. *Teknisk Ukeblad* 49/1923.
29. Kloumann 1926, p. 131.
30. Lagendijk 2008, p. 83.
31. Thue 1994/2006.
32. Cited in Thue 1994, p. 64.
33. Thue 2008.

34. Ragnar Frisch was co-winner of the first Nobel Prize in Economics in 1969 together with Jan Tinbergen. Trygve Haavelmo won the prize in 1989.
35. Søilen 1998.
36. Esping-Andersen 1990.
37. Aharoni 1981, p. 1.
38. Ibid.
39. Skjold and Thue 2007.
40. Samkjøringen 1959 and Heggenhougen 1982.
41. Gustafsson 1972.
42. Ibid., p. 339.
43. Ibid., p. 341.
44. Thue 2009.
45. Nordel annual report 1987.
46. Nilsen and Thue 2006 and Skjold and Thue 2007.
47. "Etablering av omsetningsinstitusjon for kraft og organisering av eksport og import. Rapport fra en arbeidsgruppe," Olje og energidepartementet, December 7, 1990, p. 57.
48. Skjold and Thue 2007, p. 560.
49. Ibid., Chapter 11.
50. Högelius and Kaijser 2007; without explicitly drawing this conclusion, the authors many times refer to the influence of Norwegian liberalization, see pp. 89, 100, 122, 124, 132, and 161. Arne Kaijser also shared with the author notes from several interviews conducted with Swedish actors.
51. Lilliestam 2007, p. III. Lilliestam is a researcher at the Potsdam Institute for Climate Impact Research.
52. Petersen 2006 (an unpublished manuscript). Petersen has also been kind enough to share with me notes from interviews with central Danish actors.
53. Nordel's annual reports.
54. Nordel's annual reports.
55. "Agreement (Translation) regarding operation of the interconnected Nordic power system (System Operation Agreement)": http://www.entsoe.eu/fileadmin/user_upload/_library/publications/nordic/operations/060613_entsoe_nordic_SystemOperation Agreement_EN.pdf (retrieved on May 10, 2010).
56. Examples of relevant literature are Freeman and Louçã 2001, Perez 2002, and Drechsler et al. 2009.
57. The article "Smart Roads. Smart Bridges. Smart Grids", *Wall Street Journal*, February 17, 2009, sums up some of these "smart" concepts.
58. http://intelligrid.epri.com/ (retrieved on December 14, 2009).
59. http://ec.europa.eu/energy/gas_electricity/smartgrids/taskforce_en.htm (retrieved on May 14, 2010).
60. World Bank 2005.
61. For instance, http://online.wsj.com/article/SB125663945180609871.html (retrieved on August 10, 2010).
62. The Norwegian producer's earnings are estimated by people in Statnett.
63. http://www.statkraft.com/about-statkraft/
64. http://annualreport2009.statkraft.com/activities/president/default.aspx (retrieved on May 5, 2010).
65. Thue 2009.
66. Nordiske elverksmötet 1953, report C 7, p. 1.
67. For instance, Perrow 2007, Chapter 7.
68. Doorman et al. 2004, p. 1.
69. Ibid.
70. Fadum 2009, p. 12.

9
In Case of Breakdown: Dreams and Dilemmas of a Common European Standard for Emergency Communication

Anique Hommels and Eefje Cleophas

Introduction: Transnational collaboration and critical infrastructure

Communication networks for emergency services (police, ambulance, fire brigade) are among societies' most critical infrastructure. Before the 1990s, small local or regional analog radio communication networks were used for emergency services all over Europe. Emergency services used different frequencies, standards, and operating protocols. In most cases the systems were not standardized at the national level. As a result, cross-border communication between these networks was very difficult. The (legal) regulation of cross-border cooperation among emergency services has been intensified since the 1980s. Agreements were initially characterized by local, short-term arrangements between particular villages or cities.

Since the 1980s the arrangements to facilitate emergency collaboration have become embedded in a broader context of European agreements, such as the Schengen Agreements. These changes were accompanied by attempts to develop new operational practices for cross-border emergency communication. Two technological programs were launched to improve the interoperability of emergency communication networks in Europe: short-term Schengen and long-term Schengen. Short-term Schengen referred to a plan to coordinate the radio frequencies of border regions to make direct communication possible. Long-term Schengen aimed to develop and implement a European standard for emergency communication that would ideally be used in all European countries. This technical standard, originally called Trans European Trunked Radio (later Terrestrial Trunked Radio) or Tetra, could be embedded in national networks for public safety and would also allow for cross-border communication over long distances. In this way, Tetra would enhance the quality and efficiency of the collaboration between the emergency services of one country and between different countries. The claim, of course, was that standardization would reduce our vulnerability in case of a disaster.

Infrastructure theorist Paul Edwards and others see the use of standards or other gateways to link up isolated local systems into a complex "internetwork" as a defining characteristic of modern infrastructure.[1] Standardization and gateway-building typically involve a variety of actors, and social and technical elements that need to be coordinated and connected. Furthermore, standardization researcher Tineke Egyedi distinguishes between improvised and standardized gateways.[2] In standardized gateways the interconnection between local systems takes place through a (generic) standard, a uniform solution that can be applied in other situations across time and space. In improvised gateways, by contrast, the connection between subsystems is cobbled together in an ad-hoc fashion and not reproduced in other times or places. Until the 1990s, cross-border collaboration involved such "improvised gateways." The Tetra standard (currently still in development) counts as a "standardized gateway" that aims to connect diverse (national) emergency communication (sub)systems and networks into a transnational "network of networks."[3] Our study will show, however, that a strict distinction between standardized and improvised gateways may not hold.

This chapter investigates the implications of the Tetra standard for the vulnerability of emergency communication. It is not until emergency infrastructure breaks down or runs into trouble that its dynamics, vulnerabilities, and societal importance become visible at the surface. For this reason this chapter focuses on two critical events in the recent history of cross-border emergency communication. Our first case concerns the practices of emergency communication immediately after a huge explosion in a fireworks storage facility in the Dutch city of Enschede near the German border on May 13, 2000, which destroyed considerable parts of the inner city and became inscribed as a "national disaster" in Dutch collective memory. Our second case is a test of the Tetra standard, the so-called Three Country Pilot (3CP) that took place in the border area of Germany (Aachen), Belgium (Liège), and the Netherlands (Maastricht) in 2003.

These two cases represent two so-called Euroregions – that is, microregions intent on enhancing regional economic, social, and cultural collaboration across national borders. The Euroregion around the Dutch city of Enschede and the German city of Gronau was the first official Euroregion established in 1958. The Meuse-Rhine Euroregion around the cities of Maastricht, Liège, and Aachen was established in 1976. Even in the age of globalization, regional historians tell us, such regions remain an important category of identification and economic and social activity; Europe was and still is a continent of (micro)regions, many of which cross national borders.[4] This is particularly true for emergency services, which are organized on a local or regional basis in order to respond to local or regional vulnerabilities and calamities. These two Euroregions, then, will serve as "laboratories" for studying Europe's infrastructure vulnerabilities from a regional perspective.[5] They allow us to address such questions as: How did key actors try to coordinate cross-border emergency communication? How did the Tetra standard for emergency communication change existing practices and what were its vulnerability implications? How was standardization perceived to be a solution to both the European public safety coordination problem and the vulnerability problem? And how is "Europe" constructed, materially and discursively, in regional cross-border emergency communications? We base our case studies on in-depth interviews with key actors (who are anonymized in this chapter) in addition to extensive document analysis (e.g. reports, correspondence, and minutes of meetings).

Setting the stage: Postwar developments in cross-border emergency communication

Cross-border emergency communication in our two Euroregions has long been a matter of local arrangements. After the Second World War a collaboration between the police forces of the Netherlands and Belgium gradually emerged, particularly along the Limburg borders of Belgium and the Netherlands. In 1949 this cooperation was formalized in a legal arrangement that stimulated further contact. This included joint annual meetings, common border controls, and expansion of information exchanges. In 1969 a formal Dutch, Belgian, and German police cooperation was even established, called NeBeDeAcPol. Cooperation between Dutch and German police forces took shape more slowly than the cooperation with Belgium, however. This has been explained with reference to the German occupation of the Netherlands during the Second World War, which sparked a reluctance in the Netherlands to allow German police forces to cross its borders. Before 1960, hardly any structural contact between Dutch and German police units existed. In the most southern tip of the Netherlands, the Limburg area, "spontaneous contacts" between Dutch and German police chiefs emerged after 1960. These contacts followed the increased mobility of criminals in the area around Aachen (Germany) at that time.[6] Because of better infrastructure provisions and increased car usage, transnational mobility in general increased, and the same was true for transnational criminality.[7] In consequence, an increased perception of the common need to work together in a more structural way emerged. By 1973 a telex–mobile phone connection between local police dispatch rooms in the Netherlands, Belgium, and Germany had been installed. The contacts seem to have been based on mutual "workfloor" contacts between local police officials rather than government initiatives.[8]

This changed in the late 1970s, when the Dutch Ministry of the Interior established two committees to investigate cross-border cooperation in emergency services between Belgium, the Netherlands, and Germany. However, the committees identified several obstacles to such cooperation at the level of differences in administrative structures. For instance, the Dutch administrative unit of "province" did not match the German *Länder*. The committees therefore recommended that mutual consultation on both sides of the border be improved and cooperation organized at the lower levels of public administration – that is, municipalities.[9] As a result, information exchanges between emergency services across national borders increased: Dutch and German local police units started to exchange information about officers' responsibilities, their addresses, numbers of emergency response units, maps, and so on. By 1979 a bilateral Dutch-German agreement further specified which authorities should be warned in case of an emergency. In the 1980s, further legal arrangements between the governments of the Netherlands, Belgium, and Germany detailed further procedures for requesting assistance. They also stipulated that countries receiving help from another country would not have to pay the costs.[10] In the 1970s the focus of attention shifted from the legal to the technological coordination of emergency communication. The short-term Schengen program, for instance, coordinated radio frequencies for cross-border communication on the analog radio networks of emergency services in the border region.

The idea behind all of these efforts was that cross-border cooperation would improve the responses to disasters and criminality; joining European forces would reduce society's vulnerability. Our first case of the explosion of a fireworks storage facility in the Dutch city of Enschede (close to the German border), however, demonstrates that cross-border neighbor assistance in emergency response is not easy.

Emergency communication during the Enschede disaster

If we focus on the collaboration among firefighters in the eastern Twente region of the Netherlands along the German border, where the city of Enschede is situated, we again find a history of informal and uncodified cooperation. This often focused on the interoperability of the equipment and involved "improvised gateways" to coordinate technical dissimilarities.[11] For instance, analysis of the role of technology in cross-border collaboration firefighters in this region revealed at least three crucial differences: i) the technologies used to fight fire; ii) the communication technologies on the trucks; and iii) those in the dispatch rooms. According to the director of the Twente Safety Region, in the 1970s, firefighters from Enschede joined the Germans for training and soon discovered that the water taps used to extinguish fires (which are located underground) have a different standard in the Netherlands than in Germany. And that was a problem: "We could say, let's make a nice agreement for cooperation, but if you come together and the hoses cannot be connected."[12] They pragmatically solved this problem by installing double sets of apparatus in all fire trucks in the border region, and connectors of all types to make the equipment interoperable. Improvised gateways were preferred to a more extensive system change.

Second, as mentioned above, before 2005 the analog radio networks of the German and Dutch emergency services were coordinated in the short-term Schengen program. However, radio communication in the border regions was not always successful. For instance,

> direct radio communication of Dutch ambulances with the German *Leitstelle* [dispatch room] is impossible as Germany uses different apparatus and frequencies. Furthermore, German ambulances that assist in the Netherlands cannot communicate with the Dutch ambulances. In some border regions, people exchanged mobile phones to facilitate communication, but...this is not allowed.[13]

If a Dutch ambulance crossed the German border it could communicate neither with its own dispatch room nor with the German dispatch room. Nor could the two dispatch rooms communicate with one another. These challenges to cross-border emergency collaboration, among others, also came to the fore in the case of the disaster in Enschede – as one of the things that went utterly wrong in fighting this disaster was emergency communication.

The critical event of May 13, 2000

On May 13, 2000 a huge explosion in a fireworks storage facility in the Dutch city of Enschede had disastrous consequences. What began as a small outbreak of fire escalated into a disaster. The fireworks storage facility was situated in the middle of the city and, as a result, the consequences for houses and people living in the environment of the storage facility were enormous. In total, 23 people were killed (among them four Dutch firefighters) and 950 wounded. Some 200 houses in the direct vicinity of the facility were devastated. Another 300 houses were heavily damaged and declared unfit for habitation. In total about 1250 people lost their homes as a result of the disaster.[14]

> According to the chief commander of the city fire brigade at that time, at the moment of the big explosion that Saturday afternoon at 3:30 pm, formally we were no longer in charge, we had no communication, we had nothing... Our communication devices did not work any more... We didn't have an overview of the situation... And that is what happens in all such big disasters: People are just going to do something, without any central coordination.[15]

At a certain point there was a threat of a further escalation of the disaster when a cooling installation at a large, adjacent beer factory (containing ammonia) was about to explode. The chief commander said:

> We tried to spread the message that all emergency workers had to leave the disaster area. We didn't have Tetra, everyone had their own small analog network. Moreover, everyone interpreted the message "leave the disaster area" differently. One thought "I have to go outside the fences"; someone else thought "I have to go 10 kilometers away". This shows that giving a clear piece of information to the emergency workers, to tell them what to do and what is safe, is very hard."[16]

When people in the city of Gronau across the German border warned their fire brigade after hearing the explosion, the firefighters decided spontaneously to go there and offer their assistance. Of the 17 fire brigade units present one hour after the explosion, seven were German. The German chief of the fire brigade of the neighboring Kreis Borken was notified by his own superior. He knew exactly where to go, as he followed the direction of the huge dark cloud that was rising above the city of Enschede. Close to the disaster area he contacted the Dutch fire brigade commander and took his orders. He and his team stayed the whole evening until midnight and offered assistance in the suppression of flash fires.[17]

In reports about the disaster and in the accounts of witnesses such as firefighters, one of the key vulnerabilities in the emergency response to this disaster concerned communication technologies. The communication infrastructure broke down and could not cope, nor recover quickly. Half an hour after the major explosion, all available communication channels were overloaded. This applied to the communication networks of the emergency services, the dispatch rooms, the normal telephone lines, as well as mobile phone networks.[18] There were also misunderstandings about the appropriate radio frequencies to be used for emergency communication.[19] Because of this lack of communication, emergency response in fact took place without effective central command. Emergency workers improvised to the best of their knowledge and skills, using written notes and direct, oral communication. Emergency service dispatch rooms were overloaded as well and the dispatchers lost their overview of the situation.[20]

In explanations of this vulnerability, the actors involved tend to emphasize the need for on-site *coordination*. The next sections make clear, however, that different governmental structures, laws, cultures, languages, and operational tactics also have to be coordinated to shape a more resilient infrastructure.

Coordination as a solution to vulnerability

In their 1997 study "Disaster at the Border", Hertoghs and Rambach investigate the legal aspects of cross-border emergency collaboration in Belgium, the Netherlands, and Germany.[21] They show how fundamental differences between the governmental structures of these countries shape the way in which the emergency services are organized. For instance, different definitions of what counts as a catastrophe result in different organizational structures. In German law a catastrophe is defined as a situation in which the assistance that can be provided at the local or regional level is no longer adequate. In Dutch law a catastrophe is defined as a situation that requires the coordinated assistance of different emergency services (with the level, local/regional, not specified).[22]

The director of the Twente Safety Region also emphasizes that public safety is organized differently in Germany than in the Netherlands. An important recent policy development in the Netherlands is the subdivision of the country into 25 so-called safety regions. This new policy started from the belief that cities or municipalities were not capable of bearing the full responsibility for public safety. The region, officials now believe, is the level at which disasters have to be prevented and fought. At the level of the safety region, collaboration between emergency services has intensified over the past five years. These safety regions were inscribed in national Dutch law in 2010. This is not the case in Germany, according to the director:

> The [German] fire brigade is a traditional firefighters organization, whereas in the Netherlands we have seen the development of the fire brigade into an organization for crisis management. The Dutch fire brigade had a heavy

responsibility in preparing the multidisciplinary collaboration. In Germany that was absolutely not the case. There is a big gap between the different disciplines: police and fire brigade hardly do anything together. The police has to catch criminals and the fire brigade has to help and rescue people and that hardly ever comes together.[23]

This has not changed since the disaster in Enschede, according to him.

Barriers also exist in the legal aspects of emergency work across borders. One example of a legal barrier to cross-border collaboration, until a few years ago, was the prohibition against crossing borders with sirens and flashing lights, and that the police could not cross borders when armed. These rules caused trouble in 2000, at the time of the disaster in Enschede. According to the chief commander of the city fire brigade at that time, officially the cars and trucks of the German fire brigade and Disaster Relief (Technisches Hilfswerk) were not allowed to cross the Dutch border. After the disaster in Enschede, one of the German firefighters even had to justify himself before the German government for taking the (officially unauthorized) decision to send support units to the Netherlands on his own. The Dutch chief commander suggests that, off the record, people said that if this German firefighter had not given support at that time, he would have been reprimanded as well.

Our interviewees and the Hertoghs and Rambach study emphasize that the legal differences and the dissimilarities at the level of governmental structures make it harder to coordinate transnational efforts at emergency communication. They suggest that the process is hampered by the fact that policy-makers or local emergency workers don't know whom to address to make agreements. As legal agreements and governmental support are crucial preconditions for arriving at a clear coordination of the emergency communication infrastructure, these barriers have to be overcome first – as well as cultural differences.

One of the main contrasts between the German and the Dutch fire brigades can be described in terms of the different cultural roles that the two organizations fulfill. The German "fire brigade culture" is often described as *kameradschaftlicher* (based on personal friendships),[24] while the Dutch organization is characterized by its professional profile. In Germany the fire brigade fulfills an important social role.[25] It is important in the social life of communities – for example, many youngsters start their career as firefighters in the German youth fire brigade. It is an honorable job that people do for a lifetime. Moreover, many German firefighters are unpaid volunteers, which, surprisingly or not, accounts for the much larger number of German than Dutch firemen. For cities of similar sizes, the Germans have about 2.5 to 3 times as many firefighters at their disposal as the Dutch. However, a disadvantage of the German system is that they cannot always rely on their firefighters' availability.[26]

Another cultural phenomenon crucially important to understanding the cross-border collaboration in this region is the Dutch notion of *noaberschap* – the idea that you help your neighbors whenever they need it. This idea is shared on both sides of the border and played an important role during the disaster in Enschede.

Fire trucks from Germany were not officially called to the disaster in Enschede; instead they just came when they heard about it. According to the chief commander of the city fire brigade, this can be explained by the German insistence upon the notion of neighborly assistance: "In the Netherlands we have this system of calling for assistance. In Germany, yes, in a formal sense too, but there they are very charmed by the idea of neighborly assistance, so they just came." The fact that they came unannounced was a mixed blessing according to him:

> Uncoordinated action is the worst thing that can happen ... because you lose control, people take risks, there is no communication, certainly not when they begin spontaneously and use their own communication technologies. So on the one hand, you have to be very grateful that it happens – on the other hand, it is important to coordinate this in a different way.[27]

The cultural value of *noaberschap* that is so prominent in the communities on both sides of the border has – interestingly enough – ambiguous implications for the vulnerability of emergency response. On the one hand, *noaberschap* is seen as an important asset in the reduction of vulnerability, as people can rely on being rescued by their "neighbors" when they are in trouble. On the other hand, the commander's account of the disaster in Enschede shows that it is a "mixed blessing": if the neighboring emergency services turn up without warning, this results in a situation that is very hard to coordinate – something which makes emergency response more vulnerable.

Although looking at legal differences, dissimilarities at the level of governmental structures, and cultural differences already reveals the difficulties of coordinating various aspects of emergency collaboration, an analysis of the way in which emergency services *operate* together also raises challenges to collaboration. After the disaster in Enschede, a report concluded that there are some crucial differences in the working procedures of the Dutch and German fire brigades and that this can seriously hamper collaboration: "Not all employees of the fire brigade were sufficiently aware of the organization, mandates and operational procedures of the emergency services on the other side of the border. This hampers operational collaboration in practice."[28] One general difference seems to be the existence of more strict and more explicit regulations and protocols in the Netherlands. There is also more uniformity in the way in which the Dutch fire brigade operates compared with the German fire brigade.[29]

Moreover, the operational tactics used to attack a fire differ. In the Netherlands, in most cases firefighters enter a building to go to the spot where the fire is, to fight the fire from the inside. This tactic is called "the inside attack" or "offensive firefighting." In many other countries they fight the fire from the outside, using a lot of water ("defensive firefighting"). In contrast, the Dutch "want to locate the source of the fire as precisely as possible, which is not without risk. But it is a lot more effective."[30] In Germany, firefighters have changed their tactics over the past few years and now use a combination of the inside and outside attack.[31]

Our Enschede case suggests that not only do governmental and legal barriers have to be overcome for cross-border emergency collaboration (and communication) to work; cultural, and operational practices have to be aligned as well. Its "working" seems to depend on the successful coordination of a "heterogeneous sociotechnical ensemble."[32] Stakeholders tend to perceive cross-border coordination of all of these aspects as the best way to reduce the vulnerability of emergency communication. The Euroregion played an important role in these attempts to organize cross-border emergency collaboration. Under the Euroregion flag, collaboration between emergency services intensified after the Enschede disaster. In this sense the Euroregion emerges as an important constitutive element in Europe's vulnerability governance. "Legal Europe" also played a crucial role, embodied in the Schengen Agreement that made it possible for German firefighters to cross the Dutch-German border without border controls in the first place. Furthermore, the emphasis was put on local agreements and improvising gateways between the incompatible technological tools of the two countries.

Dreaming about a common European standard for public safety

The Enschede disaster created a new sense of urgency among Dutch public safety services and the Ministry of the Interior in order to improve the effectiveness of emergency communication. Since the mid-1990s the Dutch government had been involved in the development of a European standard for emergency communication: Tetra. The Dutch version of this system, called C2000, was still under development at the time of the Enschede disaster, but there was a strong belief among government officials that Tetra/C2000 would vastly improve coordination between emergency services and across borders. Proponents of Tetra/C2000 interpreted the disaster in Enschede as an example of how emergency communication can fail if not standardized. They pushed for a speedier implementation of C2000 to enhance the quality and efficiency of emergency response.

In this section we analyze the development of the European Tetra standard as a case of European integration and fragmentation. The standard was meant to integrate Europe and to increase the safety of its citizens. As such, it was supposed to embody two key ideals of European political collaboration. As noted, the idea of improving emergency collaboration across borders in Europe was already mentioned in the Schengen Agreement of the mid-1980s. The Schengen Agreement expressed the ambition to create a pan-European network for public safety and justice, and to enhance cooperation between police services in different European countries: "Art. 44 of the Schengen Agreement contains an obligation to provide for direct radio contacts between police and custom services when operating across borders."[33] Communication technologies, preferably standardized, would be helpful in particular for the exchange of information in cases of cross-border observation and pursuit. In the late 1980s the incompatibility of systems made a direct connection between police units in border areas impossible. To overcome this, a "uniform police radio communication system using common frequencies for exclusive use by the police"[34] had to be developed. In order to achieve this goal, the use of communication standards had to be coordinated. It was expected that "standardization would lead to a better quality of the cooperation between the emergency services, increased efficiency and increased effectiveness in case of large-scale emergencies."[35] To achieve this, it was argued, a common European communication standard was needed.[36]

The European Telecommunication Standards Institute (ETSI) started developing a standard for mobile digital radio communication in 1988. In the early days this system was called the Mobile Digital Trunking Radio System. In 1992 it was renamed Tetra (Trans European Trunked Radio; later, when the global capacities of Tetra became clear, TErrestrial Trunked RAdio).[37] Initially the standard was not primarily dedicated to emergency communication. Later, due to input from the public safety sector in this standard, it became more tuned to the needs of emergency services. The main aim of the Tetra standard was to define a number of open interfaces (an air interface, Direct Mode operation and Inter System Interface are the main ones) detailed enough for different manufacturers to develop independent products based on this standard. By following the Tetra specifications, the infrastructure and terminals developed by different manufacturers were supposed to be fully interoperable. The air interface allows for communication between base stations and terminals, whereas the Direct Mode interface makes it possible to communicate in local radio networks without the support of the Tetra infrastructure. (This can be helpful in emergency situations where the Tetra infrastructure gets overloaded.) The Inter System Interface allows interoperability between two or more different networks supplied by different manufacturers (e.g. Tetra networks in different countries).[38]

A competing standard called APCO25 had already been developed in the United States. The head of the R&D department of the information and communication technology (ICT) organization of the Dutch police was informed about the developments in the United States by his British colleague. The British had been following the developments in the United States around APCO25 for some time and had become quite enthusiastic. They also knew about the new ETSI project to develop a standard for mobile radio communication for a wide market and they informed the Dutch about this. Together they agreed that the British would follow the developments in the United States and that the Dutch would focus on ETSI. Soon Britain also joined ETSI. At the same time the Netherlands became involved in ETSI RES-6, the committee in which the Tetra standard was developed.[39]

Simultaneously, a working group of the Telecom experts of the Schengen countries[40] was established. This would define the functional requirements for the ETSI standard under development. According to one of the Dutch representatives in the Schengen Telecom group, the Dutch followed the developments in the United States quite closely. However, they finally decided that they preferred a "European" standard, developed by European industry. They argued that, as Europeans, they wanted to support European not US industry. Moreover, the other countries represented in Schengen Telecom would never accept APCO25: they clearly preferred a European standard developed by ETSI.[41] According to a former high-ranking civil servant at the ministry of Justice,

> In technological matters, the Netherlands has always been a country very willing to cooperate in Europe...So, if there would be any chance of a European technological development and a European solution, you would go for that – for reasons of scale and impact.[42]

In addition to these economic and political reasons, the Dutch negotiators also expected a European standard to better address the needs within Europe. A telecom engineer for the Dutch police argues that European products take specific European circumstances into account better. For instance, in Europe, the geographical distance between big cities is much smaller than in the United States. Therefore less powerful transmitters are needed: "We need a technology that better matches the highly populated areas we have."[43] This line of reasoning is also expressed in the minutes of a Schengen Telecom meeting on June 23, 1992, in the Netherlands, where the Germans claimed that: "the advantage of [the Tetra standard] over the APCO standard is that the latter is based on the American geographical and urban reality, which is altogether different from the European situation."[44] In these reasonings, a unified "industrial Europe" is being constructed that deserves support and protection against "outside" competitors such as US industry.

This willingness to act in "the right European spirit" resulted in a hard time for Tetrapol, another standard for emergency communication under development at the time. Tetrapol was a French industry standard, developed by Matra and not acknowledged by ETSI. The Dutch did not want to limit themselves beforehand to one supplier (Matra) and therefore Tetrapol was not an option for them.[45] The discussion about the choice between Tetra and Tetrapol that originated in the early 1990s was reopened in Dutch parliamentary debates as well. MP van Heemst asked for a comparison of the costs of Tetra and Tetrapol. He raised uncomfortable questions: "What if Germany opts for Tetrapol? What shall we do then?" According to the minister, the comparison of Tetra and Tetrapol should be viewed in the context of "the praiseworthy shamelessness with which France always tries to promote its own industry."[46] The Netherlands should focus on the choices of its direct neighbors (Belgium, Germany, Luxembourg). "If the consequence of France's choice for Tetrapol is that it cannot communicate with other countries, or can only do so with extensive technical measures, Mr. Chirac should rethink his statements about the fight against international crime," according to the minister. "So far, all other countries have opted for Tetra and any government that does not do so bears a heavy responsibility",[47] he argued.

The ultimate solution thus became the creation of an "open European standard." One of the alleged advantages of an open standard was the formation of a so-called "multivendor" situation – a situation in which countries would not be dependent on one supplier or manufacturer. For the Netherlands in particular, the standard had to be "European," as the Dutch government considered itself to be a strong supporter of Europe and European integration, wanted to contribute to meeting the Schengen objectives and wished to support European industry (rather than US industry, for instance). Moreover, developing a new nationwide system on its own was not considered to be economically viable.

According to a civil servant, responsible for international relations at the Ministry of the Interior at the time,

> Yes, we knew that we were going to take a risk with the new technology and the new standard. But, okay, you decide you want to be a leader in that area or not, and someone has to be the first to implement them. And we invested a lot in spreading the Tetra standard because we found it so important to have, in Europe, a radio communication network for emergency services, comparable with GSM.[48]

In the course of 1998 it became clear that Tetra was winning ground in Europe. In October 1998 the Netherlands, Belgium, Britain, Finland, Norway, Sweden, and Portugal had opted for the Tetra standard.[49] In Germany the decision process about a nationwide network was still under way. However, the Dutch deputy minister for the interior, Gijs de Vries, still regretted the lack of consensus within Europe about the adoption of the Tetra standard. If other countries were to choose another standard, that would hinder cross-border emergency communication. Therefore he contacted the European Commission and asked it to promote the Tetra option among other European countries. The European Commission responded, however, that European public tender law does not permit an a priori choice for the Tetra standard, thereby excluding alternative technologies, such as Tetrapol. This would contradict Europe's open-market philosophy.[50] Therefore de Vries also lobbied for Tetra among other European countries and (at that time) candidate member states, such as Poland and Hungary. These countries eventually decided to opt for Tetra as well. Figure 9.1 shows that in 2009, many European countries had indeed chosen the Tetra standard, but not all of them: France, Switzerland, and the Czech Republic opted for Tetra's French competitor, Tetrapol.

In sum, the Tetra standard was promoted by countries like the Netherlands as a standard mainly because of its "European nature" – that is, produced by ETSI and supported by European rather than US industry. Standardization of emergency communication infrastructure was seen as the key to European integration and collaboration in public safety issues. France, having developed its own industry standard, was strongly criticized by the Dutch government for not acting in the right European spirit by opting for its own standard: in the Dutch view, the French government actively contributed to the technological fragmentation of Europe and made Europe more vulnerable.

The Three Countries Pilot

Having analyzed the use of improvised gateways in the case of the fireworks disaster in Enschede and the shaping of the Tetra standard, we may now ask how the implementation of the Tetra standardized gateway reshaped cross-border emergency communications. What were the implications of the Tetra standard for the vulnerability of emergency communications? We address this issue by focusing on an emergency test that took place in May and June 2003 in the border area

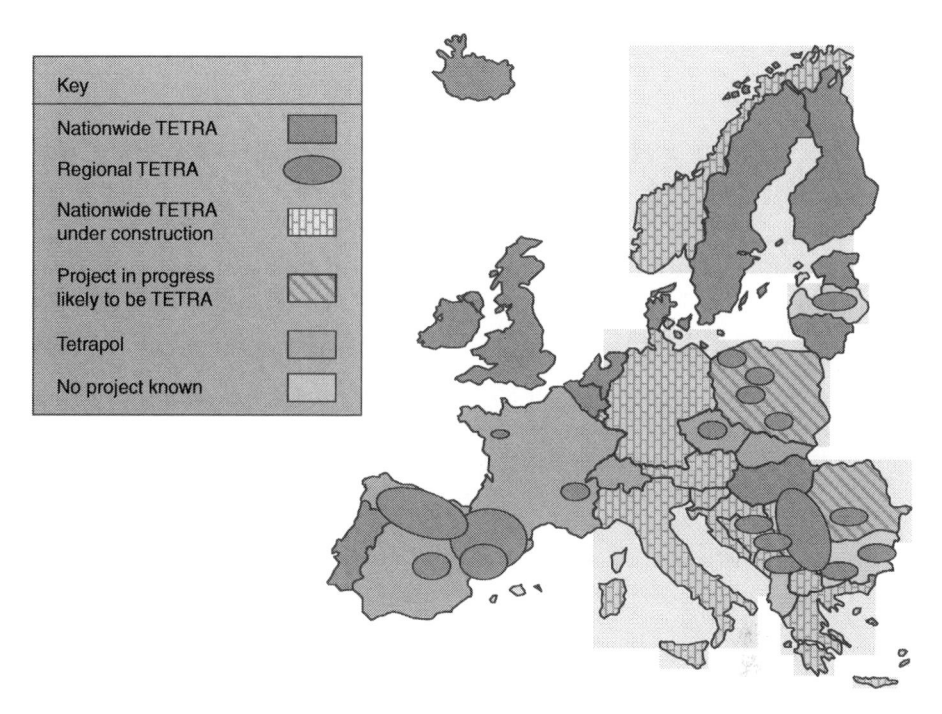

Figure 9.1 European public safety networks.
Source: Hans Borgonjen. VTS-PN 2012.

of Germany, Belgium, and the Netherlands. In this so-called Three Country Pilot (3CP) the Tetra standard was tested to investigate its operational value for cross-border emergency communication. Some 12 scenarios were tried. These included, for instance, one in which a German policeman witnessed a carjacking in Belgium, a cross-border pursuit, and a mass public demonstration at the border. Such scenarios were explained in detail and all communication was prescribed word by word, leaving little space for improvisation during the test. In essence, then, the test was a technical check of the interoperability of three independently developed Tetra systems in three different countries.

As mentioned above, the Tetra standard served as the technological basis for a number of national emergency communication networks in Europe. The Dutch C2000 network was built by KPN Getronics and Motorola. Belgium's Tetra-based network, called ASTRID, was built by Nokia. The German police used (and still use) analog radio systems but have recently decided to switch to a Tetra-based network as well. After the implementation of the C2000 network in the Netherlands, communication with Germany and Belgium had to be coordinated in a different way. According to the director of the Twente Safety Region, in that region they used the same improvisation strategy as in the past to connect the Dutch digital and the German analog networks. If a German vehicle enters the Dutch region, it can be connected to C2000, and a Dutch vehicle on German soil can be connected to its analog system. The dispatch room always acts as a kind of intermediary between the vehicles. A special arrangement has been made for the trauma helicopter from the German city of Rheine that also serves the whole Twente region.[51] Officially it is not allowed to use communication apparatus in a different country, but in this case a C2000 mobile phone was simply installed in the German trauma helicopter to make sure that it can easily be guided to the right location.[52] These examples show that, even when a standardized gateway (Tetra) is present, improvisation strategies are necessary to support cross-border emergency collaboration.

The 3CP nicely illustrates the difficulties involved in aligning different technological systems and working processes, even when a technical standard is in place. The Euroregion Maas-Rhine, where the test took place, comprises the cities of Maastricht, Aachen, and Liège and is a very densely populated area with 3.7 million inhabitants living in an area of 10,478 sq. km: "Within an area with a radius of approximately 50 km, people are living and working in three different countries with governments, laws and judicial systems of their own."[53] The region Maastricht-Aachen-Liège has a reputation when it comes to cross-border criminality. There is quite a strong influence from cross-border drug traffic. According to the 3CP project leader for the Netherlands, "For us, Holland, it [the 3CP] is very important. The three-country region is an area where there is much cross-border criminality. And of course Holland is very small and if you drive half an hour you will have visited three different countries."[54]

Before the 3CP and the development of Tetra, the emergency services of the three countries collaborated using their analog radio systems.[55] According to the Dutch 3CP project leader, cross-border communication between police and fire brigade was very limited:

> They didn't talk to each other unless they met on the spot where something was happening... There was no direct communication between the people in the field. They all had their own dispatch room and the dispatch room had to call to repeat all the communication traffic.[56]

Before the 3CP, the three countries collaborated using analog radio systems according to the radio frequency agreements made within the short-term Schengen programme.[57] In this case, too great an emphasis was placed on coordination as a means to improve collaboration and system reliability. Although the 3CP test report revealed difficulties related to the coordination of language, operational procedures, and administrative issues, the main challenge seems to have been technological coordination between the three countries.

Coordinating different technologies

When the 3CP began its preparation phase in the late 1990s, the three countries involved had made different choices about their national emergency communication systems. The Netherlands had chosen the C2000 system, based on the Tetra standard. In 1999 a consortium of Motorola and KPN/Getronics (named Tetraned) had been contracted to deliver the network. At the same time, however, the official go–no go decision for the C2000 project had not been taken yet by the Dutch government and, as we saw in the previous section, the choice for Tetra (instead of, for instance, the French Tetrapol solution) was still being debated in the Dutch parliament. In Belgium, the ASTRID network, which was also based on the Tetra standard, was under development by Finnish company Nokia. By the end of the 1990s, the German government had not yet made a choice for a standard for their new communication system. In Aachen, a temporary network based on the Tetra standard was built:

> In Germany the political issue was very important... Well, there was a discussion concerning the Aachen pilot; it was a pilot to support the decision for technology and for a country-wide network. So they were in a completely different phase of developing their Tetra standard and applying it for public safety users than we were. We were already rolling out our network in Holland, while they were still in the decision process.[58]

Although the three countries chose Tetra as the common standard for the 3CP, there were important differences between them that made it hard to integrate their Tetra networks. The 3CP shows that having a technical standard (Tetra) in place does not automatically imply that different countries can connect with different networks, even if they are based on the same official standard. One of the main reasons was that there were different suppliers in Belgium and the Netherlands (Nokia and Motorola) that were very reluctant, according to our interviewees, to openly discuss the different systems and possible solutions. One of the reasons behind this reluctance was their fear of revealing sensitive company information to their main competitors. As a member of the 3CP Coupling Group, and communication systems engineer for the German federal police, argues, "I think there was a competition in Germany to build the nationwide system and one supplier was Nokia, with EADS and Motorola the other rival. And I think they didn't want to say too much about their system to one another. This was a problem."[59] The financial and planning manager of C2000 at that time, argues along similar lines:

> Describing it [the central, common functionalities of the so-called Inter System Interface (ISI)] is no problem, making a contract of it and putting into place: a big problem. Because there were also some things going on between Motorola and Nokia – if you want to connect the systems you have to show the other one what's in your system. So it will fit. And that was the problem between the two companies.[60]

An additional technological gateway, the ISI, was needed to connect the three Tetra networks. Industry was very hesitant to invest in an ISI. The countries had to buy it before it would be developed, but they wanted to see it working before they would buy it. According to a prominent Dutch negotiator in ETSI and Schengen Telecom,

> It was a little bit of a chicken and egg problem. We said to the industry 'there should be an ISI. Develop an ISI and when you have an ISI and we know what it costs, we might buy it.' And the industry said, 'yes, but we are not going to develop an ISI when there is no contract for it, so please give us a contract first and then we will develop it.[61]

Back then, in 2007, industry was still reluctant to invest in an ISI.[62]

The financial and planning manager of C2000 relates the reluctance of the industry to work together in a project like this and its unwillingness to invest in an ISI to

> the barrier between public institutions on the one side and commercial institutions on the other side. Commercial institutions always talk about profit and money and getting a market share. Public institutions only talk about solutions for their problem: 'I can't tell my colleague on the other side of the border, fix that...' So we talk in solving problems and they talk in market shares and profits... and that clashes.[63]

Although Tetra is an "open" standard, being, in principle, supplier independent (in contrast with a "proprietary standard" developed by one supplier), this analysis shows that it still has to deal with market stakes and strategies of important suppliers.

Despite these difficulties, the telecom engineers involved in the 3CP improvised a temporary network based on Tetra that allowed for communication between emergency workers from different countries on one side of the border and between two dispatch rooms in different countries. Direct radio contact between emergency workers across the border was not possible. When writing the scenarios and also during the tests, it became clear that for a successful communication process, implementing a standardized gateway was not sufficient: different operational tactics, languages, and legal systems had to be bridged as well. Similar to the case of Enschede, many heterogeneous elements had to be coordinated to make the process work.[64] In this case, however, the main vulnerabilities in the cross-border collaboration between emergency services were attributed above all to the lack of technological interoperability. Having a (temporary) Tetra network in the three countries was not sufficient to achieve "plug and play" technological interoperability. An ISI was deemed necessary to solve this problem, but this was (and still is) impossible because of the reluctance of both industry and national governments to invest in it. According to the public safety actors involved, the lack of an interoperable network was due mainly to the unwillingness of the main suppliers to cooperate and to invest in it. Thus this case showed the clash between the ideals of cross-border cooperation, European integration and safety for citizens, and the actual practices of negotiation and deliberation, involving national and industrial stakeholders. The key vulnerability of the system thus seemed to be rooted primarily in its confrontation with unwilling or uncooperative stakeholders.

Conclusions: Europe and the vulnerability of emergency communication infrastructure

In this chapter we have discussed the shaping of emergency collaboration in the two Euroregions as an attempt to achieve heterogeneous coordination at several levels: technological, cultural, operational, tactical, and legal. These developments were embedded simultaneously in geographical scales that ranged from local collaborative arrangements among municipalities, to the involvement of international institutions (ETSI, Schengen Telecom) and legal agreements (Schengen Agreements) and standards (Tetra). We now discuss the implications of these transitions for the vulnerability of emergency communication infrastructure. Local stakeholders, we saw, have long acknowledged collaboration across national borders as an important way to reduce vulnerability to accidents and disasters. Particularly since the 1970s, local initiatives to collaborate with emergency services across the border have become more frequent. In these efforts the region (and later the formal Euroregion) became an important organizational setting for stimulating and regulating cross-border cooperation. Besides national organizational structures, such as the Dutch province and the German *Land*, the (micro)region served as a key organizational unit for the attempts to integrate emergency work. This confirms the importance of the microregion as a key unit of analysis or *laboratory* both for studying Europe's hidden integration and for understanding the governance of Europe's infrastructure vulnerabilities.

Our two cases have shown that the microregion is the unit where, according to the actors involved, coordination of emergency communication had to be organized. Stakeholders perceive "heterogeneous coordination" as a necessary step toward cross-border and transnational collaboration and toward managing critical communication infrastructure. A variety of heterogeneous elements needs to be coordinated and brought into alignment before a reliable system can be accomplished. Our cases showed two different approaches to heterogeneous coordination: the Twente case focused on improvised gateways and small-scale, local arrangements and flexibility, whereas the 3CP case studied the strategy of coordination through standardized gateways, long-term and large-scale agreements. During the Enschede disaster the communication infrastructure based on improvised gateways was evaluated as "vulnerable" by stakeholders: connections did not work and telecommunications broke down. Moreover, the rescue efforts of the German neighbors, initially seen as an important asset in reducing vulnerabilities, also suffered from coordination problems. In response, key stakeholders in the field of public safety (governments, police, fire, and ambulance officials) highlighted the need for better coordination in emergency situations, and increasingly supported the development of a joint technical standard for emergency communication.

This prioritization of a technical solution to communication and coordination vulnerabilities is clearly exemplified in the 3CP case.

A comparison of the two cases shows, however, that even technological coordination by means of a common European standard (Tetra in the 3CP case) does not guarantee the success or ease of cross-border collaboration.[65] Despite the presence of a technical standard, actors still had to improvise to connect the three country-specific networks and make cross-border communication work. These actors also found that technical standardization alone was not sufficient to make the network reliable: they also needed detailed agreements on the use of language and protocols. This apparently produced a more harmonized system, but they still had to cope with incomplete (or unfinished) technological interoperability and the lack of an ISI. Therefore we conclude that standardized solutions do not reduce the vulnerability of critical infrastructure by definition – they might very well introduce new vulnerabilities. The 3CP case showed that, ironically, the very same ideas and stakes behind Tetra (open, European, interoperable) provoked tensions on the border – tensions that resulted in reluctance among suppliers of Tetra networks and equipment to share knowledge and to make their equipment interoperable. This is a historical irony of vulnerability (see Chapter 1): attempts to reduce vulnerabilities may produce new vulnerabilities of their own later on. Such vulnerability dynamics cannot merely be derived from theory; in order to identify such vulnerabilities and study their dynamics, we need to examine empirically the historical processes in which stakeholders interpret, anticipate, and negotiate vulnerabilities and their solutions.

Our analysis has also made clear that in the discourse of politicians, public safety managers and pro-Tetra engineers, the ideal of "one public safety Europe," united by large-scale, cross-border Tetra networks, played a crucial role. Moreover, European standardization on Tetra would enhance collaboration between emergency services, thus reducing Europe's vulnerability to disasters and crime. Thus, in the reasoning of these key actors, European technological integration and vulnerability reduction through standardization were two sides of the same coin. However, our analysis of the realities of emergency collaboration using Tetra has shown that this dream has not yet come true.

Notes

1. Edwards 1998; Edwards et al. 2007; Schmidt and Werle 1998; Timmermans and Berg 2003; Timmermans and Epstein 2010.
2. Egyedi 2000.
3. Extending this line of argumentation we can say that Tetra facilitates internetworking (Edwards et al. 2007) or "vertical integration". Tetra is not a meta-generic gateway because this presumes a much higher level of integration.
4. Applegate 1999.
5. Knippenberg 2004.
6. "De politiële samenwerking in het Nederlandse, Duitse en Belgische grensgebied rond Aken. [Police cooperation in the Dutch, German and Belgian border area around Aachen] Archive Rijkspolitie in Limburg]," Report NeBeDeAcPol, 1-4-1988, 07.A20/147.
7. Fijnaut and Spapens 2005.

8. "De politiële samenwerking in het Nederlandse, Duitse en Belgische grensgebied rond Aken. [Police cooperation in the Dutch, German and Belgian border area around Aachen] Archive Rijkspolitie in Limburg]," Report NeBeDeAcPol, 1-4-1988, 07.A20/147.
9. "Samenwerking bij het bestrijden van ongevallen en rampen in grensgebieden [Collaboration in the combat of accidents and disasters in border areas]", *Brand en Brandweer* 1980 (4)1, pp. 9–10.
10. "Vraag [Questions]," *Brand en Brandweer* 1982 (6)5, p. 153.
11. Van der Vleuten and Kaijser 2006, p. 286, considers this to be the first level "of increasingly tight international infrastructural cooperation: ... purely technical coupling across national borders."
12. Interview with the director of the Twente Safety Region and chief-commander of the fire brigade at the time of the Enschede disaster, Enschede, January 22, 2008.
13. Post and Stal 2000, p. 34.
14. See the "Oosting Report" (Commissie onderzoek vuurwerkramp 2001). In this section we analyze evaluation reports written (directly) after the disaster, with a focus on the performance of emergency communication and emergency services. We also draw on interviews with two key witnesses: the commander in chief of the fire brigade of Enschede at the time of the disaster, and a German firefighter who spontaneously offered his assistance in fighting the disaster.
15. Interview with the commander in chief of the fire brigade of Enschede at the time of the disaster.
16. Idem.
17. Interview with the chief of the fire brigade of the neighboring Kreis Borken in Germany, Vreden, January 22, 2008.
18. See Commissie onderzoek vuurwerkramp 2001 and the Berghuijs report: "De toekomst van de rampenstrijding en het risicomanagement. Een evaluerende rapportage naar aanleiding van de vuurwerkramp in Enschede op 13 mei 2000, oktober 2000 [The future of fighting disasters and risk management. An evaluative report in response to the fireworks disaster in Enschede on May 13, 2000]."
19. Berghuijs report.
20. Commissie onderzoek vuurwerkramp 2001.
21. Hertoghs and Rambach 1997.
22. The way in which the emergency structure is organized in Germany can be understood in its historical context. In the First and Second World Wars, the Germans established specific services for protection against air strikes (e.g. the Sicherheit und Hilfsdienst, and later the Luftschutzdienst). Initially these services offered support at the national level but, later, regional counterparts were established. At a later stage, these services were merged again and the federal state became the prime responsible unit for their activities. Hertoghs and Rambach 1997.
23. Interview with the commander in chief of the fire brigade of Enschede at the time of the disaster [between the public safety organizations AH,EC].
24. Interview with the commander in chief of the fire brigade of Enschede at the time of the disaster. Interview with the chief of the fire brigade of the neighboring Kreis Borken in Germany.
25. In the border villages in the northern Twente region there is a lot of social contact between the fire-brigade communities of the Netherlands and Germany. They organize "firefighting matches" and exchanges.
26. Interview with the chief of the fire brigade of the neighboring Kreis Borken in Germany.
27. Interview with the commander in chief of the fire brigade of Enschede at the time of the disaster.
28. Hulpverleningsdienst Regio Twente 2000, p. 20.
29. Pater 2004.
30. Interview with the commander in chief of the fire brigade of Enschede at the time of the disaster.

31. Interview with the chief of the fire brigade of the neighboring Kreis Borken in Germany. Interview with the commander in chief of the fire brigade of Enschede at the time of the disaster. An operational similarity between the Netherlands and Germany is that they have a similar method for freeing injured people from their vehicle after an accident (by first stabilizing the victim). In the Netherlands this method is called "Kusters" while in Germany it's the "Hamburger" method. However, Robin Pater observes, "not all German emergency units work according to this method; for instance, the firemen in Nordhorn (DL) employ a method of freeing injured people from their vehicle as soon as possible on advice of the attending 'Notarzt'." Pater 2004.
32. Cf. Bijker 1995.
33. Letter by Heckmann (chairman of the Schengen Telecom group) to Oliver (chairman of STC RES 06), January 5, 1994. The Article 44 referred to here is part of the "Schengen Executive Agreement" of October 11/12, 1990. ISC Archive, box KST.2.
34. Minutes of the meeting on improving police communication in cross-border cooperation in the Schengen framework, Bonn-Bad Godesberg, July 11/12, 1989, p. 3. ISC Archive, box KICS.1. Later on the involvement was expanded in some countries to include all emergency services, including fire brigades and ambulance services.
35. Projectbureau C2000 1995.
36. A harmonization of the frequencies was also needed, but this discussion is beyond the scope of this chapter.
37. Note from German delegation to Schengen Telecom Working party, July 27, 1992. ISC Archive, box KICS.1. See the important work by Rudi Bekkers on the early development of the Tetra standard (Bekkers 2001).
38. Gray 2003.
39. Borgonjen tried to get Ginman (as a representative of a non-Schengen country) on board of the Schengen Telecom Working Group in order to act more effectively as strategic partners. Letter from R. Ginman to J.B.M. Borgonjen, September 24,1990. ISC Archive, box KICS.8. Interview with the Dutch negotiator and member of the Schengen Telecom Group.
40. Initially five countries (Belgium, the Netherlands, Luxemburg, Germany, and France) signed the Schengen Agreement (June 14, 1985). This expressed the ambition to create a pan-European network for public safety and justice, and to enhance cooperation between police services in different European countries. The Schengen Agreement was later officially adopted as EU policy in the Treaty of Amsterdam of 1997.
41. Interview with the C2000 system architect. This was also mentioned in an interview with the head of the Knowledge and Innovation Centre of the Dutch ICT Service Centre for Police, Justice and Security. A confidential note by Hans Borgonjen (September 27, 1990), since released, stated that if the European industries do not have a kind of structured debate with the European police organizations, there is "a risk that America will become dominant in the field of police communication." H. Borgonjen to Permanent Working group Schengen, September 27, 1990, p. 2, ISC Archive, box KST.1.
42. Interview with a former high-ranked civil servant at the Dutch Ministry of Justice.
43. Interview with the C2000 system architect.
44. Minutes of Schengen Telecom meeting, June 18, 1992, Bilthoven. ISC Archive, box KST.3. According to the head of the Knowledge and Innovation Centre of the Dutch ICT Service Centre for Police, Justice and Security, they were lucky that the technical experts also had technical reasons for supporting Tetra. Otherwise the political dimension would have made the debate on APCO25 versus Tetra more difficult.
45. Interviews with the head of the Knowledge and Information Centre VTS, Police Netherlands (ISC), the C2000 system architect and a civil servant at the Ministry of the Interior, responsible for international contacts police.
46. Ibid., p. 8.
47. Ibid.

48. Interview with a civil servant at the Ministry of the Interior, responsible for international contacts police.
49. Minutes of Dutch Parliament 1998–9, 25 124, no. 7. Accessed through www.parlando.nl.
50. Deputy minister of internal affairs to the parliament, February 16, 2001, 25 124, no. 19, p. 11, ISC Archive, box KIC.1. Here the European Commission plays an interesting "double role": on the one hand, it wants to promote common standards and a wide adoption of European standards; on the other hand, it does not want to stimulate this process because of their its commitment to an open market philosophy.
51. The four trauma helicopters that serve the Netherlands cannot reach Twente on time, therefore they made this agreement with Germany.
52. Interview with the chief commander of the fire department of Enschede at the time of the Enschede disaster.
53. EMRIC 2003, p. 5.
54. Interview with a former project leader of C2000 and member of the 3CP Coordination Group, The Hague, June 11, 2007.
55. It is rather unclear how intensive the cross-border collaboration actually was before the 3CP and even now. According to a Belgian police officer, "in our daily work we have a lot of trans-border cooperation with Germany, the Netherlands and Luxembourg." The financial and planning manager in the C2000 project also says that the police and fire brigades "worked a lot together in the border region." On the other hand, a former chair of the 3CP Steering Committee argues that there was "a slight lack of communication" before they started the pilot: "everybody was working a little bit on his own and simply exchanging analog radios was a common practice." Author's interviews.
56. Interview with former project leader and member of the 3CP Coordination Group.
57. Interview with two members of the 3CP Working Group Business Processes, Aachen, June 5, 2007; interview with the head of the Knowledge and Information Centre of the Dutch Police, and the chair of the 3CP Working Group Coupling, Odijk, May 15, 2007.
58. Interview with the head of the Knowledge and Information Centre of the Dutch Police and with the chair of the 3CP Working Group Coupling. Finally, in 2006, Germany chose the Tetra standard. It is very uncertain, however, when this network will be implemented there.
59. Interview with a former member of the 3CP Coupling Group, and communication systems engineer for the German federal police, Aachen, June 12, 2007.
60. Interview with a member of the 3CP Steering Committee, The Hague, June 11, 2007.
61. Interview with the head of the Knowledge and Information Centre of the Dutch Police and with the former chair of the Working Group Coupling of the 3CP.
62. See, for instance, the interview with the secretary of the 3CP Steering Committee, Swisstal, July 4, 2007:

> Of course industry wants to make money and if there is no need they will not develop anything... if four or five countries order the ISI, we will develop it, cover the development costs because there is a market... But they don't see a market now, there are not enough countries rolling out networks.

Recently (2012) an EU-funded project has started that aims to develop an ISI.
63. Interview with a member of the 3CP Steering Committee.
64. We refer to Bistra Vasileva's excellent MA thesis for a detailed discussion of the heterogeneous elements that had to be coordinated in the 3CP case. Vasileva 2007.
65. We think there are (at least) two reasons why it can be so difficult to achieve heterogeneous coordination. The first is the obduracy of existing, historically grown, sociotechnical structures and practices. When trying to build up transnational collaboration, the actors have to deal with existing operational practices, legal arrangements,

technological equipment and cultural norms that, in some cases, are hard to overcome. See Hommels 2005 for an analysis of the role of obduracy in the attempts to change urban infrastructure. The second reason is the opposite: change, or the observation that elements of heterogeneous communication infrastructure change over time and at different speeds, that makes it hard to hold the ensemble together.

Conclusion

10
Europe's Infrastructure Vulnerabilities: Comparisons and Connections

Anique Hommels, Per Högselius, Arne Kaijser, and Erik van der Vleuten

This book has investigated the historical shaping of critical transnational infrastructure in Europe, its associated vulnerabilities, and its intertwinement with broader processes of European integration and fragmentation. The chapters in the three parts have scrutinized these issues from different thematic angles and in a broad range of geographical and temporal settings. While each study has generated intriguing insights in its own right, the range of empirical cases has also set the stage for comparative and connecting observations. This final chapter sets out to harvest from our case studies by focusing on a number of such cross-cutting issues.

Differing interpretations of vulnerability

Recent constructivist research on risk and vulnerability stresses the importance of studying vulnerability not as a given, objective phenomenon that can be defined precisely and unambiguously but as something that is constantly reinterpreted, contested, and negotiated by stakeholders and analysts alike.[1] Our findings confirm that such a view of vulnerability as socially constructed is indeed crucial when seeking to grasp the making of transnational infrastructure vulnerabilities. Our case studies are largely consistent with insights from cultural studies of risk, in which the "ways in which individuals – including experts – interpret risks can be seen as an expression of socially located beliefs and world views that to a large extent stem from the individual's situated position and experiences within social hierarchies, institutions and groups."[2]

As argued in several chapters of this book, however, this is not the whole story. Rather, our results point to the importance of studying not only *vulnerability* but also *reliability* as socially and culturally interpreted and negotiated. For example, the "European blackout" of November 4, 2006, was interpreted by some as an illustration of the extreme vulnerability of electric power networks leading to a disruption of economic life in Europe. This perception went hand in hand with calls for more centralized governance and coordination of Europe's electric power grid, not least from the side of Europeanist politicians. To electricity sector spokespersons, however, the event instead suggested the high reliability of Europe's electricity supply: since the 1960s the sector had installed and continuously upgraded security measures, which had made cross-border cascading failures extremely rare and allowed fast reparation of the rare failures that did occur – in this case the sector fully restored Europe's power supply within just two hours. Overall the damage of rare and rapidly repaired breakdowns was negligible compared with the daily security gains of cross-border collaboration in the form of joint instant frequency stabilization. Importantly, this high-reliability perception, too, had implications for governance: since the sector's vulnerability management system was based on well-working decentralized response (each transmission operator repaired its own transmission area, which it knew best), European Union (EU)-level governance constituted a new threat rather than a solution to system stability.[3]

Next to such interpretations of a particular critical event, another discussion that has persisted for a longer time concerns the argument that the main problem with Europe's power grids is not its vulnerability but, on the contrary, its excess reliability; that security investments far exceed the economic costs of incidental breakdown. As emphasized in Chapter 8, the very early liberalization of the Norwegian power sector was partly intended to counteract what was seen as a huge overinvestment in excess capacity. The liberalization was indeed followed by a sharp decrease in investments in the grid, without any corresponding increase in the number of outages. This was interpreted by the proponents of liberalization as a confirmation of the old regime's excessive focus on reliability.[4] An identical discussion took place in the Netherlands a few years ago.[5] These examples illustrate how perceptions of overly vulnerable and overly reliable infrastructure may coexist in the field. It should inspire researchers to question the dominant discourse on vulnerability by simultaneously examining infrastructure vulnerability and reliability.

Our book contains many examples of situations in which actors have been forced to make choices between alternatives that imply very different kinds of vulnerabilities and thus weigh certain risks against others. The Finns discussed the pros and cons of Soviet nuclear technology, scaled up electricity imports, or increased reliance on Soviet natural gas, Polish coal or Middle Eastern oil, before concluding that Soviet nuclear technology was to be imported. But they made this choice only after tough negotiations with the Soviet nuclear establishment, in which Moscow had to give in to Finnish security demands.[6] The Greeks similarly faced a delicate choice between installing nuclear power on seismically active ground, importing electricity from beyond the Iron Curtain, or accelerating the exploitation of domestic, polluting lignite resources. In contrast with the Finns, those who regarded the risk of nuclear accidents as too great had the final say.[7] But each option was linked to a certain kind of vulnerability, and there was no objective way of deciding which of these vulnerabilities was the most problematic: nuclear accidents, pollution stemming from the burning of lignite, or import dependencies.

Our book has shown that such vulnerability trade-offs informed several lines of conflict between stakeholders. They have sometimes led to opposition between different social groups within a given country. For example, the 1995 earthquake in Greece discussed in Chapter 5 was interpreted very differently by engineers and journalists. According to the journalists, the earthquake showed the resilience of Greece's national electric power network to the devastating blackout that could have resulted from the major earthquake. For the other group, the engineers, the Greek network was resilient because of its transnational connections: "the Greek national power network proved reliable because its vulnerability was shared with the national power networks of some of Greece's neighbors." In this latter interpretation, connected countries become more resilient because vulnerabilities were shared in a transnational network.[8]

In other cases these trade-offs led to opposition between stakeholders on different sides of a border. Previous studies of risk have noted the tendency of different social groups to agree internally on certain perceptions of vulnerability and risk.[9] Yet national borders might divide them. The Finnish case illustrated different interpretations of vulnerability within the social group of engineers in Finland and the Soviet Union. According to the Soviet engineers, risks were calculable: "If calculations showed that no risk existed there was no need to build expensive backup systems and redundancy." The Finnish engineers also made calculations but did not consider this sufficient when seeking to manage nuclear risks. The two approaches clashed when transnational connections between Finland and the Soviet Union were planned, and even more when discussing the design of nuclear power plants.[10]

Such vulnerability trade-offs might set countries against international organizations. When Bulgaria negotiated its accession to the EU, the EU considered several blocks at the Kozloduy nuclear power plant unsafe and made their decommissioning a condition for Bulgarian EU membership. The International Atomic Energy Agency, however, found safety in these blocks comparable to older, Western European nuclear units that were still in operation. But the EU kept up its demand in the treaty negotiations, and the units were then actually shut down. Remarkably, when the 2009 Russian-Ukrainian gas crisis caused a disruption to Bulgaria's gas supply, the Bulgarian government suggested reopening the decommissioned blocks to counter the unexpected energy shortage. In this case, nuclear power was perceived as a potential savior rather than as a source of vulnerability. Behind these conflicting perceptions, however, lay Bulgaria's interests in retaining its position as a major electricity exporter on the Balkans.[11]

Finally, such vulnerability trade-offs may change over time. The *evolution of vulnerability perceptions over time* is a theme that has been much neglected in previous social studies of risk. Our explicit historical focus sheds light on this issue. In particular, many of the chapters demonstrate that the perceived vulnerabilities related to energy infrastructures have changed profoundly over time. For example, Bulgaria's vulnerability in the field of electricity was initially defined in terms of shortage. Having overcome the basic shortage by constructing coal-fired power plants, actors began to point out a new form of vulnerability: dependence on the Soviet Union for adequate coal supplies. The Bulgarians then sought to reduce this dependence by constructing a vast nuclear power plant. This led to a further shift in vulnerability perception, toward an emphasis on nuclear risks, and on the instability of the grid and the risk of major blackouts, which was handled by increasing storage capacity in dams and building more transmission lines.[12]

Critical events

Often, changes in the perceptions of vulnerabilities and vulnerability trade-offs have been triggered by what we call "critical events." One well-known example is the nuclear accident at Three Mile Island in April 1979, which suddenly changed the general perception of the vulnerability of nuclear power, not only in the United States but also in Western Europe. Nuclear risks, which had been abstract and hypothetical, became much more tangible, and many countries changed their nuclear policies in response. In the Balkans, two earthquakes had an even stronger role in influencing the perceptions of nuclear vulnerability. In Bulgaria a major earthquake occurred in 1977, which had its epicenter only 30 km from the nuclear power plant in Kozloduy. Even though the plant was not damaged, the government started a process to improve its safety. And in 1981 an earthquake 200 km from the site of a planned nuclear power plant led to the abandonment of nuclear power as an option in Greece. Similarly, the vulnerability of airplanes having reached their cruising altitudes was perceived as low before two actually collided over the Grand Canyon in 1956. Although this accident happened in the United States and not in Europe, it formed the point of departure for a discussion in Western Europe about air-traffic security. This led the key actors to eventually reinterpret air-traffic risks, paving the way for a new transnational infrastructure for air-traffic control.[13]

Critical events have thus been crucial to how vulnerability perceptions have changed over time, and this is why such events play a prominent role in this book. Indeed, all chapters analyze critical events of different kinds. We have found the analytical significance of critical events to be threefold: they are useful sites for analyzing the nature of infrastructure vulnerabilities; they reveal how different actors respond to crises; and they give rise to fierce discussions affecting infrastructure's further evolution. Although the critical events studied here are very diverse in terms of causes, geographical location, and impacts, they have in common that they generated new societal and political debate about how to cope with infrastructure vulnerabilities.

One interesting finding from our analysis of critical events is that many of these in fact did not affect users but "only" the systems. This underlines the importance of the distinction we made in Chapter 1 between "system vulnerability" and "user vulnerability." A telling example is the fire in the Eemshaven converter station in April 2009, which led to a sudden interruption of the electricity flow through the NorNed cable between Norway and the Netherlands, lasting for a month. No electricity consumers in either country were directly affected by this interruption because the power companies were able to avoid a blackout. However, the power companies and grid operators lost millions of euros during the month and became aware of the vulnerability of the converter stations.[14]

In the current policy debate it has been popular to distinguish between "intended" and "unintended" causes of critical events between "internal and external sources of failure"[15] and between "internal and external threats."[16] Our results, however, show that such distinctions can rarely be defined objectively. Rather, the extent to which a critical event comes to be regarded as intended or unintended (or internal or external) is the outcome of debates and negotiations – and often no agreement is reached. The European blackout discussed above, for example, has politically and popularly been interpreted as an unintended event. As we have seen, however, the way in which certain regions were disconnected from the grid and others were not followed predefined and detailed so-called emergency load-shedding plans. From this perspective the course of the blackout can be viewed as intentional. Similarly, the European gas crises of 2006 of 2009 can be interpreted either as external events caused by political decisions taken in the Kremlin power struggle with the government in Kiev, or as an internal event in which the main problem was the failure of gas managers in Russia and Ukraine to renegotiate the terms of export and transit contracts.

All in all, our book shows that the ways in which vulnerability perceptions have changed over time, not least in response to major critical events, have reshaped Europe's transnational infrastructure in decisive ways. The struggle to make certain vulnerability perceptions rather than others dominant can be seen as a struggle to influence infrastructure's future. In particular, as emphasized in much of the policy literature, reducing vulnerability is often expensive, and actors therefore often have an interest in de-emphasizing vulnerability for economic reasons. De Bruijne and Van Eeten, for example, note that infrastructure nowadays, following the liberalization and privatization of many critical infrastructures, operate "closer to the edge" than before restructurings since actors have stronger incentives to maximize their profits.[17]

As our book has shown, however, the trade-off between economy and vulnerability did not start with liberalization and privatization. Transnational integration has often been pursued for economic reasons despite anticipated increases in vulnerability. The first plans for Western European imports of natural gas from the Soviet Union and Algeria, for example, were immediately contested due to anticipated implications for energy security. Western Europe could very well have managed without natural gas from beyond the Iron Curtain and the Mediterranean, but this possibility was silenced in the debate and had to give way to the view that the risks involved were worth taking. The new vulnerabilities linked to gas imports from far away were thus accepted for the sake of economic gain.[18] All in all it is obvious, judging from the studies in this book, that minimizing vulnerability is not necessarily in every stakeholder's interest.

Coping with vulnerability

Our book has revealed a variety of strategies to cope with transnational infrastructure vulnerabilities. Some of these have aimed at reducing the risk of failures or interruptions, while others have aimed at limiting the consequences of such events. Obviously, the nature of vulnerabilities differs between, say, energy infrastructure and transport infrastructure, and some of the strategies discussed below, are not relevant to all infrastructure.

In the case of energy infrastructure, the very *creation of physical transnational links* itself has often been perceived as an effective way of reducing vulnerability. We have seen, for example, how the European electricity grid emerged out of power companies' desire to reduce the risk of blackouts and electricity shortages. Power lines across borders made it possible to share reserve power plants and storage capacity, and to help each other if needed.[19] The forging of transnational gas connections was in many cases also understood as a way to escape structural energy shortages, which was widely interpreted as the most pressing vulnerability in the field of energy in postwar Europe.[20]

Trying to *establish reliable social relations* with partners on "the other end" of the connecting lines has also been an important strategy. Early electricity cooperation among power companies in neighboring countries was often based on existing relations of trust, and in many cases the cooperation was surprisingly informal, based on gentlemen's agreements. In the case of gas imports, both parties often had to make huge investments before an exchange could commence, and here long-term contracts became an important instrument for creating reliable relations. More generally, reliable relations can be established by developing *transnational governance* of infrastructures, and this will be discussed in the next section.

However, power or gas links to neighboring countries inevitably created (inter)dependencies, and to cope with these one common strategy has been to *diversify* by building links to several countries. Greece, for example, built power lines to its three neighbors in the north: Albania, Yugoslavia, and Bulgaria. All three were ideological adversaries of Greece, but as they were also adversaries among themselves, the Greek power company felt safe that they would not simultaneously stop cooperating with it.[21] In the case of natural gas, the Western European countries that started importing gas from the Soviet Union tried to balance these imports by signing contracts for gas from other regions as well.[22]

Other chapters, however, show how *refusing connections* could also function as a strategy for reducing vulnerability. In 2006, for example, the Finnish authorities refused Russia's offer for a submarine cable from Russia to the coast of Finland. The Finnish authorities reckoned that this connection would make the national Finnish grid vulnerable to capacity overloads. It would also make the country overly dependent on the Russian energy system. This fear of dependence on Russia was deeply rooted in the history of Finnish-Russian relations. The two countries had had a very strained collaboration in the past and although the Soviet Union had been dissolved long before the new discussions about a submarine cable began, the memory of this traumatic past refused to go away.[23] In the case of natural gas, in the 1980s the US administration tried to convince Western European governments not to expand their imports of Soviet natural gas, albeit in vain.[24]

An additional strategy for coping with foreign energy dependencies has been to *create back-up capacities* of different kinds in case a major disruption should occur. In electricity systems, such capacities took the form of reserve power plants or sometimes pumped storage facilities.[25] In gas supply, underground gas-storage facilities or idle domestic gas fields played similar roles. Moreover, in the case of natural gas, importing countries could also create "virtual" backup capacity by agreeing to help each other if interruptions occurred.[26] A related kind of strategy has been to *increase substitutability* by making it easy to switch from imported to domestic fuels in power plants, for instance. In the early 1970s, for example, Bulgaria built big power plants that could use both anthracite from Ukraine and domestic lignite as fuel, thereby reducing its dependency on Soviet coal imports.[27]

The above strategies were primarily applicable to energy infrastructures, but we have also discovered strategies of a more general relevance. One such strategy was to *standardize technology or procedures* across national boundaries so as to facilitate connections and exchanges. Standardization in emergency communication served as a way to improve cooperation among policemen and fire brigades in Europe, which was seen as crucial for coping with major critical events. However, several standards competed and an all-encompassing, pan-European standard was not achieved.[28] In aviation there was a long tradition of standardization of both technology and procedures going back to the emergence of civil aviation after the First World War. But the establishment of EUROCONTROL called for a more far-reaching standardization. There was a parallel in the case of electricity, where there was also a very long tradition of technical standardization, and where the increasing importance of information and communication technologies in recent decades has called for new kinds of standardization.[29]

When standardization has not been possible to achieve, historical agents often used *gateway technologies* to counter vulnerability. Such technologies make it possible to connect systems that would otherwise not be compatible. In the case of electricity, for example, high-voltage direct current (HVDC) links emerged as a transnational gateway technology of some importance, and made it possible to connect networks that did not operate synchronously. The main motive for constructing HVDC links was rarely to reduce vulnerability, but it was an important side-effect. Hence the Nordic countries, which were connected to the Union for Co-ordination of Production and Transmission of Electricity (UCPTE) only by way of HVDC cables, proved less vulnerable to disturbances on the Continent than other regions.[30] In the case of natural gas, a problem was that Dutch and north German natural gas had a lower calorific value than Soviet and Algerian gas, which meant that they were not interchangeable. Initially it was considered necessary to build gas-merging plants as a gateway technology to transform Soviet gas to the Dutch calorific level by adding nitrogen. This would have made it possible for Dutch gas to come to the rescue in case of supply disruptions from the East. However, building such plants would have been expensive, and when high calorific gas was discovered in the North Sea, this was seen as sufficient backup so that the plans for gas-merging plants were abandoned.[31]

An important strategy for coping with the risk of a breakdown of computers or communication links has been to prepare for *fallback to manual operation*. For example, EUROCONTROL made careful preparations so that it could continue to guide airplanes if its computers failed.[32] In electricity systems, similar preparations have been made, not least to ensure that nuclear power plants can be controlled if computers fail. However, the fast development towards "smart grids" is likely to make it increasingly difficult to cope with major failures of ICT technologies supporting critical infrastructure.

Transnational governance

An important contribution of our book lies in its analysis of European integration from a critical infrastructure perspective. Many of the chapters demonstrate that the creation of transnational infrastructure has been a complicated process, involving much more than just building physical links connecting national systems. There has also been a need to develop new forms of transnational governance to coordinate flows across borders, and to cope with vulnerabilities. The organizations and actors involved in these processes were often not very well known to the general public and have not been dealt with in traditional scholarship on the political and economic integration of Europe. Our book thus contributes to the understanding of the "hidden integration" of Europe, to use a concept introduced by Schot and Misa.[33]

The forms of transnational governance have varied over time, between different kinds of infrastructure and between different parts of Europe. Chapters 2 and 3 made it possible to compare the emergence of transnational governance for electricity and natural gas in Europe. Although both systems became transcontinental in extent and came to play key roles in Europe's energy supply, there were striking differences between the two in terms of transnational governance. In the case of electricity, major power companies in neighboring and "friendly" countries established organizations like the UCPTE (1951) and Nordel (1964). These rather informal bodies became important arenas for developing guidelines, standards, and plans for building grids across borders, and for establishing conditions for electricity exchange. In Eastern Europe a somewhat more hierarchical form of organization was established in the early 1960s, in which the Moscow control center was responsible for frequency regulation and power exchanges. The rather informal character of governance, particularly in the West, is partly explained by the fact that the interdependencies were not as strong as in the case of gas. Each country had its own power plants and could be self-sufficient if needed. Electricity exchange across borders was intended to increase the efficiency and reliability of supply.

In the case of gas, by contrast, transnational governance was mainly bilateral and strongly decentralized, with hardly any international organizations of any significance, apart from a few branch organizations that functioned as arenas for community formation and knowledge exchange. On the other hand, the state was often more strongly involved in shaping the European natural gas regime than in shaping transnational electricity. Long-term contracts between exporting and importing companies, with strong direct or indirect state involvement, enabled the establishment of transnational connections. These agreements were difficult to reach and often took a long time to negotiate because of the huge diversity of interests of the actors involved. At the outset it almost appeared impossible that a highly integrated system of pipelines would actually be built across so many countries, as the countries involved often had strong ideological divergences or had been at war not long ago. The importance of long-term contracts in the transnational gas governance can largely be explained by the huge investments that were needed in both exporting and importing countries, and the huge unidirectional flows of gas. This created strong interdependencies that were strictly regulated in the contracts.

In the case of aviation, a transnational governance regime was established after the First World War based on the Paris Convention of 1919. The collision of two airplanes over the Grand Canyon in 1956 led to an ambition to expand transnational cooperation to encompass coordinated flight control for the ever denser air traffic above Western Europe. The combination of supranational governance and improved air-traffic monitoring technology – with the aspiring name EUROCONTROL – was seen as the best way to reduce this growing vulnerability in air transportation. However, EUROCONTROL's role as a European integration project soon lost momentum as a consequence of its inability to extend its operations of air traffic control beyond Belgium, Luxembourg, and the Federal Republic of Germany by 1976. Its member countries were not willing to give up control of their national airspace. This prompted the reconceptualization of EUROCONTROL's objective from being supranational to becoming international, codified in the second EUROCONTROL convention of 1981.[34]

A similar failure to achieve functioning transnational governance was seen in the case of emergency communication. Again, some of the core members of the EU failed to agree on the development of a single common European standard for emergency communication, and instead two competing standards (Tetra and Tetrapol) were allowed on the European market. Each was backed by strong industrial interests, and in the 1990s, neoliberal policies of stimulating market competition took priority over EU policies to improve public safety with a single standard for all. As a result, a very scattered and heterogeneous pattern of communication networks emerged.[35]

In the current policy debate, the EU and its predecessors are often intuitively viewed as the most important organizations contributing to the emergence and governance of Europe's transnational infrastructure.[36] The omnipresence of the EU and its institutions in media and the public debate tends to obscure the key roles played by more specialized organizations working behind the scenes in various transnational contexts, such as the UCPTE and EUROCONTROL. The role of these organizations tends to attract attention only in the case of major critical events, such as the 2006 European blackout or the collapse of air traffic in Europe following the volcanic eruption in Iceland in 2010. Historically it can be seen that this is the type of organization that has been crucial for the shaping of transnational critical infrastructure in Europe. However, they often tried to avoid media attention to their endeavors as this might induce political debates that could cause obstacles. This media aversion contributed even further to the "hidden" character of the infrastructural integration of Europe.

A porous Iron Curtain

Our book provides additional insight into the hidden integration of Europe by focusing on the "Iron Curtain". Many of the chapters in this book demonstrate that this was often neither very solid nor effective when seen through the lens of transnational critical infrastructure. From the late 1960s the Iron Curtain became increasingly "porous" for energy flows, particularly for natural gas. It was less porous in the case of electricity. NATO actively sought to and managed to prevent any far-reaching integration in electricity between East and West. There was thus hardly any perceived vulnerability in Western Europe to communist "political manipulation" of East–West electricity links; the lines were too insignificant for that, and in case of disagreement the connections could easily be disconnected without any far-reaching consequences. This was definitely not the case for natural gas supplies from the East, where NATO tried but never succeeded in exerting any notable influence on transnational system-building. Put differently, military policy objections in the case of natural gas had a much smaller impact on the transnational network geography than geological and economic aspects.

The chapters in Part II of the book analyze various ways in which small countries on both sides of the Iron Curtain maneuvered their energy connections during the Cold War. Greece belonged to the Western camp after a civil war in the late 1940s. In the 1960s and 1970s it created power links with its Cold War enemies Albania, Bulgaria, and non-aligned Yugoslavia, but not with its NATO allies, Italy and Greece. The power cooperation with the three socialist neighbors worked very smoothly, even during the years of a reactionary military dictatorship in Greece. This pragmatic and non-ideological cooperation in the field of electricity continued after the Cold War and became particularly salient after the earthquake in 1995, when a number of Greek power plants ceased functioning. In this critical situation the power company in Macedonia (or the Former Yugoslavian Republic of Macedonia, as it is called in Greece) came to the assistance of its Greek colleagues with substantial power supplies, despite the tense political relations between the two countries.[37]

Bulgaria, located on the other side of the Iron Curtain, developed very close connections with the Soviet Union, and to a lesser degree with other Council for Mutual Economic Assistance (COMECON) countries, during the 1950s and 1960s, in terms of both material flows of electricity and coal and a massive transfer of competence in the fields of nuclear-power and coal-power technologies. However, from the 1970s onwards, Bulgaria was able to become gradually more independent from the Soviet Union in terms of technological skills, and it developed suitable technologies to produce power from its domestic lignite sources. This not only decreased its dependence on energy imports from the Soviet Union but also made it possible to export electricity from the lignite-fired power plants in the southern part of the country. Therefore Bulgaria built power lines to its Cold War enemies Turkey and Greece as well as to non-aligned Yugoslavia in the mid-1970s. This power cooperation has gradually developed and made Bulgaria into the power hub of the Balkans.[38]

Finland did not belong to the socialist camp but had an "uneasy alliance" with the Soviet Union during the Cold War, not least in the field of energy. The Kremlin often exerted strong political pressure to make Finland build cross-border infrastructure or accept Soviet technologies, while Finnish actors did their utmost to increase energy self-sufficiency. This tug-of-war became particularly apparent when the Finnish state-owned power company decided to build two nuclear power plants and invited international tenders. For political reasons the company was forced to accept a bid from the Soviet Union, but after hard negotiations it managed to achieve a deal entailing that the two reactors would be fundamentally redesigned and provided with a protective shield. The forced collaboration between energy actors in the two countries prevented the development of relations of trust, and this became manifest after the fall of the Soviet Union, when Finland was no longer politically dependent. In 1995, for example, the Finnish government rejected a Russian power company's offer to build a submarine cable across the Gulf of Finland.[39]

The geography of Europe's vulnerability

In Chapter 1 we coined the term "vulnerability geography" to emphasize that different spatial configurations of vulnerability can be discerned in relation to Europe's transnational infrastructure. Indeed, our case studies have brought to attention the strikingly unequal spatial distribution of vulnerability. This can be seen in relation to Ulrich Beck's theory of the "risk society." Beck argued that politics in the developed world has now reached a stage where the main issue is not so much the distribution of wealth but rather the distribution of risk.[40] Echoing this argument, much of the infrastructure studied in this book has become so omnipresent that there is nowadays hardly any place in Europe that does not have access to electricity, natural gas, air traffic, emergency services, and the like. There is thus a reasonable equality in terms of *access* to networks. The same cannot be said when it comes to distribution of *vulnerabilities*.

In general, our research results hint at a much larger number of critical events in Eastern Europe than in Western Europe. In the West the impressive performance of large-scale infrastructure, such as electricity networks and air-traffic systems, which are so complex that they "should have failed" to a much greater extent than they actually do, prompted scholars of risk to define the responsible organizations as "high-reliability organizations."[41] The geographic reach of such organizations and systems is limited, however, as demonstrated by our book. It is not true, as suggested by Fritzon et al., that "actual experience in Europe shows that modern systems are extremely reliable, especially if historical performance is used as a measure."[42] This claim possibly holds for Western Europe. To the east of the Iron Curtain, high-reliability organizations hardly existed. Economies and societies were troubled by a more or less constant stream of blackouts, gas disruptions, air-traffic accidents, and computer failures. As in the West, some of these events may be said to have been planned, as in the case of electricity consumers being shut off in accordance with a predefined plan. But many events were not planned, and the responsible organizations completely lost control over the systems that they were set to manage. Due to the high frequency of failures, no one was surprised if an international phone call was suddenly interrupted, if the lights went out, or if there was suddenly no gas for cooking at home. Users had become accustomed to breakdowns. In short, Eastern Europe lived in a completely different vulnerability world.

Western European actors were rarely aware of the radical differences between East and West in terms of vulnerability. Had they known, for example, that the Soviet natural gas system and the attempts to expand it were in such a mess, they would hardly have been as willing to engage in far-reaching cooperation as they actually were. Lack of knowledge thus stimulated the expansion of transnational infrastructure across the Iron Curtain. When Western actors did get a detailed insight into the true functioning of the Soviet infrastructure they were shocked.

In the case of natural gas there was a direct connection between the vulnerabilities in the East and the West. While the Soviet Union did its utmost to live up to its contractual obligations vis-à-vis importers in Western Europe, the main victims of Soviet failures to produce sufficient quantities of gas were households and industries in the Soviet Union itself. Thus, while gas customers in Western Europe enjoyed steady supplies, millions of households in the Soviet Union experienced painful gas shortages, not least in wintertime.[43] On a much smaller scale there was a similar dependency across the Iron Curtain between Bulgaria and Turkey. In the early 1980s a Turkish region bordering with Bulgaria was supplied by the Bulgarian lignite power plants, which had excess capacity. However, the Bulgarian authorities adopted a very different approach to their foreign customers; when a domestic crisis of electricity supply occurred in Bulgaria in 1985, the exports across the border were cut, putting the Turkish customers in a precarious situation. As a result, Turkey decided to build new power stations to reduce its dependence on Bulgarian electricity deliveries.[44]

Vulnerability geographies, as analyzed in several chapters, were typically found to be the result of historical patterns of transnational system-building. In the 2006 "European blackout", for example, the origin of the failure in northern Germany caused lights to go out as far away as in Italy, Spain, and Portugal – and even in Morocco, Algeria, and Tunisia – but not in neighboring Denmark and Sweden. The division line between areas that were affected and those that were not coincided with the historical division between UCPTE and Nordel established in the late 1950s. Central and Eastern Europe, despite operating in parallel with the Western European grid, were also less affected by the blackout than regions located to the west of the former Iron Curtain. The political divide in Cold War Europe thus became visible again in connection with the blackout, despite the demise of the Iron Curtain nearly two decades earlier. In the case of the 2006 and 2009 European gas crises, the former Iron Curtain also seemed to reappear, as the former communist countries to the east of it were most severely affected by the disruption of gas deliveries from Russia.

It is important to note, though, that vulnerability geographies of Europe often did not correspond to the political borders of nation-states, nor to established political or military "blocs." An important result of our empirical studies is that some Western European countries have seen the reliability of their infrastructure increase thanks to well-developed physical connections with regions beyond the Iron Curtain. Moreover, actors on opposite sides of the Iron Curtain were often prepared to share vulnerabilities with one another. This is an intriguing result, especially when seen in relation to the overall Cold War tensions. It contradicts the view taken by Linnerooth-Bayer, who argues that

> if the countries involved have a history of conflict, cultural differences, or ongoing tensions, even minor potential "exports" of risk may generate widespread media coverage, societal attention, public concern and protests. The public may be especially averse to even minor risks emanating from a country they regard as hostile or untrustworthy, especially if the public views the risk producers as receiving large benefits.[45]

The closest we come to such a situation in our book is in the case of Finnish-Soviet energy relations. But in most cases we have seen the opposite. A particularly striking case was in connection with the 1995 earthquake in northern Greece, when neighboring Macedonia did not hesitate to help the Greeks.

Ironies of vulnerability

In Chapter 1 we discussed a number of ironies and paradoxes in relation to Europe's transnational infrastructure vulnerabilities. Let us conclude this book by emphasizing three ironies that have been demonstrated in several of our case studies.

One conclusion arrived at in several of our case studies is that transnational infrastructure, as it evolves over time, tends to become increasingly reliable but at the same time increasingly vulnerable. In the early career of a transnational network, failures, and breakdowns of various kinds may be frequent but (user) vulnerability may still be perceived as low, since the system is not yet firmly integrated with other user activities. Moreover, users often keep older technologies as a backup in case of breakdowns – for example, a wood stove that can replace an electric stove if needed. Once everyday activities become more deeply integrated with and dependent on the infrastructure in question, however, user tolerance is reduced and the perceived vulnerability increases – despite technical advancements and improvements. Thus in the twenty-first century users are often deeply disturbed by any blackout, gas disruption, airplane delay, or the like, no matter how brief the event. The general expectation is that infrastructure should always be available and functional, and any deviation from this availability and functionality tends to give rise to dismay.

The second irony that emerged in many of our chapters is that efforts to reduce infrastructure vulnerabilities, whether or not successful, often generate new vulnerabilities. For example, the introduction of emergency communication standards in the 1990s was intended to facilitate cross-border emergency collaboration and thereby make Europe safer in case of major accidents. However, it also generated new vulnerabilities by increasing our dependence on the suppliers of these technologies and their willingness to improve and carefully maintain the system. There are also many examples of this irony in relation to efforts to reduce vulnerabilities in electricity supply. Establishing power connections to neighboring countries for assistance in case of electricity shortage created a new risk of major blackouts cascading across borders. And the introduction of information and communication technologies for better surveillance of power grids led to greater complexity and the new risk that the information technologies themselves will break down.

A final irony is that infrastructure managers and regulators base their strategies for coping with infrastructure vulnerabilities on experiences of past critical events, but they are often totally perplexed when new, unforeseen events occur. This was illustrated in many of our historical cases, and it has also been demonstrated in a number of more recent events. For example, in April 2010 a cloud of volcanic ash originating in Iceland spread over Europe, paralyzing transnational air traffic for a period of several weeks, with disastrous consequences for many airlines as well as for travelers. Similarly, in March 2011 a devastating tsunami hit Japan and totally destroyed the huge Fukushima nuclear power plant. The most challenging task for those in charge of transnational European infrastructure might thus be to *cope with unknown and unforeseen vulnerabilities*. It remains to be seen how Europeans will respond to this challenge. International organizations such as the EU and the various sector organizations featuring in this book are sure to advance intensified cooperation between actors in different countries as one way to go ahead. But as we have seen, international cooperation does not provide any simple solution. Judging from our findings, it may even bring about new vulnerabilities. If history is to be taken as a guide, we will have to learn to live with such a constantly evolving vulnerability landscape, and, although we may wish to, we are unlikely to ever arrive at any final destination on this historical journey. What we can say with some certainty, though, is that the future calls for humility and an open mind.

Notes

1. For instance, Jasanoff 1998; Summerton and Berner 2003; Lagendijk and Van der Vleuten 2010b.
2. Summerton and Berner 2003, p. 6.
3. Chapter 3.
4. Chapter 9.
5. Van der Vleuten and Lagendijk 2010b, p. 2057.
6. Chapter 6.
7. Chapter 5.
8. Chapter 5.
9. Summerton and Berner 2003, p. 6.
10. Chapter 6.
11. Chapter 4.
12. Chapter 4.

13. Chapter 7.
14. Chapter 9.
15. Boin and McConnell 2007, p. 1.
16. Gheorghe et al. 2007, p. 8.
17. De Bruijne and van Eeten 2007, p. 20.
18. Chapter 2.
19. Chapters 3 and 4.
20. Chapter 2.
21. Chapter 5.
22. Chapter 2.
23. Chapter 6.
24. Chapter 2.
25. Chapters 4 and 9.
26. Chapter 2.
27. Chapter 4.
28. Chapter 8.
29. Chapter 9.
30. Chapter 3.
31. Chapter 2.
32. Chapter 7.
33. Misa and Schot 2005.
34. Chapter 7.
35. Chapter 8.
36. For example, Fritzon et al. 2007, p. 38.
37. Chapter 5.
38. Chapter 4.
39. Chapter 6.
40. Beck 1992.
41. Roberts 1990; La Porte and Consolini 1991; La Porte 1996; Rochlin 1996.
42. Fritzon et al. 2007, p. 37.
43. Chapter 2.
44. Chapter 4.
45. Linnerooth-Bayer 2001 p. 23f.

Acknowledgments

This book would have been almost impossible without the critical European infrastructure that it studies. It is the result of a long joint endeavor involving 17 authors at eight universities in seven different countries. Many more scholars helped us to develop key ideas and concepts during workshops in Copenhagen, Utrecht, Stockholm/Sigtuna, Helsinki, Lisbon, Athens, and Sofia. We traveled – fairly cheaply – by air and public transport. We sent thousands of emails, often with long attachments of text, and we used a common website (www.eurocrit.eu), which has had more than 75,000 visitors.

Our endeavor started in May 2006 when some of us sat down on a canal boat from Vyborg in Russia to Lappeenranta in Finland to discuss the possibility of submitting a research proposal. A call from the European Science Foundation (ESF) for a EUROCORES program named "Inventing Europe" had just opened. Its purpose was to shed new light on European history by looking at Europe through the lens of technology. Transnational infrastructure was explicitly mentioned as a crucial topic within the scope of the call. Taking inspiration from academic and non-academic discussions about "critical infrastructure" that appeared to have grown enormously in response to dramatic events such as the terror attacks in New York (2001), Madrid (2004), and London (2005), recurring Russian-Ukrainian "gas crises", and the "European blackout" (2006), we decided to focus our proposal on the common connections and shared vulnerabilities that are part and parcel of transnational infrastructure systems, and on how these vulnerabilities have been handled by different actors in the countries involved.

This proved to be a very fruitful idea. Our project, entitled "EUROCRIT: The Emergence and Governance of Critical Transnational European Infrastructures," was launched in June 2007. From the outset one of its main goals was to produce a joint book, and in October 2008 an editorial team (the undersigned) was formed to coordinate and guide the process toward its publication. The participation of scholars from different countries, with broad language skills and a deep knowledge of historical contexts, was a prerequisite for the research we pursued. All in all the project participants consulted 21 archives in 13 countries, conducted a large number of in-depth interviews, and scrutinized a vast volume of additional documentary material in different languages. Another language challenge was that none of the project participants had English as their mother tongue.

In a project of this scope there are many more people than the actual project members who have been helpful in various ways – so many, in fact, that it is impossible to name them all. We want to express our warm and sincere thanks to all academic colleagues who participated in our workshops, gave constructive feedback on our manuscripts, and presented inspiring papers of their own, and to the many people who have provided the hidden administrative infrastructure for our research: the helpful and encouraging officials at the ESF and at the national research funding agencies that have financed our work, all of the tireless administrative staff at our home universities, the knowledgeable archivists and librarians, and at a later stage the supportive editors at Palgrave Macmillan. Three anonymous peer reviewers – one for each part of the book – contributed decisively to improving our texts.

Finally we want to thank one person explicitly – Johan Schot. Without his unique and seemingly inexhaustible enthusiasm and managerial skills, the whole "Inventing Europe" program of which our project was part would not have been possible.

Per Högselius, Anique Hommels,
Arne Kaijser, Erik van der Vleuten
May 2013

Contributors

Anna Åberg holds a PhD in the history of science, technology, and environment from the Royal Institute of Technology, Stockholm, Sweden. Her thesis focuses on the attempts to introduce natural gas into Sweden in the late twentieth century. Her particular topics of interest are energy history, infrastructure history, future studies, and visions of future technologies and energy systems, particularly in popular culture.

Stathis Arapostathis is a lecturer in the Department of Philosophy and History of Science at the National and Kapodistrian University of Athens, Greece. In 2006 he finished his DPhil at Oxford University, UK (entitled "Consulting Engineers in the British Electric Light and Power Industry, c. 1880–1914"). With Graeme Gooday he co-authored the book *Patently Contestable: Electrical Technologies and Inventor Identities on Trial in Britain* (2013).

Eefje Cleophas is a PhD researcher in the Department of Technology & Society Studies, Maastricht University, the Netherlands. In her thesis she explores tensions between processes of standardization and differentation in practices of measuring, marketing, and testing car sound between 1958 and 2000.

Yiannis Garyfallos is a doctoral student in the Graduate Program in History and Philosophy of Science and Technology, National and Kapodistrian University of Athens and National Technical University of Athens, Greece. His dissertation research on the history of energy in Greece has been supported by the Greek State Fellowship Foundation.

Lars Heide is an associate professor in the Centre for Business History at the Copenhagen Business School, Denmark. He has written and published extensively on the development and application of information and other technologies in various countries and their implications for the development of society. He is the author of *Punched-Card Systems and the Early Information Explosion, 1880–1945* (2009).

Per Högselius is an associate professor in the Division of History of Science, Technology and Environment at the Royal Institute of Technology, Stockholm, Sweden. His research has focused on international aspects and in particular East–West relations in the history of science, technology, and environment. Most recently he has published *Red Gas: Russia and the Origins of European Energy Dependence* (2013). In Sweden he is also active as an author of popular history books and newspaper articles.

Anique Hommels is an associate professor in the Science, Technology & Society (MUSTS) research group, Maastricht University, the Netherlands. She completed her PhD there in science, technology, and society studies in 2001. She is the author of *Unbuilding Cities. Obduracy in Urban Sociotechnical Change* (2005). Her current research focuses on vulnerability in technological cultures, urban disasters, and standardization in emergency communication.

Ivaylo Hristov is a PhD candidate at Eindhoven University of Technology, the Netherlands, and Plovdiv University, Bulgaria. Supported by the Foundation for the History of Technology, he is currently completing a thesis project entitled "The Communist Nuclear Era: The Bulgarian Atomic Community during the Cold War."

Arne Kaijser is Professor of History of Technology at the Royal Institute of Technology, Stockholm, Sweden. His main research interests concern infrastructure, institutions, and environment in historical perspective. Together with Erik van der Vleuten he edited *Networking Europe: Transnational Infrastructures and the Shaping of Europe, 1850–2000* (2006).

Vincent Lagendijk works as a postdoc in the Department of History, Maastricht University, the Netherlands. He completed his PhD thesis at Eindhoven University of Technology in 2007. Entitled 'Electrifying Europe: The Power of Europe in the Construction of Electricity Networks,' it deals with the historical development of ideas about European unification in correlation to electricity network-building in the period 1918–2005. He is currently writing a transnational history of the Tennessee Valley Authority.

Karl-Erik Michelsen is Professor of Science, Technology and Transformation of Modern Societies at Lappeenranta University of Technology, Finland. His Finnish-language publications include *State, Technology, and Research*, *The Fifth Estate: Engineers in Finnish Society*, *Work, Production, and Efficiency*, and most recently *The Finnish Nuclear Power Program*. He is also the author of *Unknown Forest*, in English. He has written several methodological articles on the history of technology and contributed to general works on Finnish history.

Tihomir Mitev graduated in sociology from Plovdiv University, Bulgaria, in 2002. He completed his PhD thesis at the Institute of Sociology, the Bulgarian Academy of Sciences, in 2008. He is currently working as assistant professor at Plovdiv University, and his main research interests are in the field of science and technology studies, particularly actor-network theory and large technical systems. His publications include "Academic Spin-Offs as an Engine of Economic Transition in Eastern Europe. A Path-Dependent Approach," *Minerva* 48 (2010), 2, 189–217 (co-authors: Tchalakov, I. and Petrov, V.).

Ivan Tchalakov is a senior research fellow in the Technology Studies Group, Department of Sociology of Science and Education at the Institute of Sociology, Bulgarian Academy of Sciences. He is also an associate professor in the Department of Sociology, Plovdiv University, Bulgaria. He is working in the field of the historical sociology of socialism, focusing on science and technology developments in South-Eastern Europe after the Second World War.

Lars Thue is Professor of Modern History in the Centre of Business History, BI Norwegian Business School, Oslo, Norway. The history of electricity, telecommunications, and other infrastructures has been his main interests, but he has also done extensive work on other topics. He is co-editor and contributor to the book *Creating Nordic Capitalism. The Business History of a Competitive Periphery* (2008).

Aristotle Tympas is Associate Professor of the History of Technology at the National and Kapodistrian University of Athens, Greece. He specializes in the history of automation and related technologies (e.g. computing) in connection with engineering and the technology–environment relationship.

Katerina Vlantoni is a PhD student in the Department of Philosophy and History of Science, National and Kapodistrian University of Athens, Greece. She is writing her dissertation on issues concerning risk and safety in the context of transfusion medicine, focusing on debates about the introduction of biotechnology into blood screening. A recipient of a three-year dissertation scholarship "Heracleitus II," she has benefited from a predoctoral visiting fellowship to the History Office, National Institutes of Health, USA.

Erik van der Vleuten teaches at the School of Innovation Sciences, Eindhoven University of Technology, The Netherlands, and currently chairs the management committee of the pan-European research network *Tensions of Europe. Technology and the Making of Europe* (www.tensionsofeurope.eu). He studies the mutual shaping of infrastructure, societal, and environmental changes and has published on electricity, water, food, financial, and ecological networks. He co-authored, with Högselius and Kaijser, *Europe's Infrastructure Transition: Economy, War, Nature* (to be published in 2014).

Bibliography

Archives

Archives of the Finnish Foreign Ministry, Helsinki (UM)
Archives of the Finnish Ministry of Trade and Industry/Atomic Office, Helsinki (KTM/Atomitoimisto)
Austrian State Archive, Vienna (OeStA)
Bavarian State Archive, Munich (BayHStA)
Berlin Regional Archive (LAB)
Bulgarian Central State Archives, Sofia
Bulgarian Ministry of Foreign Affairs Archive, Sofia
Central State Archive of the Highest Organs of Government and Administration of Ukraine, Kiev (TsDAVO Ukrainy)
Eurocontrol Archives, Brussels
Federal German Archives, Koblenz and Berlin (BArch)
Historical Archives of the European Union, Florence
Imatran Voima Archives, Helsinki (IVOA)
ISC Archive, Odijk
Political Archive of the German Foreign Office, Berlin (PA AA)
PPC Archives, Athens
Russian State Archive of the Economy, Moscow (RGAE)
Swedegas Archive, Stockholm
Swedish Foreign Ministry Archive, Stockholm
Swedish National Archives, Stockholm (RA)
UN Archives, Geneva (UNOG)
UTCE Archives, Brussels

Interviews

Chapter 2:

Poul Nielson, Former Danish Minister of Energy, April 1, 2008
Torkel Ösgård, Former Employee of Swedegas AB, May 30, 2008
Carl-Axel Petri, Former Swedish Minister of Energy, May 29, 2008
Michael Schultz, Employee of DONG, November 11, 2008

Chapter 5:

Oved Tadjer, Former Vice Minister of Electricity, February, 2009
Nikola Todoriev, Former Minister of Electricity, April, 2004
Hristo Todoriev, Professor of Electric Power Engineering at Sofia Polytechnics (son of Nikola Todoriev), May, 2009
Nikolay Ivanov, Manager of the Power Engineering Company TOTEMA, September, 2009
Nikita Nabatov, Lasting investment Manager at Bulgarian Power Industry, September, 2009
Hristo Hristov, Deputy Director at Thermal Power Plant Mariza East 2, April, 2010
Mitko Iankov, Engineer at Kozloduy Nuclear Power Station, April, 2008

284 *Bibliography*

Chapter 7:

Hansjürgen von Villiez, First Director of Eurocontrol's Maastricht Centre, June 2008

Chapter 9:

Former High-ranked Civil Servant Dutch Ministry of Justice; June 16, 2008, Leiden (NL)
Head of the Knowledge and Information Centre VTS, Police Netherlands (ISC), May 15, 2007, Odijk (NL) by Bistra Vasileva
Telecom expert Dutch Police, Dutch Representative Schengen Telecom, System Architect C2000 May 9, 2005, Odijk; January 29, 2009, Driebergen (NL)
Chair of the 3CP Working Group Business Processes, Coordinator of the Federal Police in Eupen (Belgium), June 2008, Eupen (Belgium) by Bistra Vasileva
Chief Commander of the Fire Department of Enschede at the Time of the Enschede Disaster, Director of the Twente Safety Region, January 22, 2008, in Enschede (NL)
Member of the Project Agency 'Digitalfunk' (Digital Radio) of the German Police and Member of the 3CP Working Group Business Processes, June 5, 2007, Aachen (Germany) by Bistra Vasileva
Member of the Tetra Pilot Project Aachen and of the 3CP Working Group Business Processes, June 5, 2007, Aachen (Germany) by Bistra Vasileva
Advisor General at the Directorate-General of the Civil Security and Chair of the 3CP Steering Committee, May 14, 2007, in Brussels (Belgium) by Bistra Vasileva
Civil Servant Ministry of the Interior, Responsible for International Contacts Police; September 26, 2007, in Den Haag (NL)
Member of the 'Risk Management' Team at the 'Safety Region Twente', January 22, 2008, Enschede (NL)
Secretary of the 3CP Steering Committee, July 4, 2007, Swisstal (Germany) by Bistra Vasileva
Project Leader of C2000 and Member of the 3CP Coordination Group, June 11, 2007, The Hague (NL) by Bistra Vasileva
Financial and Planning Manager of the Dutch C2000-project and Member of the 3CP Steering Committee, June 11, 2007, The Hague (NL) by Bistra Vasileva
Chief Commander of the Fire Department of Vreden at the Time of the Enschede Disaster, January 22, 2008, in Vreden (Germany)
Member of the 3CP Coupling Group, and Communication Systems Engineer for the German Federal Police, June 12, 2007, Aachen (Germany) by Bistra Vasileva
Chief of the South-Limburg Police Corps, March 11, 2008, Maastricht (NL)
Chair of the Working Group Coupling of the Three Country Pilot, May 15, 2007, Odijk (NL) by Bistra Vasileva

Periodicals

Associated Press
BBC News
Brand en Brandweer
Economic Postman
Economy and Society (Οικονομία και Κοινωνία)
Elektrotechnische Zeitschrift
Energia-lehti
EUbusiness
EuroNuclear
Helsingin Sanomat
Independent

International Herald Tribune
Makedonia
Nea
New York Times
Nordicum
Oil and Gas Journal
ÖMV-Zeitschrift
Süddeutsche Zeitung
Tekniikan Waiheita
Tekniikka ja Talous
Teknisk Ukeblad
Wall Street Journal

Printed sources

Adas, Michael. *Machines as the Measure of Men: Science, Technology, and Ideologies of Western Dominance.* Ithaca: Cornell University Press, 1989.

Agriantoni, Christina. "I mihaniki ke i viomihania: Mia apotihimeni synantisi [Engineers and industry: An unsuccessful meeting]." In Christos Hatziiosif (ed.) *Istoria tis Elladas tou ikostou eona: 1922–1940, O Mesopolemos [History of Greece in the 20th century: 1922–1940, The Interwar Period]*, Vol. B1, 268–293. Athens: Vivliorama, 2002.

Aharoni, Yair. *The No-Risk Society.* Chatam, New Jersey: Chatham House Publishers, 1981.

Allmer, Helmuth. "Extension planning of the Austrian interconnected network with regard to HVDC back-to-back links (Report no. 2)." In *UNECE Seminar on High Voltage Direct Current Techniques,* Stockholm, May, 1985. UN doc. ser., EP/SEM.10/report no. 2. Geneva: UNECE, 1985.

Anastasiadou, Irene. *Constructing Iron Europe: Transnationalism and Railways in Interwar Europe.* Amsterdam: Amsterdam University Press, 2011.

Angelo, A.R. "Elektricitetsproduktionens centralisation." *Ingeniøren* 26 (1917): 317–325.

Antoniou, Yiannis. "Ethniko Metsovio Polytechnio: Shedia ke Mateosis [National Metsovian Technical University: Plans and cancellations]." *Theseis* 89 (October–December 2004).

Antoniou, Yiannis. *I Ellines Mihaniki: Thesmi ke Idees 1900–1940 [The Greek Engineers: Institutions and Ideas 1900–1940].* Athens: Vivliorama, 2006.

Antoniou, Yiannis. "I Efarmogi tou Systimatos Taylor stin Anikodomisi tis Korinthou (1928) ke i Idees tis Rationalization ston Elliniko Mesopolemo [The Implementation of Taylor's System in the Reconstruction of Corinth (1928) and the Ideas of Rationalization in the Greek Interwar Period]." In Vilma Hatsaoglou, Christina Agriantoni, Sokratis Anagnostou, Egli Dimoglou, Argyroula Doulgeri, Vasilis Kolonas, Emmanuel Marmaras, Ilias Beriatos, and Aggelos Siolas (eds.). *Polis tis Mesogiou meta apo Sismous [Mediterranean Cities after Earthquakes]*, 60–69. Volos: Volos Publications, 2009.

Antonopoulos-Ntomis, M. "I Asfalia ton Pirinikon Antidrastiron Ishios WWER-440/230 [The Safety of the Nuclear Power Reactor WWER-440/230]." Paper presented at the conference Kindinos Pirinikou Atihimatos: I Periptosi tou Kozlodoui, Thessaloniki, May 12, 1993.

Applegate, C. "A Europe of Regions: Reflections on the historiography of sub-national places in modern times." *American Historical Review* 104, 4 (1999): 1157–1182.

Arapostathis, Stathis, Foteini Tsaglioti, Katerina Vlantoni, Ioannis D. Margaris, Vaso Aggelopoulou, and Aristotle Tympas. "Power and resistance from nuclear plants to wind parks: A history of the Greek experience." Paper presented at the ICHOTEC 2010 Conference, Tampere, Finland.

Badenoch, Alec. "Myths of the European network. Constructions of cohesion in infrastructure maps." In Badenoch and Fickers (eds.), 2010, 47–77.

Badenoch, Alec and Andreas Fickers (eds.). *Materializing Europe: Transnational Infrastructures and the Project of Europe.* Basingstoke and New York: Palgrave Macmillan, 2010.

Baker, Peter and Susan Glasser. *Kremlin Rising: Vladimir Putin's Russia and the End of Revolution.* London: Scribner, 2005.

Batsis, Dimitris. *I Varia Viomihania stin Ellada: I Aksiopiisi ton Ploutoparagogikon Poron. To Ikonomiko Shedio gia tin Ekviomihanisi. Ekviomihanisi ke viosimotita [Heavy Industry in Greece: The Utilization of the Resources of Wealth. The Economic Plan for Industrialization, Industrialization and Viability].* Athens: Kedros, 1977.

Beck, Ulrich. *Risk Society: Towards a New Modernity.* Translated by Mark Ritter. London: Sage, 1992.

Bekkers, R. *Mobile Telecommunication Standards. GSM, UMTS, Tetra, and Ermes.* Boston and London: Artech House, 2001.

Bekkers, V.J.J.M. and Thaens, M. "Interconnected networks and the governance of risk and trust." *Information Polity* 10, 1 (2005): 37–48.

Bergin, Thomas J. and Thomas Haigh. "The commercialization of database management systems, 1969–1983." *IEEE Annals of the History of Computing* 31, 4 (2009): 26–41.

Berthelot, Y. (ed.). *Unity and Diversity in Development Ideas. Perspectives from the UN Regional Commissions.* Bloomington: Indiana University Press, 2004.

Bijker, Wiebe. *Of Bicycles, Bakelites and Bulbs. Toward a Theory of Sociotechnical Change.* Cambridge, MA: MIT Press, 1995.

Bijker, Wiebe. "The vulnerability of technological culture." In Helga Nowotny (ed.). *Cultures of Technology and the Quest for Innovation,* 52–69. New York: Berghahn Books, 2006.

Bijker, Wiebe. "Globalization and vulnerability: Challenges and opportunities for SHOT around its 50th anniversary." *Technology & Culture* 50 (2009): 600–612.

Boin, Arjen and Allan McConnell. "Editorial: Unravelling the puzzles of critical infrastructures." *Journal of Contingencies and Crisis Management* 15 (2007): 1–3.

Boin, Arjen, Magnus Ekengren, and Mark Rhinard. "Protecting the union: Analysing an emerging policy space." *Journal of European Integration* 28, 5 (2006): 405.

Boll, Georg. *Entstehung und Entwicklung des Verbundbetriebs in der deutschen Elektrizi-täts-wirt-schaft bis zum europäischen Verbund. Ein rückblick zum 20-jährgen Besiehen der Deutschen Verbundsgesellschaft e.V., Heidelberg.* Frankfurt: Verlags- u. Wirtschaftges. d. Elektrizitätswerke m.b.H., 1969.

Bondarenko, A.F., Yu.N. Kucherov, E.I. Petryaev, M.G. Portnoy, and Yu.N. Rudenko. "Entwicklungsstand und Perspektiven der Energieversorgung in der ehemaligen UdSSR." *Energiewirtschaftliche Tagesfragen* 42, 6 (1992): 384–388.

Bouneau, Christophe. "Transporter." In Maurice Lévy-Leboyer and Henri Morsel (eds.). *Histoire générale de l'électricité en France. l'Interconnexion et le marché, 1919–1946,* 777–902. Paris: Fayard, 1994.

Breulmann, H., E. Grebe, M. Lösing, W. Winter, R. Witzmann, P. Dupuis, M.P. Houry, et al. *Analysis and Damping of Inter-Area Oscillations in the UCTE/CENTREL Power System.* CIGRE Session, 2000.

Brochmann, Georg. *De store oppfinnelser. VI bind.* Oslo: Nasjonalforlaget, 1930.

Bulgarian Academy of Sciences. *Bulgarian-Soviet Relationships and Connections. Documents and Materials. Volume II, September 1944–December 1958.* Sofia: Bulgarian Academy of Sciences Press, 1981.

Bulgarian Academy of Sciences. *25 Years Bulgarian Nuclear Reactor.* Sofia: Bulgarian Science Academy Publisher, 1986.

Burgess, J. Peter. "Social values and material threat: The European programme for critical infrastructure protection." *International Journal of Critical Infrastructures* 3 (2007): 471–487.

Cahen, François and Bernard Favez. "Control of frequency and power exchanges within the framework of international interconnections (report IV.2)." In *UNIPEDE Congress of Scandinavia.* Paris: Imprimerie Chaix, 1964.

Castle, Stephen. "EU summit fails to address protectionism fears." *The Independent,* March 26, 2006.

CEEC. *Committee of European Economic Co-Operation. General Report*. Vol. 1. Washington, DC: U.S. Government Printing Office, 1947.
Chandler Jr., Alfred D. *The Visible Hand. The Mangerial Revolution in American Business*. Cambridge, MA: The Belknap Press, 1977.
Chandler Jr., Alfred D. *Scale and Scope. The Dynamics of Industrial Capitalism*. Cambridge: The Belknap Press of Harvard University Press, 1990.
Chevalier, Michel. *Système de la Méditerranée. Articles extraits du Globe*. Paris: Bureau du Globe, 1832.
Cioc, Mark. *The Rhine: An Eco-Biography, 1815–2000*. Seattle: University of Washington Press, 2002.
Cole, Laurence and Philip Ther. "Introduction: Current challenges of writing European history." *European History Quarterly* 40, 4 (2010): 581–592.
Comité Intergouvernemental créé par la Conférence de Messine. *Rapport des Chefs de Délégation aux Ministres des Affaires Etrangères*. Document Mae 120 f/56 (corrigé). Brussels: Secrétariat, 1956.
Commissie onderzoek vuurwerkramp. *De vuurwerkramp. Eindrapport*. Enschede/Den Haag, 2001.
Commission of the European Communities. "First guidelines for a common energy policy." *Bulletin of the European Communities* 12, 1/12 (1968).
Commission of the European Communities. "Single European act." *Official Journal of the European Communities* L169 (1987): 1–29.
Commission of the European Communities. *The Internal Energy Market*. COM (88) 238 final. Brussels, 1988.
Commission of the European Communities. *Critical Infrastructure Protection in the fight against terrorism. Communication from the Commission to the Council and the European Parliament*. COM (2004) 702 final. Brussels, 2004.
Commission of the European Communities. *Green Paper on a European Programme for Critical Infrastructure Protection*. Doc. COM (2005) 576 final. Brussels, 2005.
Commission of the European Communities. *Proposal for a Directive of the Council on the Identification and Designation of European Critical Infrastructure and the Assessment of the Need to Improve their Protection*. Doc. COM (2006) 787 final. Brussels, 2006.
Commission of the European Communities. *Proposal for a Directive of the European Parliament and of the Council Amending Directive 2003/54/EC Concerning Common Rules for the Internal Market in Electricity*. COM (2007) 528 final. Brussels, 2007.
Commission of the European Communities. *EU Energy in Figures 2010*. Available at http://ec.europa.eu/energy/publications/doc/statistics/part_2_energy_pocket_book_2010.pdf.
Coopersmith, Jonathan. *The Electrification of Russia, 1880–1926*. Ithaca: Cornell University Press, 1992.
Coopersmith, Jonathan. "Soviet electrification: The roads not taken." *IEEE Technology and Society Magazine* 12, 2 (Summer 1993): 13–20.
Coudenhove-Kalergi, Richard. *Paneuropa ABC*. Vienna: Paneuropa-Verlag, 1931.
Coutard, Olivier. "Urban space and the development of networks: A discussion of the 'Splintering urbanism' thesis". In Coutard et al. (eds.) 2005, 48–63.
Coutard, Olivier, Richard Hanley and Rae Zimmerman (eds.). *Sustaining Urban Networks. The Social Diffusion of Large Technical Systems*. London: Routledge, 2005.
Cutter, Susan L. *Living with Risk: The Geography of Technological Hazards*. London: Arnold, 1993.
Dahl, Klaus H. *Erdgas in Deutschland. Entwicklung und Bedeutung unter Berücksichtigung der Versorgungssicherheit und des energiepolitischen Ordnungsrahmens sowie des Umweltschutzes*. PhD dissertation, Technische Universität Clausthal, 1998.
Damian, Michael. "Nuclear power: The ambitious lessons of history." *Energy Policy*, July 1992: 596–607.
Davies, Norman. *A Modern History of Europe*. Oxford: Oxford University Press, 1996.

Davis, J.D. *Blue Gold: The Political Economy of Natural Gas.* London: Allen & Unwin, 1984.
Deaynov, D. "Obshtestvoto na mrejite kato problem predi sotsiolanalizite [The society of networks as a problem before socioanalysis]." *Sotsiologicheski problemi* 2003, 1/2.
De Bruijne, Mark. *Networked Reliability. Institutional Fragmentation and the Reliability of Service Provision in Critical Infrastructures.* PhD dissertation. Delft University of Technology, 2006.
De Bruijne, Mark and Michel van Eeten. "Systems that should have failed: Critical infrastructure protection in an institutionally fragmented environment." *Journal of Contingencies and Crisis Management* 15, 1 (2007): 18–29.
De Heem, Louis. "Expérience acquise dans le fonctionnement interconnecté du réseau belge avec les réseaux des pays voisins." In *Report to UNIPEDE Congres: Comité d'études des interconnexions internationales.* IV.1. Rome: UNIPEDE, 1952.
Directie Toezicht Energie. *Onderzoek Stroomstoring Haaksbergen: Constateringen en aanbevelingen.* Nederlandse Mededingingsautoriteit, February 2006.
Directorate General for Energy and Transport. "Moving towards improved protection. MEMO." Brussels: European Commission 2007.
Djangirov, V.A. and V.A. Barinov. *Forming Electricity Markets in the Commonwealth of Independent States and its Impact Upon Power System Development.* CIGRE Group 37: Power Systems Planning and Developent. Paris: Conférence Internationale des Grand Réseaux Electriques, 2002.
Doorman, Gerard, Gerd Kjølle, Kjetil Uhlen, Einar Ståle Huse, and Nils Fratabø. *Vulnerability of the Nordic Power System.* Trondheim: Sintef Energy Research, 2004.
Doyer, H., *Eene rijks-electriciteitsvoorziening van Nederland.* Delft, 1916.
Drechsler, Wolfgang, Rainer Kattel, Erik S. Reinert, and Carlota Perez. *Techno-economic Paradigms: Essays in Honour of Carlota Perez.* London: Anthem Press, 2009.
Edwards, P.N. "Y2K: Millennial relflections on computers as infrastructure." *History and Technology* 15 (1998): 7–29.
Edwards, P.N. "Infrastructure and modernity: Force, time, and social organization in the history of sociotechnical systems." In Th. J. Misa, Ph. Brey, A. Feenberg (eds.). *Modernity and Technology,* 185–226. MIT Press, Cambridge, 2003.
Edwards, P.N., Jackson, S.J., Bowker, G.C., and Knobel, C.P. *Understanding Infrastructure: Dynamics, Tensions and Design.* Report of a Workshop on History and Theory of Infrastructure: Lessons for New Scientific Infrastructures. Michigan: School of Information, University of Michigan, 2007.
Egan, Matthew J. "Anticipating future vulnerability: Defining characteristics of increasingly critical infrastructure-like systems," *Journal of Contingencies and Crisis Management* 15 (2007): 4–17.
Egyedi, T. "The standardised container: Gateway technologies in cargo transport." In M. Holler and E. Niskanen (eds.). *EURAS Yearbook of Standardization,* Vol. 3, 231–262. Munich: Accedo, 2000.
Eichengreen, Barry J. *The European Economy Since 1945: Coordinated Capitalism and Beyond.* Princeton: Princeton University Press, 2007.
EMRIC. *Evaluatieverslag Pilotproject Ambulance-Burenhulp Regio Aachen (D) en Zuid-Limburg. April 2002–April 2003: Euregio Maas Rijn Interventie bij Crisis.* EMRIC, 2003.
Esping-Andersen, Gøsta. *The Three Worlds of Welfare Capitalism.* Princeton, New Jersey: Princeton University Press, 1990.
Estrada, J., H.O. Bergesen, A. Moe, and A.K. Sydnes. *Natural Gas in Europe: Markets, Organisation and Politics.* London: Pinter, 1988.
Facts. *Energy and Water Resources in Norway.* Oslo: Norwegian Ministry of Petroleum and Energy, 2008.
Fadum, Hege Sveaas. *Avbruddsstatistikk 2008.* Oslo: Norges vassdrags- og energidirektorat, 2009.
Feist, J. "Status of resynchronization of the two UCTE synchronous zones." In *Power Engineering Society General Meeting, 2004.* Vol.2, 2004.

Fells, Ian. *World Energy 1923–1998 and Beyond: A Commemoration of the World Energy Council on its 75th Anniversary*. London: Atalink Projects/World Energy Council, 1998.
Fijnaut, C. and A. Spapens. *Criminaliteit en rechtshandhaving in de Euregio Maas-Rijn*. Oxford: Intersentia, 2005.
Filias, Nikos. "I Enallaktikes Piges Energias [The alternative energy soes]," *Ikonomia ke Kinonia*, February 1980: 52–60.
Fredholm, Michael. *Gazprom in Crisis*. Defence Academy of the United Kingdom: Conflict Studies Research Centre, Russian Series 06/48, October 2006.
Freeman, Chris and Francisco Louçã. *As Time Goes By. From the Industrial Revolutions to the Information Revolution*. Oxford: Oxford University Press, 2001.
Fridlund, Mats. *Shaping the Tools of Competitive Power: Government Technology Procurement in the Making of the HVDC Technology*. Research Report to the European Commission (DG XII) in the Targeted Socio-Economic Research (TSER) program under the Fourth Framework Program. Tema T Working Paper 192. Linköping: Linköping University, 1998.
Fridlund, Mats and Helmuth Maier. "The second battle of the currents: A comparative study of engineering nationalism in German and Swedish electric power, 1921–1961." *Working Papers from the Department of History of Science and Technology 96/2*. Stockholm: Royal Institute of Technology, 1996.
Fritzon, Å., K. Ljungkvist, A. Boin, and M. Rhinard. "Protecting Europe's critical infrastructures: Problems and prospects." *Journal of Contingencies and Crisis Management* 15 (2007): 30–41.
Fukuyama, Francis. "Social capital and civil society." Paper presented at the IMF Conference on Second Generation Reforms, October 1, 1999.
Furedi, Frank. *Culture of Fear Revisited: Risk Taking and the Morality of Low Expectation*, 4th edition. Continuum International Publishing Group, 2006.
Furedi, Frank. "The only thing we have to fear is fear itself." *Spiked*, April 4, 2007.
Gall, Alexander. *Das Atlantropa-Projekt: Die Geschichte einer gescheiterten Vision. Hermann Sörgel und die Absenkung des Mittelmeers*. Frankfurt: Campus Verlag, 1998.
Gall, Alexander. "Atlantropa: A technological vision of a United Europe." In Erik van der Vleuten and Arne Kaijser (eds.). *Networking Europe: Transnational Infrastructures and the Shaping of Europe, 1850–2000*, 99–128. Sagamore Beach: Science History Publications, 2006.
Gardner, Daniel. *The Science of Fear. How the Culture of Fear Manipulates Your Brain*. New York: Plume, 2009.
George, Henry. *Progress and Poverty: An Inquiry into the Cause of Industrial Depressions, and of Increase of Want with Increase of Wealth: The Remedy*. London: 1884.
Georgiev, A., M. Spirov, and M. Conev. *Energetikata v Bulgaria. Anali [Energy in Bulgaria. Annals]*. Sofia: Energoproekt Press, 1998.
Georgiev, V., A. Georgiev, and G. Pankov. *Energetikata i himiata na Bulgaria [Energetics and Chemistry in Bulgaria]*. Jambol: Hera Press, 2001.
Gheorghe, A., and D. Vamanu. On the vulnerability of critical infrastructures: "Seeing It Coming." *Journal of Critical Infrastructures* 1, no. 2–3 (2005): 216–246.
Gheorghe, A.V., M. Masera, M. Weijnen, and L. de Vries (eds.). *Critical Infrastructure at Risk: Securing the European Electric Power System*. Dordrecht: Springer, 2006.
Gheorghe, A.V., M. Masera, L. de Vries., M. Weijnen, and W. Kröger. "Critical infrastructures: The need for international risk governance." *International Journal of Critical Infrastructures* 3 (2007): 3–19.
Gicquiau, Hervé. "Les Transports d'énergie dans le Caem: Problèmes et perspectives." In *CMEA: Energy, 1980–1990. Colloquium*, 8–10 April 1981, 143–174. Newtonville, MA: Oriental Research Partners, 1981.
Giddens, Anthony. *The Constitution of Society: Outline of the Theory of Structuration*. Cambridge: Polity Press, 1984.
Gigerenzer, Gerd. "Out of the frying pan into the fire: Behavioral reactions to terrorist attacks." *Risk Analysis* 26, 2 (2006): 347–351.

Goossens, L. (ed.). "Risk and vulnerability of critical infrastructures." *Special Issue of the Journal of Risk Research* 7, 6 (2004).
Gounarakis, Petros. "I Ellinoyougoslaviki synergasia is ton tomea tou ilektrismou [The Greek-Yugoslavian collaboration in the electricity sector]", *Ikonomikos Tahydromos*, October 29, 1959: 1–6.
Graham, Steve and Simon Marvin. *Splintering Urbanism. Networked Infrastructures, Technological Mobilities, and the Urban Condition*. London: Routledge 2001.
Granadino, Ramon and Haddou Amerdoul. "Bridging the strait of Gibraltar." *Transmission & Distribution World*, July 1, 1999.
Gray, Doug. *Tetra: The Advocate's Handbook*. Cornwall: Pryntya, 2003.
Grin, J., J. Rotmans, and J. Schot (eds.). *Transitions to Sustainable Development. New Directions in the Study of Long Term Transformative Change*. Routledge Studies in Sustainability Transitions. London: Routledge, 2010.
Griniuk, R. "Centralnoto dispechersko upravlenie na Obedinenite energeticheski sistemi [Central dispatch governance of the United power systems]," *Elektroenergia* 5 (1981): 51–55.
Group of Physicists "Physics in the service of man". "Piriniki Antidrastires stin Ellada [Nuclear power reactors in Greece]." *Tehnika Hronika*, March–April 1978: 261–268.
Group of Special Contributors. "Analitiki Theorisi tou Ellinikou Energiakou Provlimatos [Analytical consideration of the Greek energy problem]," *Ikonomia ke Kinonia*, February 1980: 25–35.
Gugerli, David. *Redeströme: Zur Elektrifizierung der Schweiz, 1880–1914*. Zürich: Chronos Verlag, 1996.
Gustafson, Thane. *Crisis Amid Plenty. Soviet Energy from Brezhnev to Gorbachev*. RAND Corporation, 1989.
Gustafsson, Lars. "Framtidsvisioner om driften av det nordiske kraftsystemet." Paper presented at Elkraftsamarbete i Norden, Ronneby, Sweden, 1972.
Haaland Matlary, Janne. "Gas trade in Scandinavia: Will political factors matter?" In *Naturgasmarknader i Norden*. Gothenburg: Gothenburg University, Department of Economic History/NUPI Norsk Utenrikspolitisk Institut, 1988.
Haas, Robert. "Austausch Elektrischer Energie zwischen verschiedenen Ländern." In *Transactions of the World Power Conference, Basle Sectional Meeting*, 987–999. Basle: E. Birkhäuser & Cie., 1926.
Halacsy, Andrew A. "An early breakthrough in railway electrification." *Electronics and Power* 16, 4 (1970): 144–149.
Hall, Peter A, and David Soskice (eds.). *Varieties of Capitalism. The Institutional Foundations of Comparative Advantage*. Oxford: Oxford University Press, 2001.
Hämmerli, Bernhard and Andrea Renda. *Protecting Critical Infrastructure in the EU. CEPS Task Force Report*. Brussels: Center for European Policy Studies, 2010.
Hammons, T.J., Y. Kucherov, L. Kapolyi, Z. Bicki, M. Klawe, S. Goethe, A. Tombor, et al. "European policy on electricity Infrastructure, interconnections, and electricity exchanges." *IEEE Power Engineering Review* 18, 1 (1998): 8–21.
Hård, M. and Thomas J. Misa. *Urban Machinery: Inside Modern European Cities*. Cambridge, MA: MIT Press, 2008.
Hassan, John A., and Alan Duncan. "Integrating energy: The problems of developing an energy policy in the European communities, 1945–1980." *Journal of European Economic History* 22, 1 (1994): 159–176.
Hatziiosif, Christos. "I Elliniki Iikonomia, Pedio Mahis ke Antistasis [The greek economy, field of battle and resistance]." In Christos Hatziiosif, and Prokopis Papastratis. (eds.). *Istoria tis Elladas tou ikostou eona: B Pagkosmios Polemos, Katohi – Antistasi, 1940–1945 [History of Greece in the Twentieth Century: World War II, Occupation, Resistance, 1940–1945]*, Vol. C2, 180–217. Athens: Vivliorama, 2007a.

Hatziiosif, Christos. "Dekemvris 1944, Telos kai Arhi [December 1944, Ending and Beginning]." In Christos Hatziiosif, and Prokopis Papastratis. *Istoria tis Elladas tou ikostou eona: B Pagkosmios Polemos, Katohi – Antistasi, 1940–1945 [History of Greece in the Twentieth Century: World War II, Occupation, Resistance, 1940–1945]*, Vol. C2, 362–391. Athens: Vivliorama, 2007b.

Hatzivassiliou, Evanthis. *Greece and the Cold World: Frontline State, 1952–1967*. London: Routledge, 2006.

Hausman, William, Mira Wilkins, and Peter Hertner. *Global Electrification: Multinational Enterprise and International Finance in the History of Light and Power*. Cambridge: Cambridge University Press, 2008.

Hayes, M. "The Transmed and Maghreb projects: Gas to Europe from North Africa." In D. Victor, A. Jaffe, and M. Hayes (eds.). *Natural Gas and Geopolitics. From 1970 to 2040*. Cambridge: Cambridge University Press, 2006.

Headrick, Daniel. *The Invisible Weapon. Telecommunications and International Politics 1851–1945*. Oxford: Oxford University Press, 1991.

Hecht, Gabrielle. "A cosmogram for nuclear things." *Isis* 98 (2007): 100–108.

Hecht, Gabrielle, and Paul N. Edwards. *The Technopolitics of Cold War: Towards a Transregional Perspective*, Washington, DC: American Historical Association, 2007.

Heggenhougen, Rolv. *Samkjøringen av kraftverkene i Norge, jubileumsberetning 1932–1982*. Oslo, 1982.

Henrich-Franke, Christian. "From a supranational air authority to the founding of the Euopean Civil Aviation Conference (ECAC)." *Journal of European Integration History* 1 (2007): 69–90.

Henrich-Franke, Christian. "Mobility and European integration: Politicians, professionals and the foundation of the ECMT." *Journal of Transport History* 29 (March 2008): 64–82.

Hermansen, Ove and Bo Strøjberg. *Flyveledertjenesten 50 år*. København: Statens Luftfartsvæsen, 1986.

Hertner, Peter. "Financial strategies and adaptation to foreign markets: The German electrotechnical industry and its multinational activities, 1890s to 1939." In Alice Teichova, Maurice Lévy-Leboyer, and Helga Nussbaum (eds.). *Multinational Enterprise in Historical Perspective*, 145–159. Cambridge: Cambridge University Press, 1986.

Hertoghs, M.W.J.A. and P.H.M. Rambach. *Rampen op de grens. Een onderzoek naar de juridische aspecten van landsgrensoverschrijdende samenwerking bij rampen en zware ongevallen in het grensgebied van Nederland met België respectievelijk met Duitsland*. Maastricht: Datawyse, Universitaire Pers Maastricht, 1997.

Higgins, Polly. "Electricity: Levelling the renewables playing field." *Renewable Energy Focus* 9, 5 (2008): 42–46.

Högselius, P. *The Dynamics of Innovation in Eastern Europe. Lessons from Estonia*. Cheltenham, UK and Northampton, MA, USA: Edward Elgar, 2005.

Högselius, P. "Connecting east and west? Electricity systems in the Baltic region." In Erik van der Vleuten and Arne Kaijser (eds.). *Networking Europe. Transnational Infrastructures and the Shaping of Europe, 1850–2000*, 245–277. Sagamore Beach: Science History Publications, 2006.

Högselius, P. *Red Gas: Russia and the Origins of European Energy Dependence*. Basingstoke and New York: Palgrave Macmillan, 2013.

Högselius, Per and Arne Kaijser. *När folkhemselen blev internationell. Elavregleringen i historiskt perspektiv*. Stockholm: SNS Förlag, 2007.

Högselius, Per, Arne Kaijser, and Erik van der Vleuten. *Europe's Infrastructure Transition: Economy, War, Nature*. Basingstoke and New York: Palgrave Macmillan, in press.

Hommels, A. *Unbuilding Cities. Obduracy in Urban Sociotechnical Change*. Cambridge, MA: MIT Press, 2005.

Houry, M.P. and O. Faucon. "Defense plans: Economic solutions for improving the security of power systems." *Control Engineering Practice* 7, 5 (1999): 635–640.

Hristov, E. "Tendencii v razvitieto na demografskite procesi i naselenieto prez perioda 1965–2020g. [Tendencies in the development of demographic processes and population during the 1965–2020 period]." *Spisanie na BAN*, 2000.
Hughes, Thomas P. *Networks of Power*. Baltimore: The John Hopkins University Press, 1983.
Hulpverleningsdienst Regio Twente. *Eindrapportage INTERREG IIIA-project. Verbetering grensoverschrijdende hulpverlening 2005–2007*. Enschede: Hulpverleningsdienst Regio Twente, 2007.
Iatrides, J.O. "Perceptions of Soviet involvement in the Greek Civil War, 1945–1949." In L. Baerentzen et al. (eds.). *Studies in the History of Greek Civil War*, 225–248. Copenhagen: Museum Tusculanum, 1987.
IEA. *Annual Oil and Gas Statistics 1981/82*. Paris: OECD/International Energy Agency, 1984.
IEA. *Security of Gas Supply in Open Markets. LNG and Power at a Turning Point*. Paris: International Energy Agency, 2004.
Janáč, Jiří. *European Coasts of Bohemia. Negiotiating the Danube-Oder-Elbe Canal in a Troubled Twentieth Century*. Amsterdam: Amsterdam University Press, 2012.
Jasanoff, Sheila. "The political science of risk perception," *Reliability Engineering and System Safety* 59 (1998): 91–99.
Jasanoff, Sheila. "Citizens at risk: Cultures of modernity in the US and EU," *Science as Culture* 11, 3 (2002): 363–380.
Jensen, Lill-Ann and Alf Johansen. *I sikkerhetens tjeneste*. Oslo, 1993.
Jessen, Eike. "Surveillance radar data reduction." In Cicely Popplewell (ed.). *Information Processing 1962, Proceedings of IFIP Congress 62*. Amsterdam: North-Holland Publishing Company, 1963, pp. 219–220.
Johnson, Edward. "Early indications of a freeze: Greece, Spain and the United Nations, 1946–47." *Cold War History* 6, 1 (2006): 43–61.
Kaijser, Arne. "Controlling the grid: The development of high-tension power lines in the Nordic countries." In Arne Kaijser and Marika Hedin (eds.). *Nordic Energy Systems: Historical Perspectives and Current Issues*, 31–54. Chicago: Science History Publications, 1995.
Kaijser, Arne. "Trans-border integration of electricity and gas in the Nordic countries, 1915–1992," *Polhem: Tidskrift för teknikhistoria* 15 (1997): 4–43.
Kaijser, Arne. "Striking Bonanza. The establishment of a natural gas regime in the Netherlands." In Olivier Coutard (ed.). *Governing Large Technical Systems*. London: Routledge, 1999.
Kaijser, Arne. "Presidential address: The trail from Trail. New challenges for historians of technology." *Technology and Culture* 52 (2011): 131–142.
Kaijser, Arne, Erik van der Vleuten, Karl-Erik Michelsen, Lars Thue, and Lars Heide. *Europe Goes Critical: The Emergence and Governance of Critical Transnational European Infrastructures*. Tensions of Europe Working Document Series, 2008.
Kaiser, Wolfram and Johan Schot. *Writing the Rules for Europe: Experts, Cartels, International Organizations*. Basingstoke and New York: Palgrave Macmillan, in press.
Kampouris, G., A. Tasoulis, A. Maisis, and I. Milis. "I Autonomi Litourgia tou Diasyndedemenou Systimatos ton Valkanikon Horon [The Autonomous Operation of the Interconnected System of the Balkan Countries]," *Tehnika Hronika*, March–April 1999: 72–84.
Karydis, Panayotis G., Norman R. Tilford, Gregg E. Brandow, and James O. Jirsa. *The Central Greece Earthquakes of February–March 1981: A Reconnaissance and Engineering Report*. Washington, DC: National Academy Press, 1982.
Kaser, Michael. *COMECON: Integrated Problems of the Planned Economies*. London: Oxford University Press, 1965.
Kassimeris, Christos. "From commitment to independence: Greek foreign policy and the Western Alliance." *Orbis* 52, 3 (2008): 494–508.
Katsaloulis, Iraklis. "Istoria Tehnikon Provlepsis Sismon: I Periptosi tou VAN [History of Earthquake Prediction Techniques: The Van case]," MA Thesis, Graduate Program in the

History of Science and Technology, National and Kapodistrian University of Athens and National Technical University of Athens, 2006.
Khrushchev, Nikita S. "On peaceful coexistence." *Foreign Affairs* 38, 1 (October 1959): 1–18.
Kirzner, I. "Entrepreneurial discovery and the competitive market process: An Austrian approach." *Journal of Economic Literature* 35 (March 1997): 60–85.
Kittler, Werner. *Der internationale elektrische Energieverkehr in Europa.* Oldenburg, 1933.
Kleisl, Jean-Daniel. *Électricité suisse et Troisième Reich.* Commission Indépendante d'Experts Suisse – Seconde Guerre Mondiale. Lausanne: Chonos/Éditions Payot, 2001.
Kling, W.L. "Gekoppelde netten: Een gezamelijke verantwoordelijkheid." Inaugural speech, Technical University Delft, November 23, 1994.
Kling, W.L. "Intelligentie in Netten: Modekreet of Uitdaging?" Inaugural lecture, Eindhoven University of Technology, 2002.
Kloumann, Sigurd. "Export of electrical power from Norway." In *Verdenskraft konferencen i London 1924. Den norske nationale komites beretning.* Oslo: Grøndahl & Søns Boktrykkeri, 1926.
Knippenberg, H. "The Maas-Rhine Euroregion: A laboratory for European integration?" *Geopolitics* 9, 3 (2004): 608–626.
Kohl, Wilfrid L. "Energy policy in the communities." *Annals of the American Academy of Political and Social Science* 440, 1 (1978): 111–121.
Koliopoulos J.S. and Th. Veremis. *Greece: The Modern Sequel from 1821 to the Present.* London: C. Hurst & Co, 2007.
Kollias, I. "Epiptosis ston Elladiko horo apo Pirinika Atihimata stn Piriniko Stathmo Kozlodoui tis Voulgarias [Effects on the Greek territory by Nuclear Accidents at the Nuclear Power Plant Kozloduy of Bulgaria]." Paper presented at the conference *Kindinos Pirinikou Atihimatos: I Periptosi tou Kozlodoui,* Thessaloniki, May 12, 1993.
Kouloumpis, Evaggelos. "Piriniko Ergostasio stin Ellada: 10 Periohes Eksetastikan, Kamia Katallili [Nuclear factory in Greece: 10 locations were surveyed, none found to be appropriate]." *TEE* 2491 (June 9, 2008): 72–73.
Kouvelis, Petros T. "Skepsis epi ton ethnikon mas diekdikiseon ke epanorthoseon [Thoughts on our national claims and reparations]." *Viomihaniki Epitheorisis* 137 (March 1946): 71–76.
Kuisma, Markku. *Kylmä Sota – Kuuma Öljy. Neste, Suomi ja kaksi Eurooppaa.* Porvoo: WSOY, 1997.
La Porte, Todd. "The United States air traffic system: Increasing reliability in the midst of growth." In Thomas P. Hughes and Renate Mayntz (eds.). *The Development of Large Technical Systems.* Boulder, CO: Westview Press and Franfurt am Main, FRG: Campus Verlag, 1988.
La Porte, T.R. "High reliability organizations: Unlikely, demanding and at risk." *Journal of Contingencies and Crisis Management* 4 (1996): 60–71.
La Porte, T.R. and P.M. Consolini. "Working in practice but not in theory: theoretical challenges of high-reliability organizations". *Journal of Public Administration Research and Theory* 1991: 19–47.
Laaksonen, Lasse. *Todellisuus ja harhat: Kannaksen taistelut ja suomalaisten joukkojen tila talvisodan lopussa 1940.* Helsinki: Gummerus, 2005.
Laborie, Léonard. "A Missing Link? Telecommunications networks and European Integration 1945–1970." In Erik van der Vleuten and Arne Kaijser (eds.). *Networking Europe: Transnational Infrastructures and the Shaping of Europe, 1850–2000.* Sagamore Beach: Science History Publications, 2006.
Lagendijk, Vincent. *Electrifying Europe: The Power of Europe in the Construction of Electricity Networks.* Amsterdam: Aksant, 2008.
Lagendijk, Vincent and Frank Schipper. "Seducing the Apostate State: The Material Links of Yugoslavia to East and West, 1948–1980." Forthcoming journal article.

Lagendijk, Vincent and Erik van der Vleuten. "Electricity infrastructures." In Akira Iriye and Pierre-Yves Saunier (eds.). *The Palgrave Dictionary of Transnational History*, 315–318. Basingstoke and New York: Palgrave Macmillan, 2009.

Lambov, Todor and Ilia Borisov. *Purva atomna Kozloduyska [First Atomic in Kozloduy]*. Sofia: Partizdat, 1981.

Laporte, Georges. "Rapport sur l'Interconnexion de Réseaux de Transport d'Énergie Électrique." In *UNIPEDE Quatrième Congrès, Paris*, Vol. 2, 106–126. Brussels: Marcel Hayez, 1932.

Laurila, Erkki. *Atomienergian tekniikka ja politiikka*. Keuruu, 1967.

Laurila, Erkki. *Muistinvaraisia tarinoita*. Keuruu, 1982.

Leach, M., ed. *Re-Framing Resilience: A Symposium Report*. Vol. 13, Steps Working Paper, 2008.

League of Nations. *Proposals Put Forward by the Belgian Government for the Agenda of the Commission of Enquiry for European Union*. LoN doc. ser., C.E.U.E/3. Geneva: League of Nations, 1930.

Lebed, Sergei. "IPS/UPS Overview." Presentation at the IPS/UPS Kick-Off Meeting, Brussels, April 20, 2005.

Lee, W. Robert. *German Industry and German Industrialisation*. London: Taylor & Francis, 1991.

Legge, Joseph. *Grundsätzliches und Tatsächliches zu den Elekt-ri-zi-täts-wirt-schaften in Europa*. Dortmund: Gebrüder Lensing, 1931.

Lenin, V.I. "Report on the work of the council of people's commissars, December 22 [1920]." In *Collected Works*, Volume 31. Moscow: Progress Publishers, 1965.

Lilliestam, Johan. "The creation of the pan-Nordic electricity market." 2007. http://www.supersmartgrid.net/wp-content/uploads/2008/06/lilliestam-2007-the-creation-of-the-pan-nordic-electricity-market.pdf (retrieved on January 20, 2010).

Linnerooth-Bayer, J., Löfstedt, R.E., & Sjöstedt, G. (Eds.). (2001). *Transboundary Risk Management*. London and Sterling, VA: Earthscan Publications Ltd.

Lommers, S. *Europe – On Air. Interwar Projects for Radio Broadcasting*. Amsterdam: Amsterdam University Press, 2012.

Lubar, Steven. "Representation and power." *Technology and Culture* 36, 2 (1995): S54–S82.

Lucas, N.J.D. *Energy and the European Communities*. London: Europa Publications, 1977.

Luhmann, Niklas. *Trust and Power*. Chicester: Wiley, 1979.

Luijf, E., Nieuwenhuis, A., Klaver, M., van Eeten, M., Cruz, E. "Empirical findings on critical infrastructure dependencies in Europe." In S. Setola and S. Geretshuber (eds.). *Critical Information Infrastructure Security*, 302–310. Berlin: Springer, 2009.

Lukasik, S.J. "Vulnerabilities and failures of compex systems." *International Journal of Engineering Education* 19 (2003): 206–212.

Lundestad, Christian V. *Pleasure and Pain: Drift and Vulnerability in Software Systems*. Master's thesis, University of Oslo and Universiteit Maastricht, 2003.

Lynch, Peter and Helmuth Trischler (eds.). *Wiring Prometheus: History, Globalization and Technology*. Aarhus: Aarhus University Press, 2003.

Maier, Helmut. "Systems connected: IG Auschwitz, Kaprun, and the building of European power grids up to 1945." In Erik van der Vleuten and Arne Kaijser (eds.). *Networking Europe: Transnational Infrastructures and the Shaping of Europe, 1850–2000*, 129–158. Sagamore Beach: Science History Publications, 2006.

Makarov, Yu.V., V.I. Reshetov, A. Stroev, and I. Voropai. "Blackout prevention in the United States, Europe, and Russia." *Proceedings of the IEEE* 93, 11 (2005): 1942–1955.

Marer, P. "Prospects for integration in the Council for Mutual Economic Assistance (CMEA)." *International Organization* 30, 4 (Autumn 1976): 631–648.

Mattelart, Armand. *The Invention of Communication*. Minneapolis: University of Minnesota Press, 1996.

Maximov, I., A. Kalyadin, A. Klimov, Y. Yuvsov, O. Gerasimov, and D.B. Uglov. "Facts and figures." *International Affairs* 10, 9 (1963): 99–111.

Mazower, Mark (ed.). *After the War Was Over: Reconstructing the Family, Nation and State in Greece, 1943–1960*. Princeton: Princeton University Press, 2000.

Michel, Aloys A. and Stephen A. Klain. "Current problems of the Soviet electric power industry." *Economic Geography* 40, 3 (July 1964): 206–220.

Michelsen, Karl-Erik. *Viides Sääty. Insinöörit suomalaisessa yhteiskunnassa*. Helsinki: Tekniikan akadeemisten liitto sid, 1999.

Michelsen, Karl-Erik. "Transgressing boundaries. The Finnish nuclear power program during the Cold War." Unpublished manuscript, 2010.

Michelsen, Karl-Erik and Tuomo Särkikoski. *Suomalainen ydinvoimalaitos*. Helsinki, 2007

Mikkonen, Tuija. "Vulnerability, reliability, and security of critical Finnish-Soviet energy infrastructures." Paper presented at the EUROCRIT workshop, Stockholm, May 21–24, 2008.

Millward, Robert. *Private and Public Enterprise in Europe: Energy, Telecommunications and Transport, 1830–1990*. Cambridge: Cambridge University Press, 2005.

Milward, Alan S. *The European Rescue of the Nation State*, 2nd edition. London/New York: Routledge, 1992/2000.

Ministry of Economic Affairs. *Monitoringsrapportage leveringszekerheid elektriciteit en gas in Nederland. Letter to DG TREN*. The Hague: Ministry of Economic Affairs, 2007.

Misa, Thomas. *Leonardo to the Internet. Technology and Culture from the Renaissance to the Present*, 2nd revised edition. Baltimore: Johns Hopkins University Press, 2011.

Misa, Thomas and Johan Schot. "Inventing Europe: Technology and the hidden integration of Europe." *History and Technology* 21 (2005): 1–22.

Mladenov, Tch. and E. Dimitrov. *Urbanizaciata v Bulgaria ot osvobojdenieto do kraia na Vtorata svetovna vojna [Development of the Urbanization Process in Bulgaria from the Liberation to the end of World War Two]*. Sofia: Geografski insitut na BAN, 2009.

Morsel, Henri. "Industrie électrique et défense, en France, lors des deux conflits mondiaux." *Bulletin d'histoire de l'électricité* 23 (1994): 7–18.

Morton, David. *Power: Electric Power Technology Since 1945*. IEEE History Center. New Brunswick: New Jersey, 2000.

Moteff, John, Claudia Copeland, and John Fischer. 2003. "Critical infrastructures: What makes an infrastructure critical?" In M.T. Cogwell (ed.). *Critical Infrastructures*. New York: Nova Science Publishers.

Mousiopoulos, N. "Mathimatiki Prosomiosi tis Metaforas Atmosferikon Ripon apo tin Periohi tou Pirinikou Stathmou Kozlodoui [Mathematical Simulation of the Transfer of Atmospheric Pollutant from the Area of the Nuclear Power Plant Kozloduy]." Paper presented at the conference Kindinos Pirinikou Atihimatos: I Periptosi tou Kozlodoui, Thessaloniki, May 12, 1993.

Murray, A. and T. Grubesic (eds.). *Critical Infrastructure. Reliability and Vulnerability*. Berlin: Springer, 2007.

Myllyntaus, Timo. *Electrifying Finland. The Transfer of Technology into a Late Industrializing Economy*. London: Pinter, 1991.

Myrdal, Gunnar. "Twenty years of the United Nations economic commission for Europe." *International Organization* 22, 3 (1968): 617–628.

Neporojni, P. *Elektroproizvodstvoto v Evropejskite strain chlenki na SIV [Electricity industry in European COMECON member states]*. Moscow, 1978.

Neporojni, P. "25 godini postoianna komisia za elektrichetvoto v ramkite na SIV [25 years Standing committee on energy within Comecon]." *Energetika* 1 (1981): 5–14.

NESA. *Nordsjaellands Elektricitets og Sporvejs Aktieselskab 1902–1927*. Copenhagen: C. Ferslew & Co, 1927.

Neumann, Iver. *Russia and the Idea of Europe: A Study in Identity and International Relations*. London: Routledge, 1995.

Niesz, H. "L'échange d'énergie électrique entre pays, au point de vue économique et technique." In *Transactions of the World Power Conference, Basle, Sectional Meeting*, vol. 1. Basel: Birkhäuser & Cie, 1926.

Nikolinakos, Marios. *Ekthesi Pano sto Energiako Provlima: Simperasmata ke protaseis apo to Energiako Sinedrio tou TEE pou diorganothike ton Mai tou 1977 [Report on the Energy Problem: Conclusions and Suggestions from the Energy Conference of the TEE (Technical Chamber of Greece) held in May 1977].* Athens: TEE, 1978

Nikolinakos, Marios. "I energiaki Eksartisi tis Horas apo to Eksoteriko: I Makrohronia ke I Vrahihronia Apopsi [The external energy dependence of the country: The long-term and the short-term aspect]." *Ikonomia ke Kinonia,* February 1980: 13–24.

Nilsen, Yngve and Lars Thue. *Statens kraft 1965-2006: miljø och marked.* Oslo: Universitetsforlaget, 2006.

Noble, David F. *Forces of Production. A Social History of Industrial Automation.* New York: Alfred A. Knopf, 1984.

Nöelle, Pierre. "Beyond dependency: How to deal with Russian gas." Report to the European Council for Foreign Affairs. *Policy Brief ECFR/09* November 2009.

Nordiske elverksmötet. *Radio- och annan teleteknik i elverkens tjänst, del IV.* Stockholm, 1953.

Noutsos, Panagiotis. "I Proslipsi ton Ideologimaton tis Technocracy stin Ellada [The reception of the ideological constructs of technocrasy in greece]." In Emilios Metaksopoulos (ed.). *I Epistimes stin Kinonia [Sciences in Society],* 111–124. Athens: Gutenberg, 1988.

Noutsos, Panagiotis. "'Anikodomisi' ke 'Laokratia': To Eghirima tou Anteou ke tis 'EP-AN' ['Reconstruction' and 'people's' rule: The endeavor of Antaeus and 'EP-AN']". In *I Elliniki Kinonia kata tin Proti Metapolemiki Periodo (1945–1967) [The Greek Society during the Early Postwar Period (1945–1967)],* 371–375. Athens: Sakis Karagiorgas Foundation, 1994.

Numminen, Kalevi. *Muistelmat.* Julkaisematon käsikirjoitus, 2000.

Numminen, Kalevi and Paul Laine. *"Buyer's Participation and Well Developed Domestic Infrastructure. Keys to Successful Introduction of Nuclear Power in Small Countries."* IAEA – CN 42/43. Vienna: IAEA, 1983.

Nye, David. *Consuming Power: A Social History of American Energies.* Cambridge MA: MIT Press, 1998.

Nye, David. *When the Lights Went Out. A History of Blackouts in America.* Cambridge, MA: MIT Press, 2010.

OEEC. *Interconnected Power Systems in the USA and Western Europe: The Report of the Tecaid Mission, the Report of the Electricity Committee.* Paris, 1950.

Oliven, Oskar. "Europas Großkraftlinien: Vorschlag eines europäischen Höchst-span-nungs-netzes." *Zeitschrift des Vereines Deutscher Ingenieure* 74, 25 (June 21, 1930): 875–879.

Padgett, Stephen. "The single European energy market: The politics of realization." *Journal of Common Market Studies* 30, 1 (1992): 53–75.

Pain, Rachel. "Globalized fear? Towards an emotional geopolitics", *Progress in Human Geography* 33, 4 (2009): 466–486.

Pantelakis, Nikos. *O eksilektrismos tis Elladas: Apo tin idiotiki protovoulia sto kratiko monopolio, 1889–1956 [The Electrification of Greece: From Private Initiative to State Monopoly, 1889–1956].* Athens: Cultural Foundation of the National Bank of Greece [MIET], 1991.

Papastefanou, K. "I Simasia Enos Pirinikou Atihimatos sta Valkania [The Importance of a Nuclear Accident in the Balkans]." Paper presented at the conference Kindinos Pirinikou Atihimatos: I Periptosi tou Kozlodoui, Thessaloniki, May 12, 1993.

Papathanasiou, Ioanna. "To Kommounistiko Komma Elladas stin proklisi tis istorias [The Greek communist party against the challenge of history]." In Christos Hatziiosif and Prokopis Papastratis. (eds.). *Istoria tis Elladas tou ikostou eona: B Pagkosmios Polemos, Katohi – Antistasi, 1940–1945 [History of Greece in the 20th century: World War II, Occupation, Resistance, 1940–1945],* vol. C2, 78–151. Athens: Vivliorama, 2007.

Pappa, Elli. "Eisagogi [Foreword]." In Dimitris Batsis (ed.). *Antaios* (Journal Reprint). Athens: E.L.I.A., 2000.

Paquier, Serge and Jean-Pierre Williot (eds.). *L'industrie du gaz en Europe aux XIXe et XXe siècle. L'innovation entre marchés et collectivités publiques.* Brussels: Lang, 2005.

Pater, R. *Een goede buur.... Een onderzoek naar de operationele grensoverschrijdende brandweersamenwerking in het Twents-Duitse grensgebied.* Unpublished Thesis, Saxion Hogescholen, Enschede, 2004.

Paylor, Anne. "Datalink coverage for Europe by 2007." In. P. Butterworth-Hayes (ed.). *Forty Years of Service to European Aviation. Eurocontrol.* Brussels: Eurocontrol, 2003, pp. 71–75.

Peaceful Uses of Atomic Energy. *Proceedings of the Forth International Conference Geneva, 6–16 September 1971,* Volumes I, II, III, IV. Vienna: United Nations and the International Atomic Energy Agency, 1972.

Pegg, Carl H. *Evolution of the European Idea, 1914–1932.* Chapel Hill and London: University of North Carolina Press, 1983.

Perez, Carlota. 2002. *Technological Revolutions and Financial Capital. The Dynamics of Bubbles and Golden Ages.* Cheltenham, UK: Edward Elgar.

Permanent Committee on the Environment of the Technical Chamber of Greece. "Antidrastires ke Perivallon [Reactors and environment]." *Tehnika Hronika,* March–April 1978: 268–278.

Perrow, Charles. *Normal Accidents. Living with High-Risk Technologies.* Princeton: Princeton University Press, 1984/1999.

Perrow, Charles. *The Next Catastrophe: Reducing Our Vulnerabilities to Natural, Industrial, and Terrorist Disasters.* Princeton: Princeton University Press, 2007.

Persoz, Henri. "40 ans d'interconnexion internationale en Europe. Le rôle de l'UNIPEDE." In Monique Trédé (ed.). *Electricité et électrification dans le monde. Actes du deuxième colloque international d'histoire de l'électricité,* 293–303. Paris: Association pour l'histoire de l'électricité en France, 1992.

Persoz, Henri and Jean Remondeulaz. "Consolidating European power." *IEEE Spectrum* 29, 10 (1992): 62–65.

Petersen, Flemming. 2006. *Dansk Elforsynings liberalisering 1990–2004.* Unpublished manuscript. Copenhagen, 2006.

Petrosianc, Arkadii. *Savremennie problemi atomnaia nauka i tehnika v SSSR [Contemporary Problems of Nuclear Science and Technology in the USSR].* Moscow: Atomizdat, 1976.

Post, B. and P. Stal. *Grensoverschrijdende spoedeisende medische hulpverlening Belgie-Duitsland-Nederland.* Nijmegen: ITS, Stichting Katholieke Universiteit, 2000.

Presidential Commission on Critical Infrastructure Protection. *Critical Foundations. Protecting America's Infrastructures. The Report of the President's Commission on Critical Infrastructure Protection.* Washington, DC, October 1997.

Projectbureau C2000. *Eindrapport initiatieffase C2000, Naar een nieuw landelijk radionet voor de Brandweer, Ambulance en Politieorganisaties.* Den Haag: Ministry of the Interior, 1995.

Pugh, Emerson W., Lyle R. Johnson, and John H. Palmer. *IBM's 360 and Early 370 Systems.* Cambridge, MA: MIT Press, 1991.

Raftopoulos, Theodoros, I. "Energiaki Ikonomia ke Pige tis Energias en Elladi [Energy economy and energy sources in Greece]." *Tehnika Hronika,* January–February 1941: 28–34.

Raftopoulos, Theodoros, I. "Orthologiki Energiaki Ikonomia is tin Ellada [Rational energy economy in Greece]." *Viomihaniki Epitheorisis,* 114 (1943): 94–100 and 115 (1944): 4–13.

Raftopoulos, Theodoros, I. "I Antignomia epi tis Ethinkis Energiakis Diekdikiseos [On the disagreement regarding the national energy claim]." *Antaios* 12 (1945): 276–277.

Raftopoulos, Theodoros, I. "Ilektrike Ikomike Paranoisis ke e Limne tis Dassaritias [Electric economic misconceptions and the Desaretian lakes]." *Antaios* 17–18 (1946): 369–372.

Rambousek, H. *Die ÖMV-Aktiengesellschaft – Entstehung und Entwicklung eines nationalen Unternehmens der Mineralölindustrie.* PhD thesis, Wirtschaftsuniversität Wien, 1977.

Rathenau, Walther. *Walther Rathenau: Industrialist, Banker, Intellectual, and Politician. Notes and Diaries 1907–1922.* Revised and extended edition. Oxford: Oxford University Press, 1985.

Reintjes, J. Francis. *Numerical Control: Making a New Technology.* New York: Oxford University Press, 1991.

Resch, Ralf. *Institutioneller Wandel in einem transnationalen großtechnischen System: der Fall der europäischen Flugsicherung*. PhD dissertation, Universität Konstanz, 1994.
Riccio, Giorgio. "Rapport général du président du Comité (report no. IV)". In *UNIPEDE Congress of Scandinavia*. Paris: Imprimerie Chaix, 1964.
Roberts, Karlene H. "New challenges in organizational research: High reliability organizations." *Organization & Environment* 3, 2 (1989): 111–125.
Roberts, K.H. "Some characteristics of one type of high reliability organization." *Organisation Science* 1 (1990): 160–176.
Rochlin, Gene I. "Reliable organizations: Present research and future directions." *Journal of Contingencies and Crisis Management* 4, 2 (1996): 55–59.
RTÉ. *Memento of Power System Reliability*. Paris: Gestionnaire du Réseau de Transport d'Electricité/Département Exploitation du Système Électrique, 2004.
Rudolph, Karsten. *Wirtschaftsdiplomatie im Kalten Krieg: Die Ostpolitik der westdeutschen Grossindustrie 1945–1991*. Frankfurt: Campus Verlag, 2004.
Rüegg, Walter. *Die ersten fünfzig Jahre Kraftwerke Brusio, 1904–1954*. Bern-Bümplitz: Benteli, 1954.
Sagers, Matthew J. and Milford B. Green. "Spatial efficiency in Soviet electrical transmission." *The Geographical Review* 72, 3 (1982): 291–303.
Samkjøringen. 1959. *Samkjøringen gjennom 25 år: 1932–1957*. Oslo: Samkjøringen.
Sanne, Johan M. *Creating Safety in Air Traffic Control*. Linköping, Sweden: Linköping University, 1999.
Sanne, Johan M. "Creating trust and achieving safety in air traffic control." In Summerton, Jane and Boel Berner (eds.). *Constructing Risk and Safety in Technological Practice*. London: Routledge, 2003, pp. 140–156.
Sarewitz, D., R. Pielke, and M. Meykhah. "Vulnerability and risk: Some thoughts from a political and policy perspective." *Risk Analysis* 23, 4 (2003): 805–10.
Saunier, Pierre-Yves. "Taking up the bet on connections: A municipal contribution." *Contemporary European History* 11, 4 (2002): 507–527.
Saunier, Pierre-Yves. "Transnational/Transnationalism." In Akira Iriye and Pierre-Yves Saunier (eds.). *The Palgrave Dictionary of Transnational History*. Basingstoke and New York: Palgrave Macmillan, 2009.
Savenko, N. and M.A. Samkov. *Obiedinenie elektroenergeticheskie sistemy stran-chlenov SEV [Unification of Electro-Energetic Systems of CMEA Member States]*. Moscow: CMEA Secretariat, 1983.
Schipper, Frank. *Driving Europe. Building Europe on Roads in the 20th Century*. Amsterdam: Aksant Academic Publishers, 2008.
Schipper, Frank and Erik van der Vleuten. "Transnational infrastructure development and governance in historical perspective." *Network Industries Quarterly* 10, 3 (2008).
Schipper, Frank, Vincent Lagendijk and Irene Anastasiadou. "New connections for an old continent: Rail, road and electricity in the League of Nations' organisation for communications and transit." In Alexander Badenoch and Andreas Fickers (eds.). *Materializing Europe. Transnational Infrastructures and the Project of Europe*. Basingstoke and New York: Palgrave Macmillan, 2010.
Schmidt, Susanne. *Liberalisierung in Europa: Die Rolle der Europäischen Kommission*. Frankfurt: Campus Verlag, 1998.
Schmidt, Susanne and Raymund Werle. *Coordinating Technology. Studies in the International Standardization of Telecommunications*. Cambridge, MA: MIT Press, 1998.
Schneider, Joachim. "Bewertung von Drehstrom- und Gleichstromvarianten für Hochleistungs-fernübertragungen im Großverbund." PhD dissertation. Aachen: Rheinisch-Westfälische Technische Hochschule Aachen, 1994.
Schönholzer, Ernst. "Ein elektrowirtschaftliches Programm für Europa." *Schweizerische Technische Zeitschrift* 23 (1930): 385–397.

Schot, Johan (ed). "Building Europe on transnational infrastructures." *Mini-special issue of The Journal of Transport History* 28, 2 (2007).
Schot, Johan. "Transnational infrastructures and the origins of European integration." In Alexander Badenoch and Andreas Fickers (eds.). *Materializing Europe. Transnational Infrastructures and the Project of Europe.* Basingstoke and New York: Palgrave Macmillan, 2010.
Schot, Johan and Vincent Lagendijk. "Technocratic internationalism in the interwar years: Building Europe on motorways and electricity networks." *Journal of Modern European History* 6, 2 (2008): 196–217.
Schot, Johan and Frank Schipper. "Experts, Their Beliefs and Networks in European Transport Integration, 1945–1958." Lisbon, 2009.
Schot, Johan and Frank Schipper. "Experts and European transport integration, 1945–1958." *Journal of European Public Policy* 18 (2011): 274–293.
Schueler, Judith. "Travelling towards the 'mountain that has borne a state': The Swiss Gotthard railways." In Erik van der Vleuten and Arne Kaijser (eds.). *Networking Europe: Transnational Infrastructures and the Shaping of Europe 1850–2000,* 71–97. Sagamore Beach, MA: Science History Publications, 2006.
Schueler, Judith. "Materialising Identity. The Co-construction of the Gotthard Railway and Swiss National Identity." Amsterdam: Aksabt Academic Publishers, 2008.
Schulman, P., E. Roe, and M. van Eeten. "High reliability and the management of Critical Infrastructures." *Journal of Contingencies and Crisis Management* 12 (2004): 14–28.
Schwartz, Mischa. "Carrier-wave telephony over power lines: Early history." In *2007 IEEE Conference on the History of Electric Power,* 2007.
Scott Poole, Marshall and Andrew H. van de Ven. "Using paradox to build management and organization theories." *Academy of Management Review* 14 (1989): 562–578.
Scranton, Phil. "Writing a new history of Europe: Six considerations for discussion." Discussion paper presented at the Making Europe Workshop, Wassenaar, the Netherlands, January 2011.
Segreto, Luciano. "Stratégies militaires et intérêts économique dans l'industrie électrique italienne: Protection ou interconnexion des installations électriques, 1915–1945." *Bulletin d'histoire de l'électricité* 23 (1994): 63–82.
Semenov, Vladimir. "How Russia maintains a reliable power grid." *Transmission & Distribution World,* April 1997.
Semov, M. *Izpovedta na edna atomna centrala [Confession of a nuclear power plant],* Kozloduy NPP, 2002.
Shore, Cris. *Building Europe: The Cultural Politics of European Integration.* London: Routledge, 2000.
Sinclair, Bruce. "Engineering the golden state: Technics, politics and culture in Progressive Era California." Paper presented at the Minds and Matter Huntington Library Workshop, 2009.
Sinos, Alexandros. *I Geographiki Enotis tou Ellinikou Mesogiakou Horou, Meros B: I Ikonomiki Enotis [The Geographical Unity of the Greek Mediterranean Space, Part B: The Economic Unity].* Athens, 1946.
Skjöld, Dag Ove and Lars Thue. *Statens nett: systemutvikling i norsk elforsyning 1890–2007.* Oslo: Universitetsforlaget, 2007.
Søilen, Espen. *Fra frischianisme til keynesianisme? En studie av norsk økonomnisk politikk i lys av økonomisk teori 1945–1980.* PhD dissertation, Avdeling for samfunnskunnskap, Høyskolen i Agder, 1998.
Sokolov, D. "Power transfer 750 kV USSR – Romania – Bulgaria," *Energy,* no. 3 (1988): 13–16.
Sommer, Peter and Ian Brown. *Reducing Systemic Cybersecurity Risk.* Paris: OECD, 2011.
Sörgel, Hermann. *Atlantropa.* München: Piloty & Loehle, 1932.
Sörgel, Hermann. *Die drei grossen "A". Grossdeutschland und italienisches Imperium, die Pfeiler Atlantropas.* Munich: Piloty and Loehle, 1938.

Spirov, Mire. *History of Electrification in Bulgaria*. Sofia: Energoimpex, 1999.
Spirov, Mire, Atanas Georgiev, and Mladen Conev. *Kratka istoria na elektrifikaciata na Bulgaria [Electrification in Bulgaria. A Concise History]*. Sofia: Heron press, 1998.
Staar, R. *Communist Regimes in Eastern Europe*. Stanford: Hoover Press, 1982.
Stavropoulos, S.N. "Apantisi sto Mihaniko k. Th. Raftopoulo [Response to the Engineer Mr. Th. Raftopoulos]." *Antaios* 13–14 (1945): 303–308.
Stavropoulos, S.N. "I Limnes Dassaritias ki i Elliniki Ikonomia [The Desaretian Lakes and the Greek Economy]." *Antaios* 11 (1945): 246–247.
Stavropoulos, S.N. "I Apantisi tou k. Stavropoulou [The Response of Mr. Stavropoulos]." *Antaios* 17–18 (1946): 372–375.
Stavropoulos, S.N. "Zitimata Energiakis Ikonomias: I Apantisi tou k. S. Stavropoulou [Energy economy issues: The response of Mr. S. Stavropoulos]." *Antaios* 19–20 (1946): 415–416.
Steklov, V.Y. *Electrification in the USSR*. Translated by David Skvirsky. Moscow: Foreign Languages Publishing House, 1960.
Storm, E. "Regionalism in history, 1890–1945: The cultural approach." *European History Quarterly* 33, 2 (2003): 251–265.
SUDEL. "SUDEL, 1964–1984." Ente Nazionale per l'Energia Elettrica, 1984.
Summerton, Jane. "Power plays: The politics of interlinking systems." In Olivier Coutard (ed.). *The Governance of Large Technical Systems*, 93–113. London: Routledge, 1999.
Summerton, Jane and Boel Berner (eds.). *Constructing Risk and Safety in Technological Practice*. London: Routledge, 2003.
Tarr, J. and G. Dupuy. *Technology and the Rise of the Networked City in Europe and America*. Philadelphia: Temple University Press, 1988.
Thiry, J. *Interconnections and Electric Power Exchanges in Europe: Options and Prospects*. Committee on Energy: Working Party on Electric Power. Geneva: UNECE, 1994.
Thue, Lars. "Den politiske kraften. Fredrik Vog og historien om norsk krafteksport." In K. Endresen (ed.). *Vår vidunderlige vannkraft. Fredrik Vogt og norsk vannkraftutbygging*. Oslo: Universitetsforlaget, 1992.
Thue, Lars. "Barriers to trade: The history of the Norwegian power export." In R.P. Amdam and E. Lange (eds.). *Crossing the Borders. Studies in Norwegian Business History*. Oslo: Universitetsforlaget, 1994.
Thue, Lars. *Statens kraft 1890–1947. Kraftutbygging og samfunnsutvikling*. Oslo: Universitetsforlaget, 2006 (1994).
Thue, Lars. *For egen kraft. Kraftkommunene og det norske kraftregimet 1887–2003*. Oslo: Abstrakt, 2003.
Thue, Lars. "Norway: A resource-based and democratic capitalism." In S. Fellman, M. Iversen, H. Sjögren and L. Thue (eds.). *Creating Nordic capitalism: The Business History of a Competitive Periphery*. Basingstoke: Palgrave Macmillan, 2008.
Thue, Lars. *A Culture of Continuous Improvements. The evolving triangle of men, models and machines in Statkraft*. Oslo: Norwegian School of Management, 2009.
Timmermans, S. and M. Berg. *The Gold Standard. The Challenge of Evidence-Based Medicine and Standardization in Health Care*. Philadelphia: Temple University Press, 2003.
Timmermans, S. and S. Epstein, S. "A world of standards but not a standards world: Toward a sociology of standards and standardization." *Annual Review of Sociology* 36 (2010): 69–89.
Todoriev, Nikola. "Bulgarskata energijna politika [Bulgarian energy policy]." *Energetika* 6 (1988): 5–11.
Todoriev, Nikola, Scheli Benatova, Atanas Georgiev, Milen Bankov et al. *Elektrifikaciata na Bulgaria [The Electricity Industry in Bulgaria]*. Sofia: Izdatelstvo Tehnika, 1982.
Todoriev, Nikola et al. *Durjavata i durjavnika. Spomeni za Todor Zhivkov [State and the statesmen. Memoirs to Todor Zhivkov]*. Sofia: Phenomen 21 Publishing House, 2001)
Tooley, T. Hunt. "German political violence and the border plebiscite in Upper Silesia, 1919–1921." *Central European History* 21, 1 (March 1988): 56–98.

Trieb, Franz and Hans Müller-Steinhagen. "Europe–Middle East–North Africa cooperation for sustainable electricity and water." *Sustainability Science* 2, 2 (October 1, 2007): 205–219.
Troebst, Stefan. "Introduction: What's in a historical region? A teutonic perspective." *European Review of History: Revue europeenne d'histoire* 10, 2 (2003): 173–188.
Tsotsoros, Stathis. *Energia ke anaptiksi sti metapolemiki periodo: I Dimosia Epihirisi Ilektrismou 1950–1992 [Energy and Development in the Postwar Period: The Public Power Corporation 1950–1992]*. Athens: KNE/EIE, 1995.
Tyrrell, Ian. "American exceptionalism in an age of international history," *American Historical Review* 96, 4 (1991): 1031–1055.
Tyrrell, Ian. "Reflections on the transnational turn in United States history: Theory and practice." *Journal of Global History* 3 (November 2009): 453–474.
UCPTE. *Rapport annuel 1951–1952*. Paris: UCPTE, 1952.
UCPTE. "Resultaten van Afschakelproeven die in de Westeuropese Netten Werden Uitgevoerd ter Bepaling van voor de Regeling van Belang zijnde Grootheden." In *U.C.P.T.E. Jaarverslag*, 1958–1959, 130–138. Heidelberg: UCPTE, 1959.
UCPTE. "Maatregelen ter Bevordering van een zo Constant Mogelijke Frequentie en Voorzorgsmaatregelen bij Sterk Dalende Frequentie." In *U.C.P.T.E. Kwartaalbericht*, I-1965, VII–IX. Vienna: UCPTE, 1965.
UCPTE. "Maatregelen ter Beperking van de Kans op en de Omvang van Grote Storingen in Gekoppeld Verband." In *U.C.P.T.E. Kwartaalbericht*, IV-1966, 5–19. Paris: UCPTE, 1966.
UCPTE. *U.C.P.T.E. 1951–1971: 20 ans d'activitè*. Rome: UCPTE, 1971.
UCPTE. *1951–1976: 25 jaar UCPTE*. Arnhem: UCPTE, 1976.
UCPTE. *Rapport annuel 1976–1977*. Arnhem: UCPTE, 1978.
UCPTE. "Uitval van eenheden - simulatieberekeningen over de invloed op vermogentransporten in het UCPTE-koppelnet." In *UCPTE Kwartaalbericht IV-1986*, 39–43. Heidelberg: UCPTE, 1986.
UCPTE. "De toetsing van het UCPTE-net tijdens de storing in Frankrijk op 12 januari 1987." In *UCPTE Kwartaalbericht II-1987*, 5–9. Heidelberg: UCPTE, 1987.
UCPTE. "Measures to counteract major disruptions in interconnected operation and to re-establish normal operating conditons." In *Half-yearly Report*, I-1990, 19–35. Arnhem: UCPTE, 1990.
UCPTE. "Protokoll über die Sitzung der UCPTE-ad-hoc-Gruppe Ost-West-Verbund am 30. Mai 1994 in Wien." UCTE Archives, Brussels.
UCPTE. *Rapport annuel 1997*. Madrid: UCPTE, 1998.
UCTE. "Electric system reliability in the context of market liberalisation." In *Annual Report 2000*, 23–33. Brussels: UCTE, 2001.
UCTE. *UCTE System Adequacy Forecast 2003–2005*. Brussels: UCTE, 2002.
UCTE. *Annual Report 2004*. Brussels: UCTE, 2004.
UCTE. *Final Report of the Investigation Committee on the 28 September 2003 Blackout in Italy*. Brussels: UCTE, 2004.
UCTE. *Operational Handbook*. Brussels: UCTE, 2004.
UCTE. *Final Report System Disturbance on 4 November 2006*. Brussels: UCTE, 2007.
UNECE. *The Situation and Prospects of Europe's Electric Power Supply Industry in 1961/62*. Geneva: United Nations, 1963.
UNECE. *The Situation and Future Prospects of Europe's Electric Power Supply Industry in 1962/63*. Geneva: United Nations, 1964.
UNIPEDE. *Compte-rendu du Xe Congrès International, London 1955*. Paris: Imprimerie Chaix, 1955.
United Nations. *United Nations 2005 Energy Statistics Yearbook*. New York: United Nations, 2006.
United States General Accounting Office. *Information Security. Computer Attacks at Department of Defense Pose Increasing Risks*. Report GAO/AIMD-96-84. GAO, 1996.

United States Strategic Bombing Survey. *The United States Strategic Bombing Survey: Over-All Report (European War)*. Washington, DC: Government Printing Office, 1945.

Urwin, Derek W. *The Community of Europe: History of European Integration since 1945*. London: Longman, 1995.

Valden, Sotiris. *Antistathmistiko Emporio: I Anaptiksi tou stis Diethnis Synallages ke I Thesi tis Elladas [Barter Trade: Its Development in International Transactions and the Position of Greece]*. Athens: Sakkoulas Publications, 1985.

Valden, Sotiris. *Ellada ke Anatolikes Hores 1950–1967: Ikonomikes Shesis ke Politiki [Greece and Eastern Countries 1950–1967: Economic Relations and Politics]*. Athens: Odysseas: Mediterranean Studies Foundation, 1991.

Valden. Sotiris. *Ellada-Yiougoslavia: Genisi ke Ekseliksi mias Krisis [Greece-Yugoslavia: Birth and Development of a Crisis]*. Athens: Themelio, 1991.

Valden. Sotiris. *Parateri Eteri: Elliniki Diktatoria, Komounistika Kathestota ke Valkania (1967–1974) [Odd Partners: Greek Dictatorship, Communistic Regimes and the Balkans (1967–1974)]*. Athens: Polis, 2009.

Van der Vleuten, Erik. "Electrifying Denmark: A symmetrical history of central and decentral electricity supply until 1970." PhD dissertation, University of Aarhus, 1998.

Van der Vleuten, Erik. "Constructing centralised electricity supply in Denmark and the Netherlands." *Centaurus* 41 (1999): 3–36.

Van der Vleuten, Erik. "Understanding network societies: Two decades of large technical systems studies." In Erik van der Vleuten and Arne Kaijser (eds.). *Networking Europe. Transnational Infrastructures and the Shaping of Europe, 1850–2000*, 279–314. Sagamore Beach, MA: Science History Publications, 2006.

Van der Vleuten, Erik. "Toward a transnational history of technology: Meaning, promises, pitfalls." *Technology and Culture* 49, 4 (2008): 974–994.

Van der Vleuten, Erik and Per Högselius. "Resisting change: The transnational dynamics of energy regimes in historical perspective." In G. Verbong and D. Loorbach (eds.). *Governing the Energy Transition: Reality, Illusion or Necessity?* London: Routledge, 2012.

Van der Vleuten, Erik and Arne Kaijser. "Networking Europe." *History and Technology* 21, 1 (2005): 41–48.

Van der Vleuten, Erik and Arne Kaijser. *Networking Europe: Transnational Infrastructures and the Shaping of Europe, 1850–2000*. Sagamore Beach: Science History Publications, 2006.

Van der Vleuten, Erik and Vincent Lagendijk. "Transnational infrastructure vulnerability: The historical shaping of the 2006 European 'Blackout'." *Energy Policy* 38 (2010a): 2042–2052.

Van der Vleuten, Erik and Vincent Lagendijk. "European blackout and the historical dynamics of transnational electricity governance." *Energy Policy* 38 (2010b): 2053–2062.

Van der Vleuten, Erik, Irene Anastasiadou, Vincent Lagendijk, and Frank Schipper. "Europe's system builders: The contested shaping of transnational road, electricity and rail networks." *Contemporary European History* 16, 3 (2007): 321–348.

Van Laak, Dirk. "Der Begriff 'Infrastruktur' und was er vor seiner Erfindung besagte." *Archiv für Begriffsgeschichte* 41 (1999): 280–299.

Vanneste, Alex. *Kroniek van een dorp in oorlog. Neerpelt 1914–1918. Het dagelijks leven, de spionage en de elektrische draadversperring an de Belgisch-Nederlandse grens tijdens de Eerste Wereldoorlog*. Deurne: Universitas, 1998.

Varaschin, Denis. *Etats et électricité en Europe occidentale. Habilitation à diriger des recherches*. Habilitation, Université Pierre-Mendes-France: Grenoble III, 1997.

Vasileva, B. *Heterogeneous Standardization: Reconstructing the "working" and "successful" implementation of TETRA-based technologies in the European Three-Country Pilot*. Unpublished MA Thesis, Maastricht University, 2007.

Venikov, V.A., Ja.N. Luginsky, V.A. Semenov, and S.A. Sovalov. "General philosophy of emergency control in the UPG of the USSR." *International Journal of Electrical Power & Energy Systems* 11, 1 (January 1989): 19–26.

Verbong, Geert. "Dutch power relations: From German occupation to the French connection." In Erik van der Vleuten and Arne Kaijser (eds.). *Networking Europe: Transnational Infrastructures and the Shaping of Europe,* 217–244. Sagamore Beach, MA: Science History Publications, 2006.

Verbong, Geert and Derk Loorbach (eds.). *Governing the Energy Transition: Reality, Illusion or necessity?* London and New York: Routledge, 2012.

Victor, N. and Victor, D. "Bypassing Ukraine: Exporting Russian gas to Poland and Germany." In D. Victor, A. Jaffe, and M. Hayes (eds.). *Natural Gas and Geopolitics. From 1970 to 2040.* Cambridge: Cambridge University Press, 2006.

Viel, Georges. "Etude d'un reseau 400.000 volts." *Revue generale de l'electricité* 28 (1930): 729–744.

Villiez, H. von. *The Process of Integration of European Air Traffic Services.* Aachen: Technische Hochschule Achen, 1987.

Voigt, Wolfgang. *Atlantropa: Weltbauen am Mittelmeer. Ein Architektentraum der Moderne.* Hamburg: Dölling und Galitz, 1998.

Voropai, N.I., I. Reshetov, and N. Efimov. "Organization principles of emergency control of electric power systems in a market environment." *IEEE Powertech* 2005: 1–8.

Voskopoulos, George. "The Greek-Bulgarian relations in post-Cold War era: Contributing to the stability and development in southeastern Europe." *Mediterranean Quarterly* 19, 4 (2008): 68–80.

Vovos, Nikolaos. "Diasyndesis Systimaton Ilektrikis Energias me Synehes Revma [Interconnections of Electric Energy Systems Using Direct Current]." In TEE, *Seminario: Systimata Ilektrikis Energias, Athina 6–10 Martiou 1989, tefhos 2, kefaleo 15 [Seminar: Systems of Electric Energy, Athens, March 6–10, 1989, issue 2, chapter 15].* Athens: TEE, 1989.

Vovos, N., G. Georgantzis, E. Dialynas, N. Koskolos, E. Leonidaki, A. Marinakis, G. Sakkas, N. Frydas, and N. Hatziargyriou. "Meleti tou Rolou tis Elladas stin Ilektrika Diasyndedemeni Evropi [Study of the Role of Greece in the Electrically Interconnected Europe]." *Tehnika Hronika* 1 (January–February 1999): 127–139.

Wallace, W. *The Dynamics of European Integration.* London and New York: Pinter, 1990.

Whitley, Richard. *Divergent Capitalisms. The Social Structuring and Change of Business Systems.* Oxford: Oxford University Press, 1999.

Williams, Elisa. "Climate of fear." *Forbes Magazine,* February 4, 2002.

Wilson, K. and J. van der Dussen. *The History of the Idea of Europe,* 2nd edition. London and New York: Routledge, 1995.

Winner, Langdon. *The Whale and the Reactor. A Search for Limits in an Age of High Technology.* Chicago: University of Chicago Press, 1986.

Wistoft, B., J. Thorndahl, and F. Petersen. *Elektricitetensaarhundrede. Dansk elforsyningshistorie.* Copenhagen: DEF, 1992.

Woodhouse, C.M. *The Struggle for Greece, 1941–1949.* London: C. Hurst & Co. Publishers, 2002.

Working Group 37.12. *The Extension of Synchronous Electric Systems: Advantages and Drawbacks.* CIGRE Group 37: Power Systems Planning and Developent. 1994 CIGRE Session Papers. Paris: Conférence Internationale des Grand Réseaux Electriques, 1994.

World Bank. "Building regional power pools: A toolkit." *Energy, Transport, and Water Department,* The World Bank Group, 2005.

Zobaa, A. "Status of international interconnections in North Africa." In *Power Engineering Society General Meeting,* Vol. 2, 1401–1403. IEEE, 2004.

Index

Abel, Hans, 219–20
AEG, 66, 123–4
AEG-Telefunken, 197–8, 207
Aeronautical Fixed Telecommunication Network, 206, 208
Africa, 4, 7, 14–15, 24–5, 32, 47, 54–5, 62, 70, 72–3, 86, 96, 274
  see also North Africa; South Africa
Aharoni, Yair, 227
air-traffic control, 9, 11–12, 16, 187–90, 191–209, 210–1, 273
  and analog radar, 188–9, 192, 194–5, 197–8, 200, 204–5, 207–9
  and computer-mediated radar, 196–201
  and Eurocontrol, *see* Eurocontrol
  and "flight strip board," 193, 195
  and high-speed craft, 191–2, 195
  and Maastricht Control Centre, 188–9, 196–205, 208–9, 211
  and technology and organizations, 201–7
  and "way points," 193
Air Traffic Flow Management, 189, 204–6, 208–9
Algeria, 24, 32, 36–7, 40–3, 46–9, 52, 55–6, 58, 62, 86, 267, 269, 274
Allied powers, 75, 98, 196–7
Alps, 8, 14, 73
alternating current (AC), 65–6, 85, 174
American Society of Mechanical Engineers (ASME), 125
America Online, 3
analog systems, 188–9, 192, 194–5, 197–8, 200, 204–5, 207–9, 217–18, 224, 227–8, 239, 242–3, 251–2, 259
Antaeus (journal), 164
APCO25 (US), 248–9, 258
APX (Netherlands), 233
Arctic Circle, 24
Areva (France), 214
ASEA (Sweden), 85, 123–4, 155
ASME, *see* American Society of Mechanical Engineers
ASTRID (Belgium), 251, 253
"Atlantropa project," 70–6
"Atoms for Peace" program (1955), 121–2

Austria, 24–5, 27–9, 32–4, 36, 40–5, 49–50, 54, 56–7, 65–6, 68, 79–80, 84, 86, 134, 156, 173–4
Averianov, Sergei, 109

BaltEnergo, 108
Baltic Sea, 27, 37, 59, 72
Baltic States, 9, 49, 54, 73, 81, 86, 110, 112, 115
Batsis, Dimitris, 164
Bavaria, 12, 33–5, 42, 48, 54, 57, 59, 61, 65, 86
Bayerngas, 48, 59
Beck, Ulrich, 10, 273
Belarus, 49, 53–4, 60
Belemeken-Sestrimo cascade, 147, 155
Belene nuclear power plant (Bulgaria), 147, 155
Belgium, 29–31, 40–1, 43, 63, 67–8, 74–5, 79, 91, 134, 145, 188, 195–204, 208, 240–2, 244, 249–51, 253, 258–9, 271
Beniger, James, 219, 224
Berlin Wall, 33, 39, 50
Bignami, Paolo, 64–5
blackouts, 4, 6, 11, 14–16, 24–6, 62–5, 86–96, 97, 100, 139, 148, 157–8, 174–5, 188, 213, 223, 236, 263–8, 271, 273–6, 278
  cascading (1987), 86–8
  European (4/11) (2006), 4, 11, 25–6, 62–5, 86–8, 90, 93–6, 97, 263–4, 267, 271, 274, 278
  Italian (2003), 4, 11, 90, 93–4
  Netherlands (2005), 95
  New York City (1965), 11, 89
  New York City (1977), 11
Black Sea, 14, 131, 139–41
Bobov Doll thermal power plant (1971–1975) (Bulgaria), 141
Bosporus, 14
Bozhinov, Todor, 131
Brandt, Willy, 34–5
"Bratstvo" (Brotherhood) pipeline, 30, 33
Brochmann, Georg, 188, 222–3
Brown Boveri (Switzerland), 68

Bulgaria, 12, 14, 16, 36, 85–6, 89, 105–7, 131–53, 154–6, 274
  as Balkan power hub, 131–53
  and coal, 139–42
  and communism (1944–65), 133–7
  and the electrical communist revolution, 134–7
  and electricity shortages, 137–9
  and grid optimization, 145–7
  and hydropower, 147–8
  and nuclear relations, 142–3
  and "People's Court" (*Naroden sad*), 133
  and "socialist industrialization," 134
  and technological choices, 136–7
  and transnational connections, 137–9, 143–5, 148–51
Bulgarian Communist Party, 106, 133–6, 139, 148, 151–3, 155
Burgbacher, Fritz, 52

C2000 (Netherlands), 247, 251–4, 258–9
capitalism, 30–3, 54, 82, 105, 132, 142, 166, 218
Carter, Jimmy, 39
Castberg, Johan, 221
Caucasus, 38, 51, 81
Central Air Traffic Flow Management Unit, 189, 209
Central Electricity Board, 67
Central Europe, 5, 27, 53–4, 58, 81, 225
Central Intelligence Agency (CIA), 3, 39
centralization, 80, 85, 113, 133, 137, 202, 204–5, 209, 222–4, 228–9, 231–2, 263
CENTREL, 86
Cerna Voda nuclear plant (Romania), 177–9
Chaira (Bulgarian hydropower plant), 145, 147–8, 153, 155
Channel Tunnel, 9, 13
chemical industry, 9, 17, 30, 152, 155
Chernobyl catastrophe, 109, 155, 183
Chevalier, Michel, 8
China, 9, 34, 108
Christian Democrats, 34–5
Citibank hacking, 3
climate change, 6, 94, 216, 232–7
Clinton, Bill, 3
coal, 12, 29, 32–4, 48, 51, 53, 57–8, 65–8, 73–4, 86, 91, 106, 121, 124, 131–2, 139–42, 146–7, 154, 159–61, 169, 221–2, 264–5, 269, 272
  see also lignite

Cold War, 5, 15, 23–5, 27–57, 60, 82, 86, 105, 112–14, 120–2, 126–7, 132, 151, 153, 158, 164–7, 169, 196, 207, 272–5
COMECON, *see* Council for Mutual Economic Assistance
Commission for Electric Power Exchange and Utilization of the Hydropower Potential of the Danube, 138
communism, 25, 30, 34, 39–40, 49, 53–4, 80, 89–90, 105, 112–13, 133–7, 139, 148, 151–3, 155, 163–5, 167, 272, 274
Companie Générale de Télégraphie Sans Fil (CSF) (France), 198
Confederation of Danish Industries, 221–2
Continental European Synchronized Area, 13
Continuation War of 1941–4, 113
Coordinating Committee for the Development of the Interconnection of the Electric Network of the Balkans countries, 167
Copeland, William, 112, 237
Council for Mutual Economic Assistance (COMECON), 25, 30, 54–5, 85, 88, 125, 131–3, 137–9, 144–5, 148–51, 153, 155–6, 272
"critical events," 3, 11, 25, 44, 52–3, 60, 62–3, 125, 128, 177–9, 213, 240, 243–4, 264, 266–7, 269, 271, 273, 276
critical infrastructure (CI) vulnerabilities, 3–17, 263–76
  and ambivalence, 9–12
  and back-up capacities, 268–9
  and "connectedness," 15–16
  defined, 5–6
  and "Europe," 12–15
  findings on, 263–76
  geography of, 273–5; *see also* "vulnerability geography"
  and the Iron Curtain, 272–3
  ironies of, 275–6
  and manual operation, 269
  and "the myth of the network," 8
  and the paradox of infrastructure, 7–9
  and physical transnational link, 268
  and public policy, 267–70
  and "refusing connections," 268
  and reliable social relations, 268
  and standardization, 269
  strategies for, 267–70
  and substitutability, 269
  and transnational governance, 268, 270–1
  and vulnerabilities, 3–17, 263–76

306 Index

Croatia, 4, 58
crossborder infrastructure, 15–16, 33, 243, 245, 251, 254
Cuban Missile Crisis (1962), 33
cybersecurity, 4–5, 10, 189, 218, 222, 236
Czechoslovakia, 29–30, 34, 36, 43, 49, 53, 58–9, 65–6, 85–6, 99, 134, 138–9, 146, 156
Czech Republic, 27, 250

Danube River border, 14–15, 138
DATABUS, 228
decentralization, 62–3, 67, 74, 79–80, 84, 87–90, 92–3, 208–9, 231, 233, 264, 270
Denmark, 4, 37, 44–5, 48, 50, 59, 65–6, 68, 84–5, 91–2, 93, 122, 145, 148–9, 168, 219–22, 224–5, 229–32, 234, 274
Desaretian lakes, 160–5
Dimitrov, Borislav, 145
Dimitrov, Georgi, 133, 136
Directorate-General for Energy (EU), 91
Disaster Relief (*Technisches Hilfswerk*), 245
diversification, 36, 47–51, 55, 91, 132, 153, 268
Donbass industrial region, 29, 139
"Druzhba" (Friendship) oil pipeline system, 30, 58

earthquakes, 10, 107, 145–6, 155, 157–8, 177, 180, 227, 265–6, 272, 275
Eastern Europe, 16, 24–5, 30, 32, 34, 39–40, 46, 49, 53–4, 76, 85–6, 89, 132, 135, 138, 143, 155, 189, 206, 270, 273–4
EBASCO, 182–3
ECAC, *see* European Civil Aviation Conference
Edwards, Paul, 187, 239–40
EEC, *see* European Economic Community
Eesti Energia A/S, 110
EEX (Germany), 233
EFI, *see* Norwegian Electric Power Research Institute
"EFI's Multi-area Power-market Simulator" (EMPS) (*Samkjøringsmodellen*), 228–9, 235
Eisenhower, Dwight, 121
Electrical Industry Act (1948), 135
"Electric Curtain," 15, 25–6, 85–6
Électricité de France, 75
electricity, 6–16, 23–5, 26–7, 37, 40, 46, 51, 58, 62–97, 105–7, 109–10, 112, 114–23, 127, 131–82, 187–90, 213–37, 263–76
  and Bulgaria, *see* Bulgaria
  and communist revolution, 134–7
  and Greece, *see* Greece
  and interdependencies, *see* electricity interdependencies
  and Norway, *see* Norway
Electricity Act, 67
electricity interdependencies (European), 62–96
  and alliances, 83–6
  contours of, 62–3
  and cross-border collaboration, 65–70
  and the EU, 91–5
  and high-reliability organizations, 87–90
  and internationalism, 64–5
  and the invention of vulnerability, 90
  and mesoregional collaboration models, 79–83
  and Pan-Europe, 70–5
  and power grid governance, 62–3
Elektrobank, 68
Elektrochemische Werke, 65
Elprom (Bulgaria), 135, 155
emergency communication, 239–56
  and coordination, 244–7
  and critical infrastructure, 239–41, 255–6
  and the Enschede disaster, 242–4
  and postwar developments, 241–2
  and public safety standard, 247–50
  and technology, 252–5
  and Three Country Pilot (3CP), 250–2
EMPS, *see* "EFI's Multi-area Power-market Simulator"
Energohydroproject (Bulgaria), 135
Energoobedinenie (Bulgaria), 135
Energoproekt (Bulgaria), 139–40, 143, 147
Energostroy (Bulgaria), 135
Energy Act (1990) (Norway), 230–1
energy, *see* coal; electricity; hydroelectricity; natural gas; nuclear power; oil; thermal energy
ENI (Italy), 32–4, 39, 42, 45, 55, 58
Enschede fireworks disaster (2000), 188
Enso community, 116–20, 129
Enso-Gutzeit pulp and paper mill, 116–17
ENTSO-E, *see* European Network of Transmission System Operators for Electricity
Ernst & Young, 15
Essent Netwerk BV, 95
Esso, 34–5, 43, 45
Estlink project, 110–11, 127
Estonia, 58, 81, 86, 110–11, 115, 129

ETSI, *see* European Telecommunication Standards Institute
EU, *see* European Union
EURATOM, *see* European Atomic Energy Community
Eurelectric, 68
Eurocontrol, 14, 188–9, 191–209, 269–71
 and "Air Traffic Flow Management," 189, 204–6, 208–9
 and Central Flow Management Unit, 205–6
 and Central Route Charges Office, 202
 establishing, 195–6
 and Permanent Commission, 196, 203
European Atomic Energy Community (EURATOM), 30, 91, 196
European blackout (4/11) (2006), 4, 11, 17, 25–6, 62–5, 86–8, 90, 93–7, 263–4, 267, 271, 274, 278
European Broadcasting Union (1950), 14
European Civil Aviation Conference (ECAC), 205–6, 212
European Coal and Steel Community, 30, 91, 196
European Commission, 3, 6, 13, 15, 26, 62, 80–1, 90–4, 214, 229, 234, 250, 259
European Conference for Ministers of Transport (1954), 14
European Conference for Post and Telecommunications (1957), 14
"European Constitution," 94
European Economic Community (EEC), 30, 34, 45–7, 52–4, 91, 196
European Network of Transmission System Operators for Electricity (ENTSO-E), 63, 83, 90, 94, 171, 173, 234
European Regulators Group for Electricity and Gas (2003), 234
European Telecommunication Standards Institute (ETSI), 188, 248–50, 254–5
ETSI RES–6, 248
European Union, 3, 6, 23, 27, 62, 74, 92, 112, 152, 158, 215, 264

fear, culture of, 10, 18
fertilizers, 3, 30
Financial Times, 214–15
Finland, 16, 35–7, 40–1, 44, 48, 51, 54, 58–9, 66, 86, 105–29, 230–2, 235, 250, 253, 264–5, 268, 273, 275
 and Soviet border, 108–28
First Komsomolska thermal power plant, 140

Former Yugoslav Republic of Macedonia (FYROM), 158–9, 172, 174, 177, 179
Foss, Alexander, 221–2
France, 4–6, 8, 13–14, 25, 29–32, 34–6, 39–43, 46–7, 50, 55–6, 63–8, 72–3, 75, 79–80, 83, 85, 90–1, 94, 142, 148–9, 177, 195–6, 198, 202–3, 206, 214, 249–50, 253, 258
Franco-Iberian Union for Coordination and Transport of Electricity (UFIPTE), 83
Freiberger, Heinrich, 79
French Association of Electricians, 72–3
Friends of the Supergrid, 214
Frisch, Ragnar, 226–7, 238
Fuchs, Martin, 94

"gas crises," 4, 14–16, 25, 27–8, 53, 265, 267, 274, 278
gas fields, 5, 28–30, 32–3, 36–7, 49–50, 53, 58, 269
Gaz de France, 39, 46, 55
Gazprom, 27
General Atomics (US), 123
General Electric (Britain), 123
General Electric (Canada), 123–4
George, Henry, 221
German Democratic Republic (GDR), 35–6, 43, 50, 85, 99, 138, 143
Gibraltar Dam, 72
globalization, 5, 8, 38, 240
GOELRO, *see* State Commission for Electrification of Russia
Gotthard railway line, 8–9
Gounarakis, Petros, 165
Grand Canyon mid–air collision (1956), 191–3, 266, 271
Great Britain, 9, 12–13, 29, 32, 37, 40–1, 67, 72–3, 76, 85–6, 91, 123–4, 134, 159, 161, 163, 176, 181, 196, 198, 202, 214, 234, 248, 250
Great Depression, 74
Great Nordic Telegraph Company, 9
Great Northeast Blackout (1965), 11, 89
Greece, 12, 16, 54, 58, 80–1, 84, 86, 105–7, 141, 144–5, 150–1, 153, 157–80, 180–3, 264–6, 268, 272, 275
 and the Balkans, 159–65
 and Eastern bloc imports (1953–66), 169
 and electricity exchanges (1953–2009), 170–1
 and electricity exchanges (1961–2009), 172–3
 and energy flows, 157–80

308 Index

Greece – *continued*
  and frequency drops (1997), 179
  and infrastructure, 165–74
  and nuclear power, 175–80
  and projected electricity network (1943–44), 161
  and sources of electricity (1961–2005), 159
Greek Civil War (1946–9), 163
Greek Society for the Scientific Organization of Work (*Elliniki Eteria Epistimonikis Organoseos tis Ergasias*), 162
Groningen gas field (Netherlands), 30, 34, 44–7
Guidelines for a Common Energy Policy, 91
Gustafsson, Lars, 228

Haaksbergen incident (2005), 95–6
Haavelmo, Trygve, 226, 238
hackers, 3, 10, 96, 236
Hanko naval base (Finland), 117
Hassi R'Mel (Algerian gas field), 32
Hecht, Gabrielle, 177, 183
Heide, Lars, 189, 191–209
Henry George Association, 221
Hertoghs, M.W.J.A., 244–5, 257
Heyerdahl, Fritjof, 219–20
H-gas, 47–8
"hidden integration" of Europe, 14–15, 25, 28, 51–2, 54–5, 63, 158–9, 180, 196, 255, 270–2, 279
"high-reliability organizations," 25, 78–9, 87–90, 92–4, 199, 263–4, 273
High Reliability Theory, 11
High Voltage Center (Greece), 157
high-voltage direct-current (HVDC), 174, 182, 269
Hitler, Adolf, 118
Hughes, Thomas, 18, 97, 224
Hungary, 14–15, 27, 33–4, 36, 58, 65–6, 85–6, 99, 134, 138, 250
Hveding, Vidkunn, 231
hydroelectricity, 64–5, 70, 116–17, 134, 155, 157–67, 178

IAEA, *see* International Atomic Energy Agency
IATA, *see* International Air Transport Association
Iberian peninsula, 25, 83–6
ICAO, *see* International Civil Aviation Organization
Iceland, 189, 206, 209, 215, 237, 271, 276

ICT, *see* information and communication technology
Imatran Voima, 116, 119, 123–6
industrialization, 106, 121, 133–7, 151, 161, 164, 219, 226
information and communication technology (ICT), 5, 10, 12, 15–17, 187–90, 213, 217, 219, 226, 228–9, 232–3, 235–6, 248, 258, 270
Information Revolution, 219
infrastructure, the term, 7–9
  see also critical infrastructure and vulnerabilities
integrated power system (IPS), 132, 138–9, 148–9
interdependency, 6, 13, 15–16, 25, 28, 41, 58, 63–4, 70, 74, 79, 86, 95–6, 144, 187, 219, 222, 270–1
Interexecutive Working Group on Energy in 1961, 91
International Air Transport Association (IATA), 195
International Atomic Energy Agency (IAEA), 122, 124, 143, 151
International Civil Aviation Organization (ICAO), 189, 195, 205
International Convention relating to Cooperation for the Safety of Air Navigation, Eurocontrol (1960), 195
International Council on Large Electric Systems, 68
International Electrotechnical Commission, 68
International Labour Organization, 5, 74
international power highways, 74
International Power Program, 80
International Union of Producers and Distributors of Electrical Power, 68
"internetwork", 187, 190, 239–40, 256
Inter System Interface (ISI), 248, 253–4, 256, 259
invasion of Afghanistan (1979), 39
Iran, 37–9, 46, 108, 146
Ireland, 12, 40, 91–2, 96, 201–4, 206
Iron Curtain, 15–16, 24–6, 28, 30, 32, 34, 37, 51, 54–6, 85–6, 105–6, 119, 121–2, 126, 141, 153, 167, 174, 264, 267, 272–5
  as "Electric Curtain," 15, 25–6, 85–6
  as "porous," 272–3
Iron Lady line, 116
IRT–1000 (nuclear reactor) (Bulgaria), 142
ISI, *see* Inter System Interface

Italian blackout (2003), 4, 11, 90, 93–4
Italy, 4–5, 11, 24–5, 29, 32–42, 45, 47, 50, 54–6, 58, 62, 64–6, 68, 71–3, 75, 79, 81, 84, 90–1, 93–4, 158–9, 166–7, 169, 172, 174, 272, 274

Japan, 9, 37, 146–7, 155, 276

Kairyakov, Lyubomir, 133
Kalburov, Marin, 133
Karamanlis, Konstantinos, 167–8, 175
Karelian Isthmus, 114–16, 118–19
Karlsruhe (West Germany), 134, 201–3, 208
Kazakhstan, 88
Kekkonen, Urho, 126
Khomeini, Ayatollah, 39
Khrushchev, Nikita, 81–2
Kiesinger, Kurt Georg, 34
Kindle, Fred, 110
Kitsikis, Nikos, 163–4
Klingenberg, Georg, 66–7
Kloumann, Sigurd, 224–5
Knudsen, Gunnar, 221–2
Kortunov, Alexei, 49–50
Kostov, Traycho, 133, 136
Kosygin, Alexei, 126
Kotilainen, A.E., 117
Kouvelis, Petros, 162–3, 181
Kozani Energy Center (Greece), 157
Kozloduy nuclear power plant, 143–8, 151–2, 155, 177–80, 183, 265–6
KPN Getronics, 251–2

Large Technical Systems, 8, 18, 218–19
Latvenergo, 110
Latvia, 49, 53, 61, 86, 115
Laurila, Erkki, 122–3
League of Nations, 5, 64, 66, 68, 74, 180
Lehtonen, Heikki, 123
Leningrad Nuclear Power Plant, 109
Lenin, Vladimir, 67, 127, 134
L-gas, 47–8
Libya, 32, 40, 47
Lietuvos Energija, 110
lignite, 12, 66, 106–7, 131, 140–4, 146–7, 150, 154, 159, 161, 166, 264, 269, 272, 274
liquefied natural gas (LNG), 32, 37, 42, 47
Lithuania, 53, 66, 86, 115
LNG, *see* liquefied natural gas
load factor management, 50
Lubar, Steven, 179
Lukesch, Rudolf, 33, 57

Maastricht Control Centre, 188–9, 196–205, 208–9, 211
Maastricht Treaty (1992), 6
Macedonia, 151, 157–60, 162, 165–8, 177, 179, 272, 275
Macedonia (newspaper), 157, 165
Madrid train bombings (2004), 3–4
Mannesmann (Germany), 33
Mao Zedong, 34
Maritsa East (Bulgaria), 131, 140–2, 145–8, 150, 153–4
Marshall Plan (1947–1951), 79–80, 99
Mashinoexport (Soviet Union), 119
Matra (France), 249
media, 5, 8, 10, 15, 27, 33, 35, 63, 108, 124, 191, 271, 275
Mediterranean, 14, 25, 32, 34, 37–9, 51, 56, 70, 73, 96, 162, 267
Merlin, André, 63
mesoregions, 75–84, 94, 96, 187
Mikkelsen, Bård, 215, 235
"Millennium Bug," 5
Misa, Thomas J., 180, 270
Mobile Digital Trunking Radio System, 248
modernization, 10, 39, 92, 109, 133, 163, 221
Moldova, 36, 81–2
Molotov, Vyacheslav, 27, 115–16, 118
"Molotov-Ribbentrop Pipeline," 27
Morocco, 4, 32, 62, 86, 274
Moscow Energy Institute (Moskovskii Energeticheskii Institut), 143
Moscow Power Institute, 88–9
Motorola, 251–3
Muslims, 105
Myrdal, Gunnar, 13

Nabatov, Nikita, 139, 154
NAM (Netherlands), 43–4
National Bank of Greece, 160–1
NATO, *see* North Atlantic Treaty Organization
natural gas, 5, 16, 23–58, 85–6, 105, 113, 150, 264, 267–70, 272–4
  in Cold War Europe, 24, 27–57
  and contractual arrangements, 43–4
  and diversification, 47–9
  and domestic reserves, 49–51
  and "earth gas," 29, 58
  as "energy weapon," 25
  Europe through the lens of, 51–7
  and formation of transnational links, 30–6

natural gas – *continued*
  and gas fields, 5, 28–30, 32–3, 36–7, 49–50, 53, 58, 269
  and interruptible customers, 51
  late rise of, 29–30
  and liberalization, 58
  and "linguistic divide," 29, 58
  and mesoregions (figure), 56
  and North Sea gas, 36–7
  and partnerships, 41–3
  and political turbulence, 37–41
  and regions of Europe, 54–7
  and role of state, 44–5
  and transcontinental network, 29–51
  and transnational governance, 41–51
  and vulnerability geography, 52
  and vulnerability in practice, 46–7
Nazi Germany, 70–2, 75, 78, 225
NeBeDeAcPol, 241
neoliberalism, 8, 24, 88, 158, 217–18, 229–31, 271
NESA, 65, 70
Netherlands, 6, 13, 24, 28–32, 34, 36–7, 40–1, 43–9, 54, 56, 58, 61, 66, 68, 75, 79, 91, 94–6, 145, 188, 195–7, 199–202, 204, 213–14, 224–5, 233–4, 240–55, 257–9, 264, 266, 269
Netz, E.ON, 62
New York City blackouts, 11, 89
  1965, 11, 89
  1977, 11
Nielsen, Poul, 45
noaberschap (helping your neighbors), 245–6
Nöelle, Pierre, 112
Nokia, 251, 253
NORDEL, 84–5, 89–90, 92, 226, 228–32, 234, 270
"Nordism," 221
Nord Pool, 231, 233, 235
Nord Stream pipeline, 27
No-Risk Society, 227
Normal Accident Theory, 9, 11
NorNed cable, 213–14, 231, 234, 266
North Africa, 4, 24–5, 32, 47, 54–5, 62, 72–3, 86, 96
North Atlantic Treaty Organization (NATO), 7–8, 30, 33, 39, 52, 54, 57, 86, 105, 112, 196, 200, 203, 272
North Sea, 36–7, 41, 46–8, 54–6, 61–2, 72, 214–16, 233–4, 269

Norway, 24–5, 37, 40–1, 46–9, 59, 62, 66, 73, 96, 187–9, 213–37, 250, 264, 266
  as "battery of Europe," 214–16
  and climate, 232–4
  and complexity, 218–19
  and digital shift, 227–8
  and hydropower, 219–21
  and liberalization, 229–32
  and power export, 224–5
  and profit, 234–5
  and risk, 235–7
  and "risk-free" electricity supply, 225–7
  and "swing producers," 215
  and technocratic hubris, 222–4
  and transnational connections, 216–18, 221–2
Norwegian Electric Power Research Institute (EFI), 228–9
Norwegian Power Pool, 222–4, 227–9, 231
Norwegian Watercourses and Electricity Administration (NVE), 221
Norwegian Water Resources and Electricity Directorate, 225
nuclear power, 9–10, 12, 16–17, 36–7, 58, 72, 75, 106–7, 108–10, 114, 121–7, 131–3, 137, 139, 142–8, 151–5, 159, 168, 175–80, 183, 264–6, 270, 272–3, 276
NVE, *see* Norwegian Watercourses and Electricity Administration

Obama, Barack, 233
OECD, *see* Organisation for Economic Co-operation and Development
oil, 23, 25–6, 29–30, 33–4, 36–7, 42, 44–6, 51, 53, 57–8, 91, 112, 121, 124, 159–61, 169, 264
Oliven, Oskar, 73–4, 77, 225
ÖMV (Austria), 33–4, 39, 42, 45, 49–50
"Oosting Report," 257
OPEC, *see* Organization of the Petroleum Exporting Countries
"Orange Revolution" (Ukraine), 27
Organisation for Economic Co-operation and Development (OECD), 5, 10, 99, 122
Organization of the Petroleum Exporting Countries (OPEC), 25, 36–7, 45
Ottoman Empire, 105, 107

Panagiotopoulos, George, 168
"Panropa project," 70
Papandreou, George, 163, 181

Paris Peace Conference, 162
"peak shaving," 50
Pekkarinen, Mauri, 110
Perrow, Charles, 9–11
Petri, Carl-Axel, 45
Pezopoulos, Georgios, 165
Plesser, Norbert, 45
Plessey Radar (UK), 198
Poland, 27, 29–30, 35, 39, 43, 49, 54–5, 58, 60, 63, 65–6, 85–6, 99, 115, 121, 138, 154, 156, 250, 264
Popov, Konstantin, 139, 142, 154
Portugal, 4, 13, 41, 62–3, 67–8, 73, 80–1, 83, 91–2, 95, 168, 250, 274
PPC, *see* Public Power Corporation
Prague Central Dispatch Organization, 85, 90
Prespa Transboundary Park, 159–60, 180–1
Prodi, Romano, 62–3
Protocol of Agreement on Energy Policy, 91
Public Power Corporation (PPC) (Greece), 157–9, 165–78, 180, 183
Putin, Vladimir, 113

Raftopoulos, Th. I., 160–4, 180
railways, 4, 7–8, 14, 16, 50, 65, 73, 97, 144, 192
Rallis, George, 168, 175
Rambach, P.H.M., 244–5, 257
Rask, Tauno, 123
Raunio, Helena, 110–11
Reagan, Ronald, 39
Red Army, 115, 117–18
Resch, Ralf, 192
Rheinisch-Westfälisches Elektrizitätswerk (RWE), 65, 71, 141
Rhinard, M., 273
Ribbentrop, Joachim von, 115
"risk society," 273
road accidents, 10
Romania, 14–15, 29, 36, 58, 62–3, 85–6, 105–7, 138–9, 143–6, 148, 153, 177–9
Rosenergoatom, 108
Rouhiala Power Company, 117
Royal Air Force (RAF), 98
Ruhrgas, 34, 39, 44, 48, 55, 58–9
Russia, 4–5, 9, 12, 27–57, 67, 69, 73, 81–2, 86, 105, 108–28, 143, 168, 174, 206, 212, 265, 267–8, 273–4
  see also Soviet Union
Russian United Power System, 81
RWE, *see* Rheinisch-Westfälisches Elektrizitätswerk

Sahara, 29, 32, 48, 72
Sakelarov, Manol, 133, 136
Sanne, Johan M., 192, 210
Scandinavia, 13, 25–6, 28, 36, 55, 73, 76, 84, 86, 114, 122, 236
Schedl, Otto, 33–5, 57, 59
Schengen Agreement, 239, 242, 247–9, 252–5, 258
Schönholzer, Ernst, 73–4, 78
Schot, Johan, 19, 96, 180–1, 270, 279
Second General Plan for Electrification, 136
Semenov, Vladimir, 88
September 11, 2001, 4, 10
Seven July London bombings (2005), 3–4
Seven Year Plan, 81
Shannon (Ireland), 202–3, 208, 211
Shell, 34, 43, 45
Siberia, 5, 24, 29, 33, 40, 53, 58, 81–2, 96, 124–5
Siemens & Halske, 197, 207
Siemens, 123, 214
Sikorski, Radosław, 27
Single European Act, 91–2
Sinos, Alexander, 162
Slovakia, 27
Smith, Bengt, 228
SNAM Progetti, 58
socialism, 67, 82–3, 85–6, 108, 115, 132–4, 138, 150, 152, 166–9, 226, 272–3
Sonatrach, 32
Sörgel, Hermann, 70, 72–4, 76
South Africa, 177
Soviet Union, 9, 12, 23–5, 27–59, 61, 67, 74, 81–2, 85–90, 98, 105–7, 108–29, 131–54, 166–9, 177, 181, 189, 193, 209, 212, 264–5, 267–9, 272–5
  and the Balkan Power hub, 131–53
  and the Finnish border, 108–28
  and natural gas in Europe, 24–5, 27–57
Spaak Report, 91
Spain, 5, 32, 40–1, 58, 62, 75, 80–1, 83, 86, 91–2, 142, 146, 155, 168, 274
Staar, Richard, 136
Stalin, Joseph, 119, 136
standardization, 188–9, 201, 219, 239–40, 247–8, 250, 252, 254–6, 269
Standing Commission on Electric Power (Postoiannaia komisia po elektroenergia), 138
Standing Commission on Energy, 148, 150, 156
State Commission for Electrification of Russia (GOELRO), 67, 81

Statkraft (Norway), 215, 227, 230, 234–5
Statnett (Norway), 214–16, 223, 230–1, 234, 238
Stylianidis, Kleonymos, 166
submarine electricity cables, 4, 62, 84, 86, 108, 110, 127, 268, 273
SUDEL, 84
Summerton, Jane, 174
supergrid, 72, 76, 82, 95, 214–16, 233–4
Sweden, 4, 36–7, 41, 44–5, 48, 50–1, 59, 65–6, 93, 115, 119, 122, 126, 160, 218, 220–2, 225–6, 229–32, 234–5, 250, 274
Swedish Natural Gas Committee, 38
Switzerland, 3, 40–1, 64–6, 68, 73, 75, 79–80, 90, 126, 134, 166, 174, 206, 250
Sydgas (South Gas) pipeline, 37, 45
"system vulnerability," 11, 23, 41, 266
Szép, András, 108, 110

Taylorism, 162
Tchorbadjiiski, Ivan, 140, 154
Technical Chamber of Greece, 163–4, 175, 182
Technical University of Sofia (Sofiiska Politehnika), 137, 141–2
Technopromexport, 125–6
Tekniikka ja talous (Technology and Economy), 110–11
TEN-E programme, 46
TenneT (Netherlands), 213
Teploenergoproekt, 139–40
terrorist attacks, 3–4, 6–7, 10, 18, 94, 97, 236
  see also September 11, 2001; Seven July London bombings; Madrid train bombings
Tetra, 188–9, 239–40, 243, 247–56, 258–9, 271
Tetraned, 252–3
Tetrapol, 249–51, 253, 271
Thalweg, 65
thermal energy, 65, 73, 75, 79, 110, 121, 131–3, 136–7, 139–41, 144–8, 151, 153–4, 159, 170–1
Third Communist International, 136
Third Industrial Revolution, 219, 229
Three Country Pilot (3CP), 240, 250–6, 259
  Coupling Group, 253, 259
  Steering Committee, 259
Three Mile Island accident (1979), 176, 266
Thyssengas (Germany), 44
Thyssen (Germany), 33
Tito, Josip Broz, 136, 181
Todoriev, Hristo, 141, 150

Todoriev, Nikola, 140–2, 154
Todt, Fritz, 75
Toshiba, 147, 155
TOTEMA, 155
Trans-European Pipeline, 33
Trans-European Synchronously Interconnected System, *see* Continental European Synchronized Area
Trans-Mediterranean Pipeline, 32
transmission system operator (TSO), 63, 94, 213, 215, 232–4, 236
Treaties of Rome (1957), 91
Treaty on European Union (1992), 92
Treaty of Lisbon, 94
TSO, *see* transmission system operator
Tunisia, 32, 62–3, 86, 95, 274
Turkey, 12–13, 38–9, 58, 86, 105–6, 150–1, 153, 156, 158, 167–9, 172, 176, 181, 272, 274
Twente Safety Region, 242, 244, 251–2, 255, 257, 259
Tzantzareni substation (Romania), 148

UCPTE, *see* Union for the Coordination of Production and Transport of Electricity
UCTE Operational Handbook, 93
UCTE, *see* Union for the Coordination of Transport of Electricity
UFIPTE, *see* Franco-Iberian Union for Coordination and Transport of Electricity
UKAEA, 124
UK Cabinet Office, 15
Ukraine, 4, 27–30, 36, 49, 53–4, 58, 73, 81–2, 85–6, 131, 138–40, 148, 206, 265, 267, 269, 278
UNECE, *see* UN Economic Commission for Europe
UN Economic Commission for Europe (UNECE), 13–14, 32, 46, 60, 150, 167
UNESCO, 160
Union for the Coordination of Production and Transport of Electricity (UCPTE), 75–6, 79–90, 92–3, 150, 174, 269–71, 274
Union for the Coordination of Transport of Electricity (UCTE), 63, 90, 92–4, 97
Union of Greek Nuclear Scientists, 175–6
United Power Systems (Russia), 81, 108–11, 127
Urals, 14, 81–2, 96
US Air Force, 3, 98
US Department of Justice, 3

US Electric Power Research Institute (EPRI), 233
"user vulnerability," 11, 23, 41, 154, 266, 275
US Federal Aviation Act of 1985, 191
US Federal Aviation Administration, 191
US Federal Aviation Agency, 197

Vallinkoski rapids, 116–19
Varna, 139–40, 148
Vattenfall (Sweden), 225, 228
Vereinigung Deutscher Elektrizitätswerke, 79
"vertical integration," 187, 217–18, 235, 256
Viel, George, 72–3, 77
VÖEST (Austria), 33
Vogt, Fredrik, 225–6
von Villiez, Hansjürgen, 208
Vries, Gijs de, 250
VTT Technical Research Center of Finland, 109
"vulnerability geography," 12, 51–3, 96, 273–4
Vuoksi River, 114, 116, 118–20, 127
VVER reactor, 125–7

WAMS, *see* Wide Area Monitoring Systems
Watercourse Act (1887), 220
waterfalls, 166, 215, 218–20
"waterfall speculators" (*fossespekulanter*), 220
Water Syndicates Act (1948), 135

West Germany, 24–5, 31, 35–6, 40, 43–5, 52, 58–9, 86, 124, 126, 142, 154, 195–200, 202, 208
Westinghouse (US), 123–4
Wide Area Monitoring Systems (WAMS), 93
Williot, Jean-Pierre, 58
Winter War of 1939–40, 113
"Working Group for Expansion of Norway's Electricity" (Arbeitsgemeinschaft für den Elektrizitätsausbau Norwegens), 225
World Bank, 155, 233
World Energy Council, 68–9, 77
World Power Conference, 68, 73, 224–5
World Trade Center (NYC), 4
World War I, 65, 67, 75, 218, 220, 269, 271
World War II, 9, 29, 54, 67–8, 79, 82, 106, 113–14, 142, 159, 196, 207, 218, 241, 257
World Wetlands Day, 159–60

Yamal pipeline, 39
Yeltsin, Boris, 108
YUGEL (Yugoslavia), 167
Yugoslavia, 33–6, 58, 84, 86, 99, 105–7, 136, 144, 150–4, 156, 158–60, 163, 165–9, 172, 174, 176–7, 179, 181, 209, 268, 272

Zapolyarnoe (gas field), 58
Zhivkov, Todor, 141, 154

Printed in Great Britain
by Amazon